Curious Future Insight

Ulrich A. K. Betz
Editor

Curious Future Insight

Science for a Better Tomorrow

Springer

Editor
Ulrich A. K. Betz
Innovation
Merck KGaA
Darmstadt, Hessen, Germany

ISBN 978-3-031-41780-1 ISBN 978-3-031-41781-8 (eBook)
https://doi.org/10.1007/978-3-031-41781-8

This Springer imprint is published by the registered company Springer Nature Switzerland AG
The registered company address is: Gewerbestrasse 11, 6330 Cham, Switzerland

Preface

(Written with the chat mode of Microsoft Bing search based on ChatGPT from OpenAI)

The Curious2022 Future Insight Conference was a remarkable event that brought together some of the world's brightest scientists, including several Nobel laureates, to present their work and explore the future of science and technology. The conference took place from July 12 to 14, 2022 in Darmstadt, Germany, as a hybrid event allowing on-site and online participation.

The conference covered a wide range of topics such as health, nutrition, synthetic biology, materials, energy, digitalization, artificial intelligence, machine learning, robotics, mobility, space flight, secrets of the human mind, and new ways of working together. Each topic was addressed by keynote speakers who share their insights on their respective topics in more depth in this book. The conference also featured panel discussions, workshops, and poster sessions. The conference also featured a special session on the Future Insight Prize, a prestigious award that recognizes visionary research projects that aim to solve the challenges of today and enable the dreams of a better tomorrow.

The author of this book is Ulrich A.K. Betz, a senior executive with 30 years of leadership experience in the pharmaceutical and chemical industry. He is the initiator and manager of the conference. In this book, he provides a comprehensive summary of the conference highlights, insights, and learnings. He also shares his personal reflections on how science and technology can shape a better future for humanity.

This book is intended for scientists who are interested in interdisciplinary dialogue and collaboration. It offers an overview of cutting-edge research across various fields and disciplines. It also showcases inspiring examples of how science can address global challenges such as climate change, pandemics, food security, and resource scarcity.

Science and technology are essential for the benefit of humanity. They enable us to understand our world better, to improve our lives and well-being, to create new opportunities and solutions. They also inspire us to dream big and to pursue our passions.

Conferences like Curious2022 are important for fostering a culture of curiosity, creativity, and collaboration among scientists. They provide a platform for sharing ideas, learning from each other building networks. They also create a positive outlook for the future by highlighting the potential of science and technology to make a difference.

The Curious2022 Future Insight Conference was more than just a scientific meeting. It was a celebration of curiosity, creativity, and collaboration. It was an opportunity to exchange ideas, network with peers, and learn from experts. It was an invitation to imagine what is possible with science and technology.

This book aims to capture some of that spirit and share it with you.

Darmstadt, Germany
March 2023

Ulrich A. K. Betz

Contents

Part I

United by Science for a Better Tomorrow

From a 350th Anniversary to a Global Movement: United by Science for a Better Tomorrow

Ulrich A. K. Betz

When Merck celebrated its 350th anniversary as the world's oldest pharmaceutical and chemical company back in 2018, the establishment of the Curious—Future Insight Conference was a major milestone. Along with the inauguration of the annual upto one million euro Merck Future Insight Prize, it underlined the company's commitment and ambition to help advance science and technology for the benefit of humanity. in collaboration with the scientific community worldwide.

An outstanding highlight was the selection and announcement of "pandemic preparedness" as the first topic of the one million euro Merck Future Insight Prize in July 2018, the prize then was given out in summer 2019, and the devastating COVID-19 pandemic broke out end of the same year. In addition to the theme of pandemics, another important theme was highlighted at the Curious2018—Future Insight Conference that later unfortunately received unexpected urgency: war. The Darmstadt Science Declaration—Make Science Not War, was rolled-out at Curious2018 as a global call to all nations, organizations, and societies to invest more resources in the further advancement of science and technology to solve humanities' biggest problems and to help avoid global conflict. With the invasion of Ukraine by Russia and the horrible war that resulted in February 2022, and with the terrorist attack of Hamas on Israel in October 2023, we now live in another world and the term "Zeitenwende" has been used. Seeing the global issues we have to solve such as climate change, disease, and global pollution, we cannot afford to waste scarce resources on fighting each other.

Readers are cordially invited to sign the Make Science Not War declaration and support the cause at http://make-science-not-war.org. Please share this link in your respective networks to help collect signatures globally.

The original version of this chapter was previously published non-open access. A Correction to this chapter is available at https://doi.org/10.1007/978-3-031-41781-8_21

U. A. K. Betz (✉)
Innovation, Merck KGaA, Darmstadt, Hessen, Germany
e-mail: Ulrich.betz@merckgroup.com

U. A. Betz (ed.), *Curious Future Insight*,
https://doi.org/10.1007/978-3-031-41781-8_1

From the beginning on it was planned to run the Curious—Future Insight conference each second year, making 2020 the date for the next planned edition after its inauguration in 2018. The COVID-19 pandemic, however, made this impossible, and the initially planned 3-day 2020 conference was reduced to an afternoon online event featuring the awarding of the 2020 Merck Future Insight Prize, the 2020 Johann Anton Merck award, and the 2020 Nature Spin-off Prize. Everybody was then eager to see the full Curious—Future Insight Conference being conducted in the following year 2021. But the pandemic had different plans, and due to the infection waves still running over the planet, it turned out that also the Curious 2021—Future Insight Conference had to be postponed toward Curious 2022. In 2021 instead online Future Insight Days were conducted with a portfolio of scientific prizes and laudatory speeches. Then finally, 4 years after the inaugural event, in July 2022 we at last saw again a full Curious 2022—Future Insight Conference taking place! The event with the motto "United by science for a better tomorrow" went live as a hybrid event allowing for on-site and online participation and broke all records of excitement and engagement. The agenda running in two parallel streams featured keynote presentations from >80 speakers, among them 9 Nobel Laureates, >2000 attendees from 77 countries, 320 academic institutions, and 160 companies. Twenty-three exhibitors showed their innovations in the frame of an exhibition, a panel discussion organized by AAAS/Science educated the attendees on "Can AI save the world?", an ignite session gave visibility to raising young scientists and their start-up companies, an ask-me-all circle allowed direct interaction with keynote speakers in a smaller round, 23 workshops triggered discussions and co-creation, a dream board collected visions of visitors on how science and technology can contribute to a bright future, and an opportunity board allowed for direct interaction and posting of collaboration and business opportunities. The conference also saw the hand-over of several awards such as the one million euro Merck Future Insight Prize on the topic of "CO2 to fuel conversion" to Tobias Erb from the Max-Planck-Institute in Marburg, Germany, and the Johann Anton Merck Award for groundbreaking oncology research to Stephen Jackson from the University of Cambridge. In addition, the Nature "Science in Shorts" award for science communication, the Keeling Curve Prize for reduction of greenhouse gas emissions, the ScienceFluencer award for young talent in science communication, and the Junior Visions Prize for children painting their views of a great positive future shaped by science and technology were handed over.

The full agenda can be seen here (Fig. 1):

The dream board continues to be available online, and readers are actively encouraged to share their dreams and wishes on how science and technology can help to shape a bright future and what should be invented that has not yet been invented and which products should be developed. The board entries will be used to inspire future topics of the Merck Future Insight Prize to honor and enable outstanding achievements in science and technology that help to realize these visionary dream products important for humanity.

The dream board can be accessed here: https://www.curiousfutureinsight.org/dreamboard/

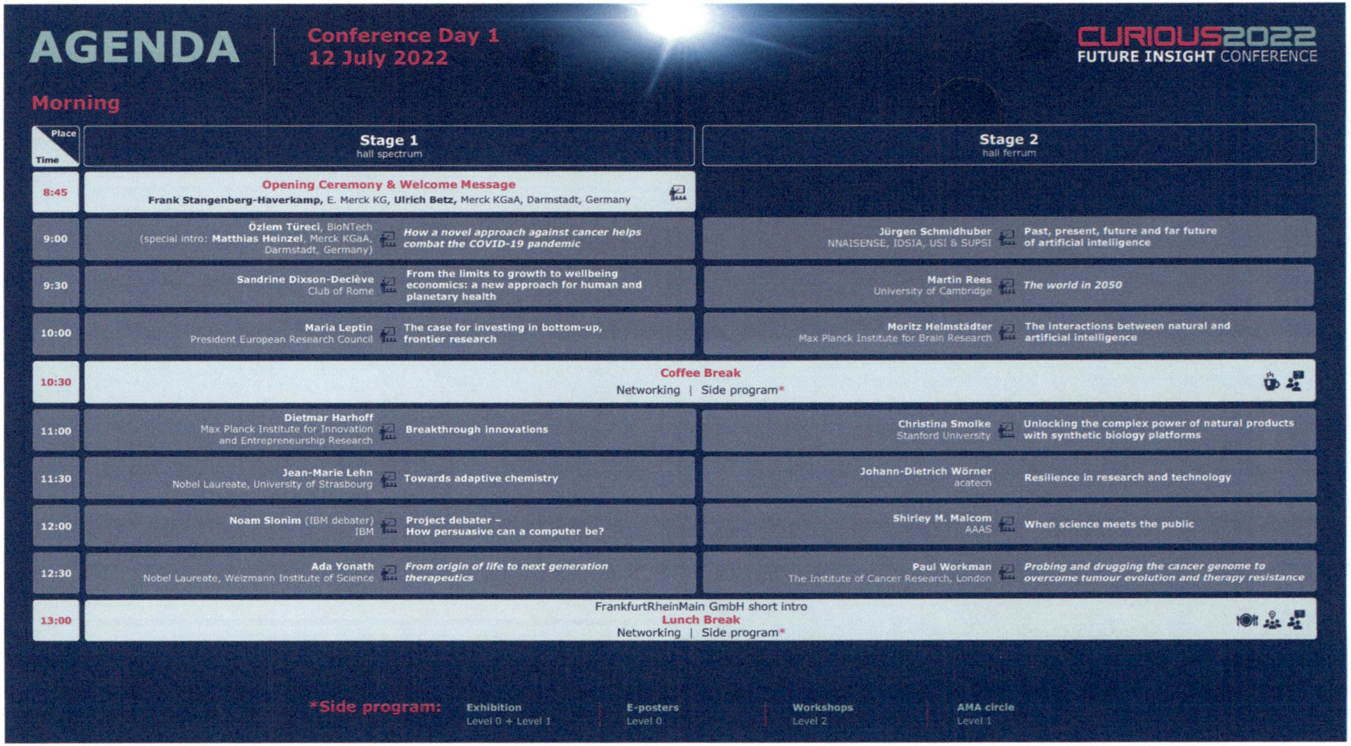

Fig. 1 Agenda of the Curious2022—Future Insight Conference

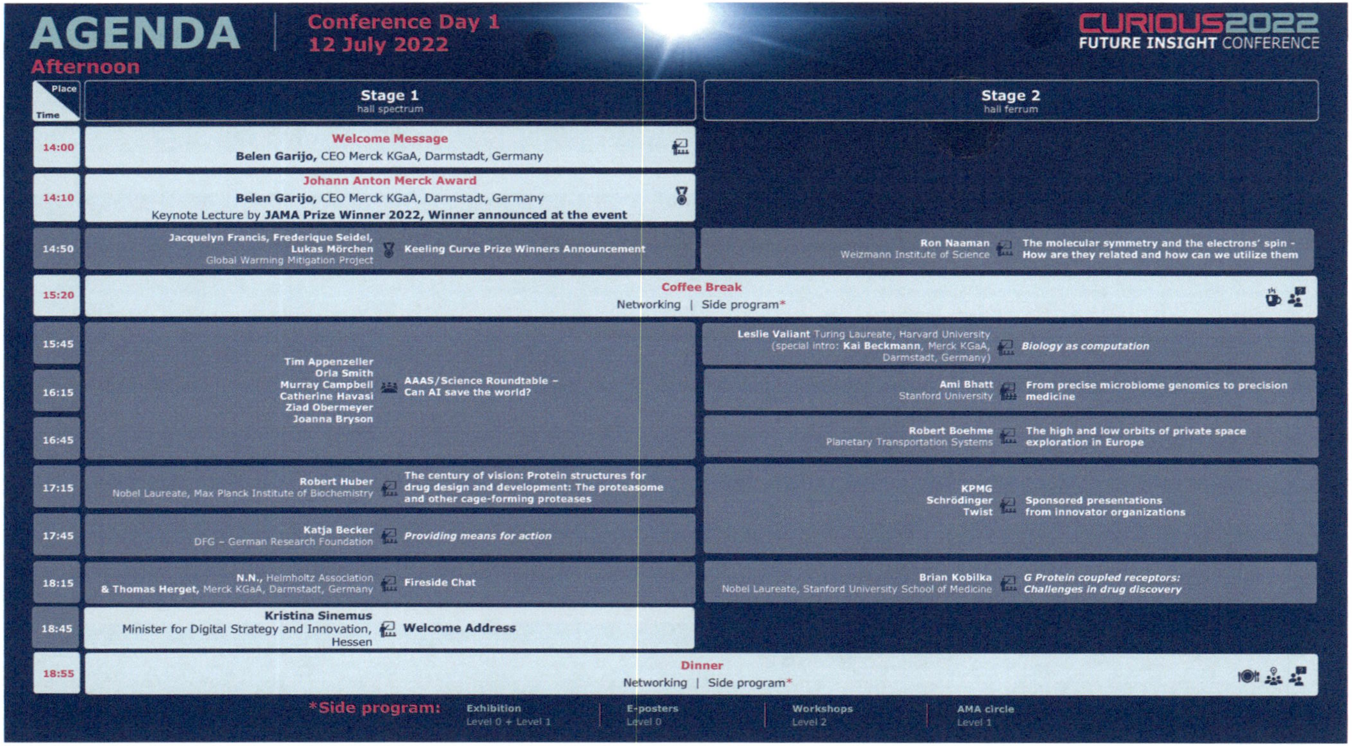

Fig. 1 (continued)

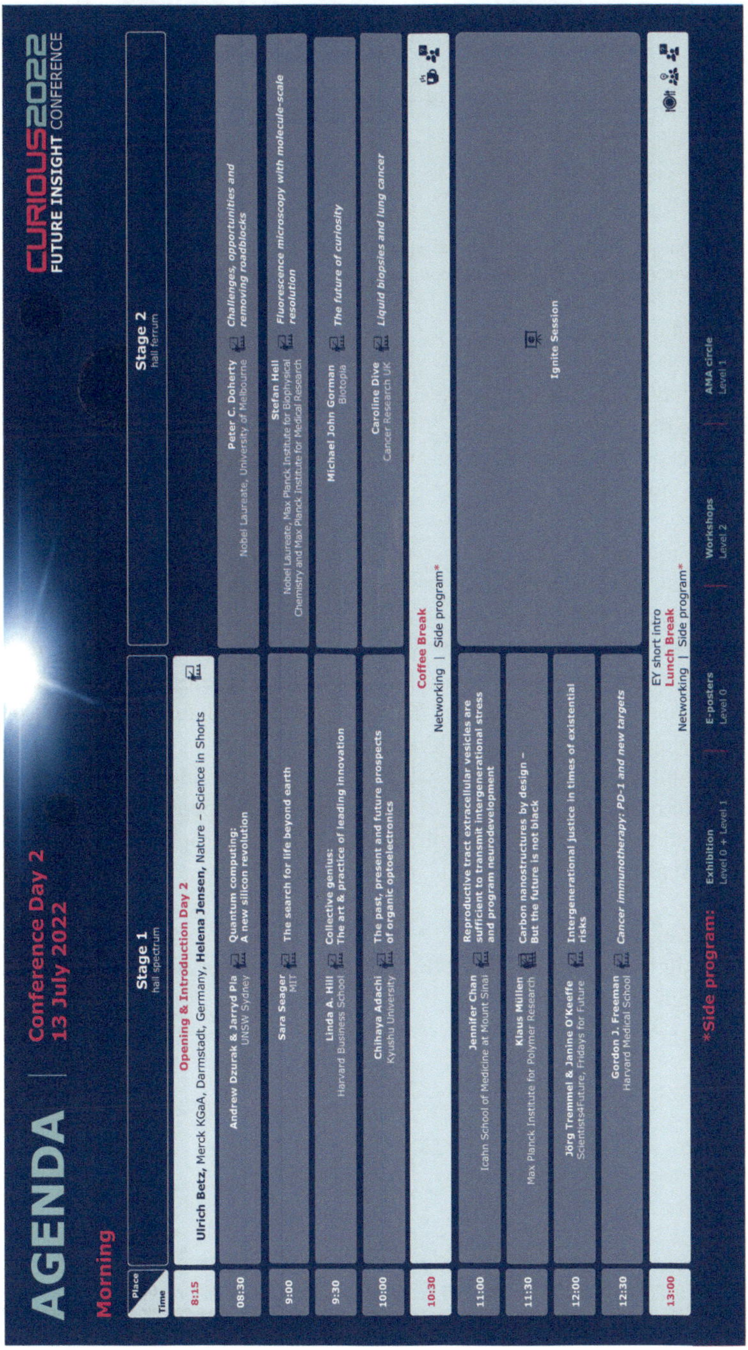

Fig. 1 (continued)

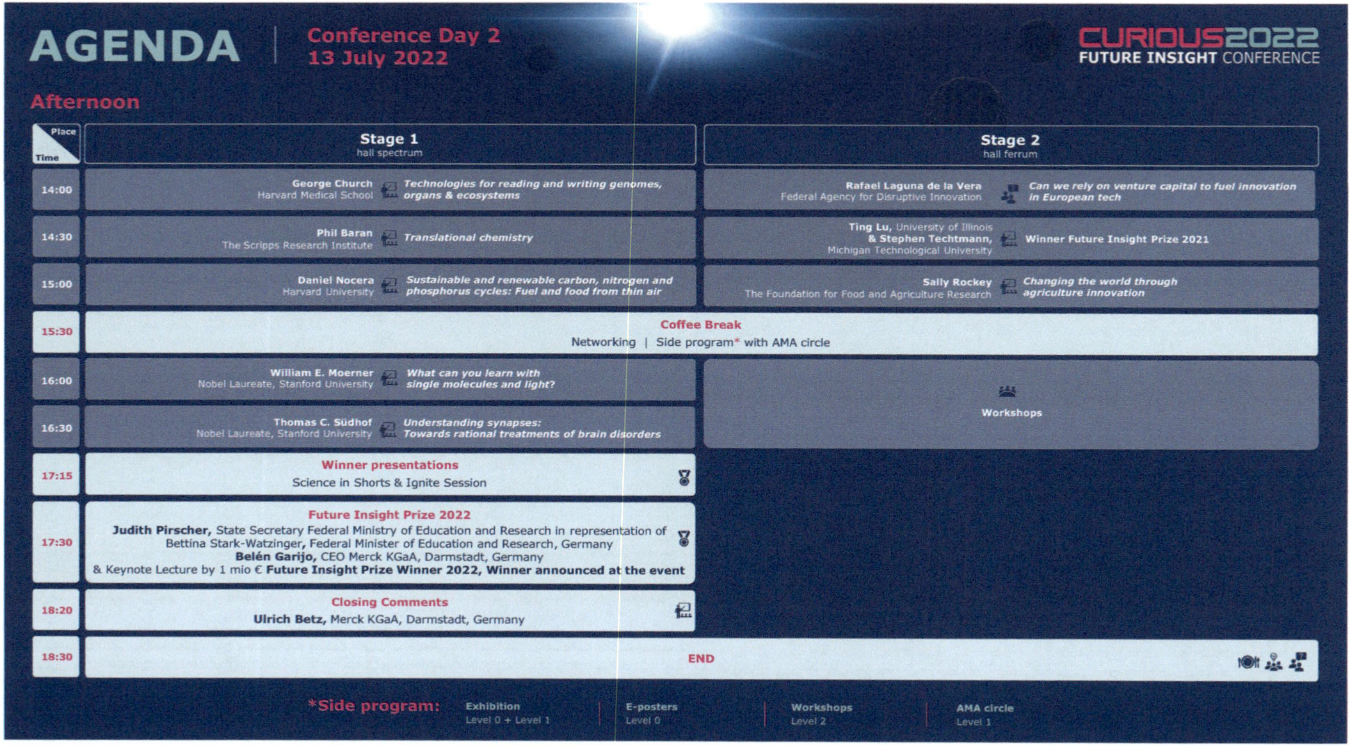

Fig. 1 (continued)

Fig. 1 (continued)

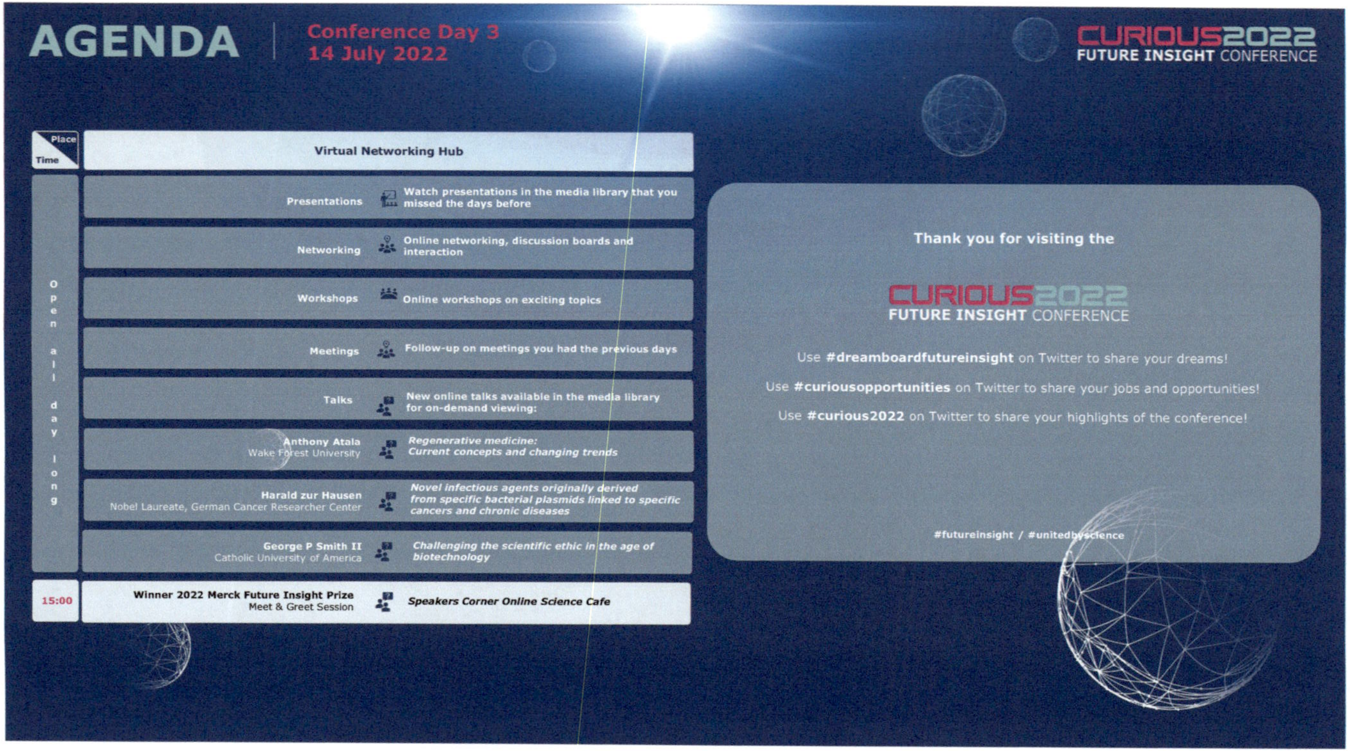

Fig. 1 (continued)

Fig. 1 (continued)

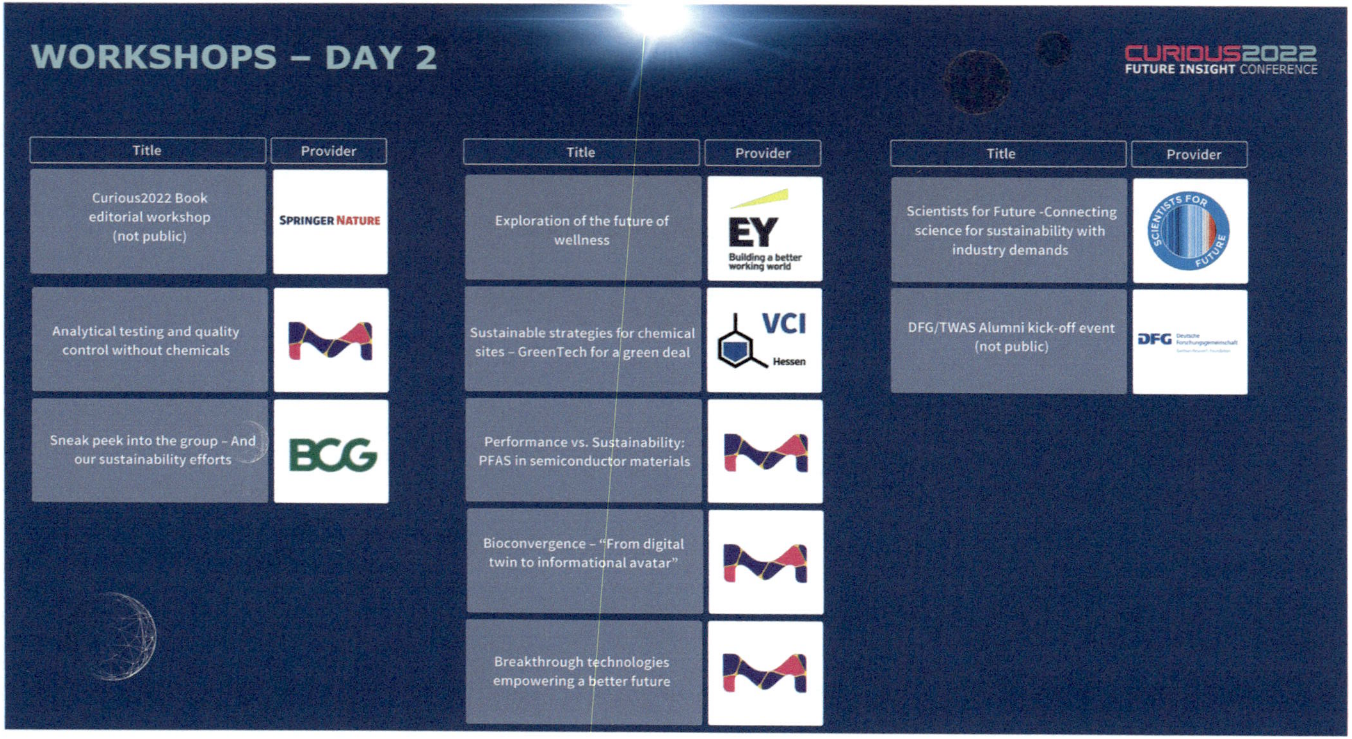

Fig. 1 (continued)

Fig. 1 (continued)

For the Curious2022—Future Insight Conference, a network of >150 partners could be assembled that joined forces in partnering with, conducting, and promoting the conference. Among this network are renowned organizations such as AAAS/ Science, the German Research Foundation (DFG), the Max-Planck-Society, Helmholtz Society, Leibniz Society, Fraunhofer Society, EMBO, Springer Nature and many others. Over the years also the German, Hessen, and EU governments have supported the event with keynotes and messages from two German Federal Ministers of Education and Research, the German Federal Minister of Economic Affairs and Energy, the German Federal Minister of Health, the Hessian Minister of Higher Education, Research and the Arts, and last but not least the European Commissioner for Innovation, Research, Culture, Education and Youth.

In total 15 Nobel Laureates spoke at the Curious—Future Insight Conferences so far: Ada Yonath, Benjamin List, Brian Kobilka, Bruce Beutler, Emmanuelle Charpentier, Frances Arnold, Fraser Stoddart, Harald zur Hausen, Jean-Marie Lehn, Joachim Frank, Peter Doherty, Robert Huber, Stefan Hell, Thomas Südhof, and William Moerner; another one, Robert H. Grubbs thankfully had accepted to speak but then passed away before being able to come. We commemorated him at the closing ceremony of Curious2022. Benjamin List, Emmanuelle Charpentier, and Frances Arnold received the Nobel Prize after they spoke at the Curious—Future Insight Conference!

Examples of other highly renowned speakers that gave a presentation at Curious—Future Insight are Craig Venter, Daniel Zajfman, Dietmar Harhoff, George Church, George Whitesides, Gordon Freeman, Johann-Dietrich Wörner, Katja Becker, Klaus Schwab, Linda Hill, Martin Rees, Paul Workman, Phil Baran, Rafael Laguna de la Vera, Renée Mauborgne, Sandrine Dixson-Declève, Sara Seager, Tom Knight, and Otmar Wiestler.

How much attending Curious—Future Insight inspires participants can best be seen from key quotes gathered at or after the event:

"The best conference I have ever attended."

"This highlights the key role of science for humanity."

"I will never see such a line-up of speakers anymore in my entire life."

"This was the Woodstock of science."

"The ambitious look far ahead is unique."

"Such days let us hope for the world."

"It is events like these that keep me inspired and excited for the future."

"Brilliant work and inspiring talks – walked away feeling very hopeful about the wonderful world of innovation and continued investment in the next generation of scientists."

"They address problems that affect the whole mankind and honor scientists who find creative solutions to tackle these problems, not only debating what is going wrong but what can be done to counter steer."

"They provide hope for a better future based on progress in science and technology."

"The conference is great as it gives insight into a broad variety of extremely relevant areas of science and business."

"Incredible networking opportunity."

"The conference stimulates people to think about the future, which we all want to be based on peace, no arms, freedom, good health, good education, great jobs, harmony and prosperity for all."

"Probably the next step up in quality is the Nobel Prize ceremony."

"I am at a lot of conferences, and I can say that this was by far the most interesting, energetic and diverse I have witnessed."

"The Curious – Future Insight Conference is the place to meet the people that shape the global future of science & technology."

So, the question is, what is special about the Curious—Future Insight Conference and what explains its magic?

The Curious—Future Insight Conference brings together some of the world's brightest scientists and innovators to solve the challenges of today and enable the dreams of a better tomorrow. It covers a broad range of topics such as health, life sciences, nutrition, material sciences, digitalization, AI, energy, mobility, space flight, robotics, the secrets of the human mind, and new ways of working together. The conference, which is now one of the world's leading gatherings on science, technology, and innovation, was initiated by Merck, the world's oldest pharmaceutical and chemical company, at the occasion of its 350th anniversary in 2018. The event triggered the birth of a global movement in support of science, technology, and innovation and now the non-profit Future Insight e.V. association, which is open for organizations and individuals from all over the world to join, carries the flame, and organizes this event biannually together with hundreds of partners.

Participating to and visibility at the event as a partner provides multiple benefits such as the opportunity to (a) position you and your organization as a leading innovator and be able to work with some of the world's best scientists and most accomplished entrepreneurs, (b) create and sustain enabling partnerships with top global science and technology players, (c) learn early on about key game changing developments in science and technology shaping our future, (d) access breakthrough ideas and top talent, (e) motivate and inspire your employees with a positive future-oriented mindset, and finally (e) join a global movement to create a bright and peaceful future.

In the end, there are three main differentiators vs. other conferences:

1. Interdisciplinary coverage of all top hot areas of science explained by leaders and pioneers in the field with deep dive science talks
2. Top science presented in a fancy way, with a cool stage, huge futuristic screen, etc. in a kind of "science fiction" atmosphere
3. Emotionally engaging, with a positive outlook to a bright future, combined with a call to action, united by science for a better tomorrow

In the future it is planned to continue to organize the Curious—Future Insight Conference as a hybrid event, allowing for physical on-site and virtual online

Fig. 2 Logo of the Future Insight e.V.

participation. While the conference clearly is an event with global attraction and visibility, it is planned to keep the venue of the Curious—Future Insight Conference in Germany and to circulate between different cities in the Rhine-Main-Neckar innovation hotspot region.

As already mentioned above, at the closing of the Curious 2022—Future Insight Conference in July 2022, the Future Insight e.V. non-profit organization was founded which will from now on organize future editions of the Curious—Future Insight Conference.

The association's goal is to unite like-minded people in its mission to support the further advancement of science and technology for the benefit of humanity creating a bright and peaceful future (Fig. 2).

Science and technology are incredibly powerful forces in the advancement of humanity, but they have their limitations. Science, for example, forever stays silent on the most important question in life: "Why do we live and what should we do?" Science is a double-edged sword that can be used for the good and the bad alike. And, unless we define ourselves as puppets without free will entirely pulled on the strings of natural laws with cause and effect, it is clear that there must be another dimension, guiding our thinking, the world of ethics, mind, spirit, and consciousness.

Embracing this concept will be of increasing importance in the light of future technological developments. Already 2022 has seen amazing progress in the field of artificial intelligence, such as the release of the chatbot ChatGPT on November 30, 2022, by the company Open AI and other large language models. ChatGPT is able to have "intelligent?" conversations with humans where it is hard to tell if a living human is on the other side or not, basically passing the famous Turing test. It can write summaries, essays, articles, and computer programs. We can be sure that further enhancements of AI will soon lead to disrupting shifts in the labor market and even the definition of what is human, the entire concept of humanism, might come under pressure. We will not be able to create a bright and peaceful future without a clear guiding ethical framework embracing the mind/spiritual dimension.

This is also reflected by the Future Insight logo. The association's logo symbolizes the amalgamation of the iron rule of science with the golden rule of ethics. The iron rule of science can be summarized as "Show evidence or counterevidence for hypotheses via observable facts from experiment or nature." The golden rule of ethics can be summarized as "Do to others as you want them do to you and also think of future generations." Together they are inspired by the three eternal fundamental principles of truth, love, and hope.

The association is rapidly growing and is currently in its first year accepting new members with no fees involved. Readers are encouraged to take this unique opportunity to become a member in an influential global network and be involved in the further organization of the Curious—Future Insight Conference at https://www.curiousfutureinsight.org/about-us/.

United by science for a better tomorrow.

Advancing Human Progress as a Twenty-First-Century Science and Technology Pioneer

Belén Garijo, Matthias Heinzel, Peter Guenter, Kai Beckmann, and Marcus Kuhnert

Since the emergence of the COVID-19 pandemic, society has been under tremendous stress. Geopolitical, environmental, and social forces have been converging at an unprecedented pace to make our world more complex and uncertain. Within this challenging environment, science has been one beacon of hope across borders, cultures, and markets.

From the first day of the pandemic, the global scientific community responded with speed, urgency, and a singular purpose. Thanks to countless scientific breakthroughs across R&D, manufacturing, and supply, humanity gained access to essential vaccines, therapeutics, and other solutions in record time. Merck KGaA (hereafter referred to as "Merck") has been proud to serve as an enabling partner and reliable supplier during this challenging period, collaborating with more than 80 different COVID-19 vaccine developers, plus many other customers for their therapeutic and diagnostic programs [1].

This pandemic contribution by Merck reflects our proud history of harnessing the power of science and technology as a force for good. Since being established in Darmstadt, Germany, in 1668, Merck has navigated its way through many wars, recessions, and other challenging periods of geopolitical or economic crisis. Through 13 generations of Merck family ownership, we have evolved time and time again to become a highly resilient and globally diversified organization with a multi-generational perspective on value creation.

As of 2022, more than 66,000 Merck employees across 66 countries worldwide shared the common purpose of being curious minds that advance human progress.

The original version of this chapter was previously published non-open access. A Correction to this chapter is available at https://doi.org/10.1007/978-3-031-41781-8_21

Marcus Kuhnert stepped down as Chief Financial Officer and a Member of the Executive Board on June 30, 2023.

B. Garijo (✉) · M. Heinzel · P. Guenter · K. Beckmann · M. Kuhnert
Merck KGaA, Darmstadt, Germany

U. A. Betz (ed.), *Curious Future Insight*,
https://doi.org/10.1007/978-3-031-41781-8_2

Thanks to our business structure and portfolio, robust culture, and proud history of performance, Merck today holds leading positions across three large and fast-growing markets at the core of tomorrow's reality.

In Life Science, we are a diversified industry leader providing scientists and researchers with lab materials, technologies, and services to make research and biotech production simpler, faster, and safer. In Healthcare, we are a global specialty innovator delivering personalized treatments for severe diseases and enabling people to achieve their dream of becoming parents. And within Electronics, we are a leading player uniquely positioned to serve and enable the electronics industry to cater to the ongoing data explosion.

Each of these three business sectors offers a differentiated, well-balanced portfolio to generate long-term value for target customers or patients. The forthcoming sections of this chapter will demonstrate some of the contributions we are making across life science, healthcare, and electronics.

However, the true potential of Merck to be a science and technology pioneer of the twenty-first century becomes fully apparent when the combinatorial synergies that exist across each sector are unlocked under a unified growth strategy. This multi-disciplinary approach reflects a significant new scientific megatrend called bioconvergence that we also wish to discuss during this chapter.

Most of all, as the executive leaders of Merck, we want this opening chapter to highlight initiatives where we, together with various partners, are striving to increase the speed, impact, and resilience of scientific discovery.

We believe that the COVID-19 pandemic will ultimately become a historic inflection point for humanity. It is our responsibility to ensure that we at Merck, together with other scientists, policymakers, and healthcare professionals worldwide, seize this moment to aim higher for customers, patients, and society. Suppose, for example, that we allowed the momentum of scientific research to slow. In that case, we might miss pivotal opportunities to address critical issues such as climate change, future pandemics, and the supply of water, food, and other scarce resources.

It must be our collective goal to foster an open, collaborative environment that allows curious scientific minds to outperform in an increasingly complex and connected world.

1 Reflections on the Curious 2022 Future Insight Conference

Countless scientific events and conferences occur around the world each year. Most focus on a particular discipline, technology, or field of research. However, until 2018, no global interdisciplinary event existed to leverage the potential of science and technology for a better tomorrow. There are so many challenges facing society and our planet today which are only likely to be solved with a bolder, more unified approach to scientific discovery. We at Merck recognized a need to help fill this void.

The inaugural Curious 2018 Future Insight Conference, held in Darmstadt, Germany, brought together many of the world's leading scientists and other curious minds that shared Merck's ambition. Over 2 days, dozens of speakers and 1000 participants united in identifying and helping enable solutions where science and technology could address current and emerging needs for society. The success of the Conference saw it quickly gain a reputation as a leading international gathering on the future of science and technology.

To reflect the spirit of the Conference and commemorate the 350th anniversary of Merck in 2018, we announced our intention to award a prize of up to one million euros annually for pioneering innovation over the next 35 years. The Future Insight Prize would recognize critical scientific breakthroughs in health, nutrition, and energy. Unlike other notable scientific awards, such as the Nobel Prize, which rewards outstanding past achievements, we wanted the Future Insight Prize to be forward-looking. The Future Insight Prize seeks to identify promising scientific champions early and provide funding and recognition to help bring groundbreaking innovations to market faster.

The first focus area selected for our Future Insight Prize was pandemic preparedness. Declaring this as the topic 18 months before the arrival of the COVID-19 pandemic demonstrated the visionary nature of the Prize and its intention to address some of the most significant challenges facing humanity.

The inaugural 2019 recipients for pandemic preparedness were Dr. Pardis Sabeti of the Broad Institute of M.I.T. and Harvard and Dr. James Crowe of Vanderbilt University Medical Center. The Prize helped them accelerate essential areas of their scientific research, such as the development of diagnostic assays and technologies for genomic sequencing, testing, and surveillance of pathogens. Both winners went on to make significant contributions during the global response to the COVID-19 pandemic.

While the pandemic made it challenging to come together in person, we continued to award the Future Insight Prize virtually. In 2020, Stephan Sieber from the Technical University of Munich received it for his research to prevent another potential health crisis—the growing emergence of multidrug-resistant bacteria. And in 2021, it was shared by Ting Lu from the University of Illinois and Steve Techtmann from Michigan Technological University to create technologies that convert plastic waste into edible food.

In July 2022, Merck hosted the second Curious—Future Insight Conference. More than 2500 participants from around the world joined over 70 speakers, including 8 Nobel laureates, to share insights, discuss, and run workshops. Each part of the agenda reviewed how the latest science and emerging technologies can help us better address unmet global needs in human health, sustainable nutrition, synthetic biology, smart materials, renewable energy, digitalization, and mobility.

We also announced the winner of the 2022 Future Insight Prize in the category of CO2 conversion. Tobias Erb from the Max Planck Institute for Terrestrial Microbiology in Germany received the Prize for his impressive efforts to re-engineer photosynthesis to develop more efficient solutions to capture and convert carbon dioxide into a range of sustainable products such as renewable fuels.

The 2023 Future Insight Prize will focus on developing an early warning system to prevent future pandemics. Our goal will be to accelerate the development of a technology or process that can constantly monitor all known viruses and bacteria circulating worldwide, including mutations of concern, to detect potential threats before it becomes too late.

The third Curious Future Insight Conference is scheduled to occur in Germany from July 10 to 11, 2024. To reflect the growing international stature of the Conference, and its support by many leading organizations, the 2024 event will be organized by a newly formed non-for-profit association called Future Insight e.V. Merck is pleased to be the founding member. We invite other public and private organizations that share our commitment to accelerating the impact of science and technology to join us in preparing for this important event.

1.1 Accelerating the Speed, Impact, and Resilience of Science

The Conference and the Future Insight Prize are only two examples of how we at Merck live up to our purpose of being curious minds that advance human progress. In 2020, we decided to direct this curiosity toward the global health of scientific research itself. Merck commissioned a report to examine the state of research today. It also reviewed options for how the scientific community can further boost productivity moving forward. A full analysis of existing literature was conducted together with novel econometric analysis. Dozens of leading scientists from around the world were also interviewed, while thousands of others were surveyed.

When the final report was published in 2021, it found that productivity levels across various scientific fields and regions varied based on a range of factors, including scientific complexity, the level of pressure to publish, sources of funding, and how parties choose to collaborate or outsource. However, the report identified several structural issues that represent challenges to future research productivity.

Overall, 85% of scientists surveyed found research was becoming increasingly complex. With scientific teams increasingly being larger in size and representing a broad array of specializations and different partners, such complexity could serve as a drag on productivity. Furthermore, 74% said a trend toward shorter funding cycles discouraged opportunities for research in unexplored areas. To become more resilient, the global scientific community must move beyond short-termism to support more long-term projects of significant importance to society's future health and well-being. The report also underlined the need for private and private organizations to build long-term partnerships.

How the international scientific community came together from the early days of the COVID-19 pandemic onward represents a viable framework for our way forward. By collaborating in a spirit of open innovation across borders and disciplines, groundbreaking solutions were developed, scaled up, approved, and supplied quickly and effectively.

Such pioneering achievements during the pandemic should give us confidence that the scientific community can build upon this recent momentum to help achieve other important societal goals. Indeed, the public increasingly expects science and scientists to lead the way. Global indexes for trust indicate that scientists are now the most trusted societal group [2].

Approximately 90% of people also say they trust science and expect the scientific community to deliver solutions that can minimize the effects of climate change and make the world more sustainable. A majority also expect science to provide cures and better treatments for chronic diseases and cancer while improving access to quality healthcare worldwide [3].

It is the responsibility of the global scientific community to earn this societal trust by pushing the boundaries of science and technology in ways that help people live longer, healthier, and more sustainably than ever before. New and emerging healthcare and biotech technologies put us in a solid position to help prevent and predict disease and personalize medicines. Furthermore, data and digital technologies point to a future of instant access to limitless information, where we have the potential to interconnect anything with anyone.

Merck is well positioned to make an important contribution toward fulfilling these and other societal aspirations by operating not only within but across its business sectors as a diversified science and technology leader.

1.2 Bioconvergence: A Multidisciplinary Approach to Scientific Discovery

The history of modern science can be characterized by a silo-based approach to research, with disciplines such as physics or biotechnology operating mainly in isolation. However, if scientific research continues to operate largely in separate siloes over the coming decades, it may increase the risk of technological stagnation at a time when transformational breakthroughs are vital. For example, today's generation of computers may reach capacity in processing power and energy consumption. Traditional healthcare models have also struggled to meet industry demand for patient care solutions that are more personalized, preventative, and predictive.

Bioconvergence represents a significant opportunity to generate a new wave of scientific breakthroughs that help ensure innovation remains a powerful engine for societal growth. It seeks to bring together a mix of technologies, processes, and expertise across digital, biotech, and material science fields. Success requires aligning these diverse competencies under multidisciplinary projects to generate novel functionality or application outcomes that can disrupt existing markets or create new ones.

Merck is well-placed to help realize such disruptive opportunities given our expertise and capabilities across three complementary business sectors, plus our strong network of partnerships. We are pleased to provide the following examples of how bioconvergence is already helping Merck and our partners to transform healthcare.

While the megatrend of bioconvergence represents a significant opportunity to redefine value chains across many industries, participants must accept the ethical and regulatory responsibilities that come with using such a broad mix of technologies in previously unexplored ways. Here, Merck is once again demonstrating our accountable, transparent leadership. First, we have adopted a Code of Digital Ethics [4] that lists the core principles guiding our activities and decision-making processes. And we are forming joint panels of independent experts with different perspectives across biotech and digital to help us determine the right path forward.

1.3 Accelerating Our Science and Technology Leadership

Our company adopted a new long-term ambition in 2021 to further accelerate our global science and technology leadership. Looking forward, we aspire to be the world's leading science and technology pioneer of the twenty-first century. We intend to fulfill this ambition through bold investments, organizational realignments, and strategic partnerships that concentrate growth in high-priority areas such as process solutions, new healthcare products, and semiconductor solutions.

To develop pioneering solutions that have a positive societal impact and foster organic growth, we continuously review many transformative technologies beyond our existing portfolio. In parallel, we constantly explore how synergies within and across our business sectors can generate additional value for our business and stakeholders. These efforts are conducted via internal incubation, partnerships, strategic investments, and collaborations with academia and other research organizations under an open innovation mindset.

As a values-based company with a multi-generational mindset, Merck will remain accountable for being a global leader that addresses long-term challenges to human health and well-being. Sustainability is one core area of focus where Merck is well positioned to create significant value for customers, as well as current and future generations of society.

From an internal business perspective, we introduced three new strategic goals for sustainability in 2021. First, we will advance human progress for more than one billion people globally through sustainable science and technology by 2030. Second, we will integrate sustainability into all our value chains by 2030. And third, we will achieve climate neutrality and significantly reduce our resource consumption by 2040. We remain on track to achieve these targets [5].

Furthermore, we believe we can amplify our impact and extend our competitive advantage as a sustainable innovator providing differentiated solutions that enable customers to meet or beat their own targets. It's our goal to integrate sustainability into every part of the value chain, from R&D to commercial supply, across every market in which we participate. In Life Science and its portfolio of 300,000 products, for example, our SMASH strategy is helping to reduce packaging, achieve zero deforestation, shift to sustainable plastics, and maximize recycling. And in Electronics, sustainable innovations include a new generation of greener solvent alternatives and the development of gas solutions with low global warming potential (G.M.P.) for semiconductors.

1.4 Our Ongoing Commitment to Open Innovation

Today, Merck is a globally diversified science and technology company with leading positions across life science, healthcare, and electronics. Despite such strong internal competencies and momentum, it is essential that we continue to foster long-term relationships with other industry leaders, start-ups, and academia that share our purpose and values.

Our long-standing commitment to open innovation is reflected in how we are addressing unmet global health challenges affecting millions of people around the world. An open innovation framework is helping us to develop innovative treatments for infectious diseases such as malaria. Here, we share access to our proprietary compound library for drug discovery activities to identify new potentially life-changing drugs.

We also engage non-profit organizations and academia in developed and developing countries to improve the health of underserved populations in low- and middle-income countries. One such collaboration is with the World Health Organization to treat schistosomiasis in endemic African countries. As of 2022, we had provided 1.5 billion tablets to WHO to support this initiative. The Lancet confirmed these efforts had reduced the prevalence of this neglected tropical disease by 60% between 2000 and 2019. Through such open innovation collaborations, we remain on track to eliminate this disease as a public health problem by 2030 [6].

Below, we are pleased to provide additional examples of where Merck is accelerating its leadership as a twenty-first-century pioneer for science and technology.

2 Building on a History of Success to Drive Further Innovation and Market Within the Life Science Arena

From 1890 onward, Merck has manufactured lactic acid biotechnologically, with the process taking place at an industrial scale in a dedicated "fermentation site" [7]. This feat marks an early record of our pioneering role in biotechnology. Our historic successes since then, and even pre-dating this time, have resulted from ongoing innovation leadership within Merck and within companies that have become part of the Merck portfolio over the years. That rich history has built on itself to strengthen our innovation, including manufacturing the first highly pure ATP (adenosine triphosphate)—a critical biologic "energy" molecule [8]; creating the first biosafety protocols for polio vaccines in the early 1950s [9]; and launching the first single-use bioreactor [10].

In 2015, Merck solidified itself as a life science powerhouse through the acquisition of Sigma-Aldrich, creating a laboratory supply leader offering over 300,000 products to global customers in research, pharmaceutical manufacturing, and diagnostic and testing labs. This followed the acquisition of Millipore Corporation in 2010, which pioneered the use of membrane technologies in hundreds of applications, importantly in bioprocessing [11]. To position ourselves for continued innovation and success, we reorganized our Life Science business sector in early 2022 to better serve the evolving needs of our global customers. Since then, we have continued taking steps to realize our vision of a world in which our innovative products, services, and digital offerings positively impact life and health through science.

In working to realize this vision, we are helping to transform the life science landscape on multiple fronts. Next-generation monoclonal antibodies (mAbs), cell and gene therapies (CGTs), and mRNA-based therapies are only a few of the many areas in which we are helping to overcome twenty-first-century healthcare challenges. However, these areas exemplify our ability to innovate, streamline, and simplify while enabling the highest-quality therapies with accelerated times to market.

2.1 At the Forefront of the Therapeutic Antibody Revolution

Since the approval of the first mAb therapy (Orthoclone OKT3®) in 1986, mAbs have transformed the lives of patients with diverse cancers, a wide variety of autoimmune diseases, and a growing number of more common diseases, such as age-related macular degeneration. mAb-based therapies are designed to bind to individual proteins on the cell surface or in the circulatory system with incredibly high specificity. This binding can destroy disease-causing proteins, inactivate disease-related pathways, or catalyze the destruction of diseased cells. More than 100 mAb therapies have already been approved by the US Food and Drug Administration, and nearly 600 clinical trials of mAbs are ongoing [12]. This robust, growing portfolio of commercial mAb therapies is the result of continued innovation in target discovery and next-generation antibody technologies (Fig. 1).

In the three-plus decades since the approval of Orthoclone OKT3®, a great deal of effort, investment, and innovation has been focused on establishing reliable, robust platform processes for manufacturing mAb therapies. Today's key challenge in bringing the benefits of mAb therapies to the patients who need them is to continue optimizing product quantity and quality while accelerating time to market. Overcoming this challenge requires a modular approach to mAb manufacturing, utilizing templates that eliminate the need to recreate the production process for each new mAb.

Today, Merck's antibody templates support the rapid development of new mAb expression constructs, radically reducing the time and cost of producing mAbs for clinical trials and commercial use. We innovate along the entire process, from cell line engineering and media, to single use systems, intensified processing, automation, and downstream processes (Fig. 2).

Our goal is to catalyze significant advances in efficiency, purity, scale, and cost for producing these lifesaving therapies by streamlining and simplifying the entire mAb manufacturing continuum. Our BioContinuum Platform™ was designed with exactly this in mind, and, as part of the platform, our Bio4C™ Software Suite comprises multiple products and services, including our in-process analytical

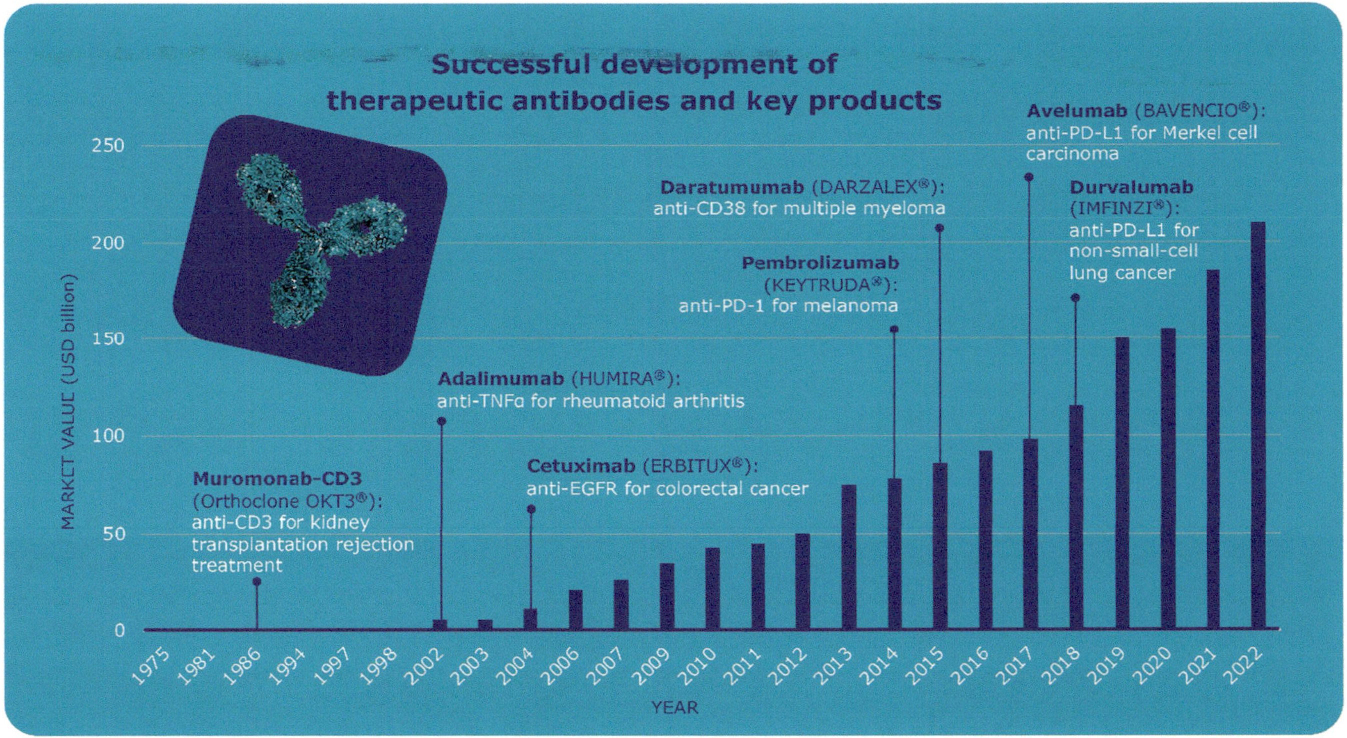

Fig. 1 The explosion of mAbs as therapies, showing market growth and key product highlights over time (Data combined from: https://doi.org/10.1186/s12929-019-0592-z; https://www.antibodysociety.org/resources/approved-antibodies/; https://www.grandviewresearch.com/industry-analysis/monoclonal-antibodies-market; various industry reports)

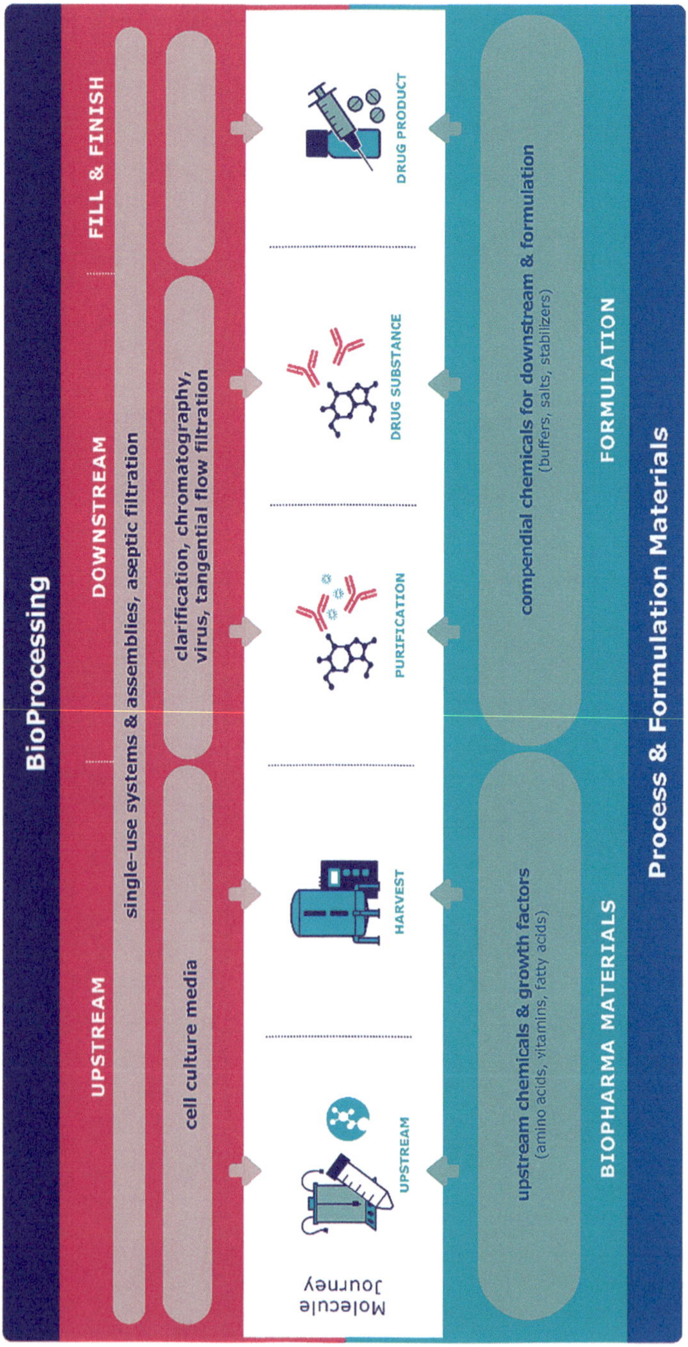

Fig. 2 Addressing the full value chain, from upstream and downstream processes to drug product

technologies (PAT) Portfolio, which brings the capabilities of the quality control (QC) lab to the manufacturing floor on the path to real-time release. Collectively, these offerings help our customers produce the highest quality mAbs and biologic molecules possible. Our continued innovation in intensification, perfusion, and digitalization of mAb manufacturing with breakthrough products and technologies are catalyzing significant advances.

Examples of our advanced mAb manufacturing products include the Eshmuno® CP-FT chromatography resin, a first-in-class technology that simplifies the purification of mAbs; our Cellicon™ Cell Retention Device and Cell Retention System, which provides an integrated solution for perfusion bioreactors run as intensified batch or intensified seed train, and serves as the beginning of a platform of products that will scale to 2000 liters and provide continuous production; and the Breez™ micro-bioreactor platform technology, which we acquired from Erbi Biosystems in 2022 to complete our full-range offering of bioreactors [13], from two milliliters to 2000 liters, with rapid lab-scale process development. Together with the rest of our portfolio, these products help to reduce the time, cost, and complexity of delivering safe and effective therapies to patients.

As we work to enable more efficient mAb manufacturing that meets the needs of patients, drug developers, and regulatory authorities, we continue to support innovation in mAb technology. Our industry-leading capabilities in mAbs and small molecule drugs have provided a natural pathway for Merck to enable new antibody-drug conjugate (ADC) therapies that are further realizing the therapeutic potential of mAb technology. As "magic bullets" that precisely deliver high-potency cell-killing agents to cancer cells while protecting healthy tissue, ADCs have already demonstrated their ability to improve patient outcomes in cancer indications with previously unmet needs. Continued ADC innovation offers the promise of effectively and more safely treating additional cancers for which outcomes remain suboptimal.

A key ADC-enabling focus for Merck is innovating novel linkers that release cell-killing drugs in targeted cells, at the right time and at an optimal dose. We also support our customers' innovations with critical products and services, including novel payloads, highly efficient chemical processes and synthesis methods, highly potent active pharmaceutical ingredients (HP-APIs), as well as reagents and services for ADC research and development. We have leveraged our services and products to advance multiple ADCs within the healthcare portfolio of Merck, including M9140, an anti-CEACAM5 ADC, and M1231, a bispecific ADC targeting MUC1xEGFR, both of which are being evaluated in Phase 1a clinical trials in patients with solid tumors.

Many other ADCs in development throughout the biopharmaceutical industry utilize Merck innovations, and we continue to take a templated approach to our enabling technologies to ensure that manufacturing processes remain as simple as possible, even as therapeutic complexity increases. This is also fundamental to our approach to innovation across other emerging therapeutic modalities.

2.2 Enabling Cell and Gene Therapy Innovation

Cell and gene therapies (CGTs) were initially envisioned as transformative therapies for rare diseases that resulted from the dysfunction of a single gene. They have achieved this goal in a growing number of disease indications, including blood and immune system disorders, neuromuscular disease, and inherited retinal disease. In these indications, relatively small quantities of viral vectors have been sufficient to achieve therapeutic benefits, which have also been sufficient to support the early clinical development of CGTs in other indications. However, as CGTs show promise in a growing number of more common indications, including cancer, age-related macular degeneration, and Parkinson's disease, there is an urgent need for increased viral vector manufacturing capacity to support clinical development and commercialization.

As with mAb manufacturing, templated approaches are essential for meeting viral vector production demand. We are applying the expertise amassed through our ongoing mAb innovation efforts to advance the discovery and development of novel gene therapies. Our approach leverages our established capabilities in viral vector templates to enable novel cell lines, cell culture media, and filtration technologies that will support innovation in the development and manufacture of viral gene therapies (VGTs) while streamlining, simplifying, and speeding-up their production.

In addition to supporting our customers' VGT innovation, we also are a leading VGT contract manufacturer. Our commitment to meeting this customer demand is reflected in the €100 million expansion of our Carlsbad, California, US facility for viral vector manufacturing to support large-scale commercial and industrial manufacturing for viral and gene therapy. As the second expansion of this contract testing, development, and manufacturing organization (CTDMO) facility, the investment doubles our VGT capacity [14].

Our VirusExpress® 293 Adeno-Associated Virus (AAV) Production Platform positioned Merck as one of the first CTDMOs and technology developers to provide a full viral vector manufacturing offering, including AAV, Lentiviral vectors, CTDMO, and process development services and solutions [15]. This platform enables biopharmaceutical companies to increase the speed-to-clinical manufacturing of VGTs while reducing their process development time and costs.

2.3 Finding the Silver Lining of the COVID-19 Pandemic to Leapfrog Our mRNA Capabilities

While it was not the origin of our mRNA offerings, the COVID-19 pandemic necessitated a rapid acceleration of our capabilities, products, and services in this space to support the development of mRNA vaccines. mRNA-based therapies are intended to promote or reduce the expression of disease-related proteins. Well before COVID-19 was the subject of daily headlines in news outlets around the world, we had established a robust suite of products and services to enable these therapies. For mRNA vaccines, the goal is to express proteins that stimulate the immune system to provide protection against a specific infectious agent. As history has shown, worldwide crises accelerate innovation, and we were well positioned in these areas to innovate rapidly in response to customer and global health needs and to build an end-to-end value chain of services and products for mRNA.

We did that by building on our strong foundation and specialization in GMP-scalable processes that yield high-quality lipids for pharmaceutical applications and gene therapy. Lipids are the most commonly used delivery method for RNA therapeutics and vaccines, and lipid nanoparticles (LNPs) played a game-changing role in the COVID-19 mRNA vaccines and have the potential to be applied in many other disease areas. Amid the urgency of the pandemic, we leveraged our portfolio, which includes all the raw materials required for mRNA LNP formulation development, to provide BioNTech and other key players with the lipids that enabled their life-saving COVID-19 vaccines [16]. With a truly innovative approach, we produced these lipids using a wholly novel method of producing synthetic cholesterol that provided a simple way to manufacture this critical raw material used to encapsulate and protect the mRNA contained in these vaccines (Fig. 3).

Unleashing the full power of this novel technology requires robust, consistent processes to make, purify, and then formulate the mRNA. Further, realizing the potential of mRNA-based therapies beyond vaccines and infectious diseases will require innovating novel nucleic acid design and manufacturing capabilities, as well as formulations. Integrating mRNA design and delivery strategies is essential for supporting the delivery and expression of therapeutic mRNA molecules in target cells while minimizing off-target expression that can lead to adverse effects. With this need—and our ability to meet it—established, we bolstered our offerings to be highly specialized in manufacturing synthetic lipids and mRNA for pharmaceutical applications. We also made additional investments in our existing RNA assets, including a €130 million expansion at our site in Molsheim, France, which strengthened our manufacturing capabilities for single-use assemblies, a key technology to produce COVID-19 vaccines and other lifesaving therapies [17].

As part of our commitment to furthering our mRNA offerings, we took the bold step to expand support across the full value chain, advancing our services with strategic investments in new capabilities and assets that enable this novel therapeutic class. We acquired AmpTec, a leading mRNA CDMO, in early 2021 to expand our mRNA capabilities for vaccines, therapeutics, and diagnostics [18]. Their differentiated polymerase chain reaction (PCR)-based technology has shown to have advantages over other technologies for mRNA manufacturing, and its diagnostic business, which is focused on producing customized long RNAs and DNAs for in vitro diagnostics, complemented our existing portfolio. Importantly, we are also leveraging this acquired expertise to develop analytical capabilities for mRNA and a sequencing panel for QC of mRNA, both of which should help address the lack of release criteria for mRNA used in therapeutics.

We also acquired Exelead in 2022, a full-service CDMO focusing on LNP formulations, enhancing our mRNA and lipid manufacturing capabilities and enabling us to provide customers with unique and integrated solutions across the

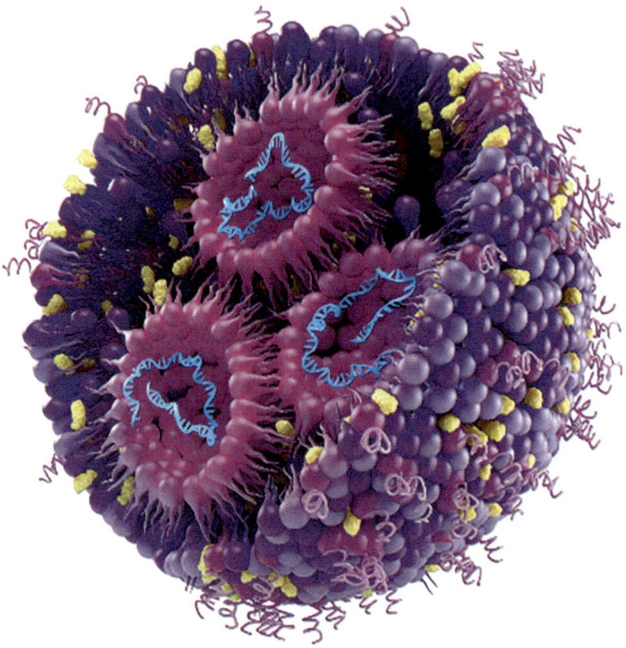

Fig. 3 Diagram of a lipid nanoparticle

mRNA value chain [19]. Merck will continue innovating and investing in resources and technologies that our customers need to realize the potential of RNA-based medicines across diverse indications with unmet needs.

2.4 Ensuring Quality Products Across Multiple Therapeutic Modalities

Regardless of how well any given therapy is designed and optimized, its ability to treat disease safely and effectively is heavily dependent on manufacturing processes that yield potent, highly pure, and high-quality products at the correct specifications.

We continue to expand our portfolio of novel microbiology and release testing products and services to support VGT manufacturing. Our Blazar® platform, which received R&D 100 and CPhI Awards in 2020 for reducing testing times by 80%, provides researchers with accurate and highly sensitive viral detection in just days compared with months [20, 21]. Ensuring the removal of replication-competent viruses used in gene and cell therapy manufacturing is essential for producing safe and effective therapies.

Our portfolio of products and services also supports new approaches for analytical development for AAV and other viral vector types, and next-generation sequencing approaches that enable rapid and efficient genetic stability testing and detection of adventitious agents (such as viruses introduced via raw materials and helper viruses that are used to manufacture VGTs). Our contract testing services are also changing the paradigm for evaluating the safety and efficacy of ADCs and other next-generation mAb-based therapies, including bi-specific and multi-specific mAbs and proteolysis-targeting chimeras (PROTACs).

2.5 Committed to Meeting Our Customers' Twenty-First-Century Life Science Challenges

We recognize that our continued ability to innovate—and to support our customers in their own groundbreaking work—requires organizing our business to address the challenges of today, strengthen our resilience, and capitalize on unmet market needs or emerging opportunities. In early 2022, we rolled out a differentiated operating model that positions us even more firmly as an enabling partner for customers across their value chains [22]. With our business units of Life Science Services, Process Solutions, and Science and Lab Solutions, as well as a dedicated Integrated Supply Chain Operations organization, we provide our customers with broad, technical expertise across all aspects of the life science industry. In parallel, we can maintain the focused customer service and tailored offerings that allow us to meet each customer's holistic needs.

Recognizing that biologics, such as next-generation mAbs and gene and cell therapies, will only continue to take an increasing role in the treatment of diverse diseases, we are also making significant investments in building out our capacity to support the development of these important new products. This includes the opening of a viral clearance (VC) laboratory as part of the first building phase of our new €29 million China Biologics Testing Center, meeting the double-digit increase in demand for VC testing services in China [23]; the opening of our new facility in Martillac, France, to boost our CTDMO capacity and services [24]; and our €290 million investment in biosafety testing capacity at our Rockville, Maryland, US facility [25].

We also continue to enhance our resources to support small-molecule drug development, with the opening of a new €59 million expansion of our facility near Madison, Wisconsin, US, that doubles our production capacity for HP-APIs, making us one of the largest manufacturers globally for these compounds that play a critical role in cancer therapies [26]. With all these investments, our continued focus is on geographic distribution, diversifying, and de-risking our supply chain as we ensure our footprint brings us to be closer to our customers around the world.

Addressing the challenges of today and preparing to innovate solutions for tomorrow, we have established a Technology Office and Technology Pioneering Group to guide our continued investment in the future technologies that will drive life science growth and innovation. We also have launched an "entrepreneurs in residence" program to explore innovative technologies with cross-sector teams, bringing together the expertise and resources of Merck's Life Science and

Healthcare business sectors to enable our customers to ask and answer wholly new questions in biology and medicine.

One of these projects, which may also be extended to include Merck's Electronics business sector, is developing a platform for Artificial Intelligence for drug discovery (AIDD) that will help significantly accelerate the discovery of new and better drug candidates and reduce the timelines for delivering new therapies. Another is focused on cell-specific mRNA delivery and expanded use of RNA-based medicines. Enabled by the extensive Merck expertise and capabilities in mRNA synthesis, LNP synthesis, formulation, and targeted delivery, this project will leverage AI to develop "smarter" LNPs that can more effectively target different tissue types, including difficult-to-reach biological targets in various disease areas. We expect the development of these enhanced mRNA LNP delivery platforms will open new treatment possibilities for patients relevant to our Healthcare sector and our Life Science pharma and biotech customers.

The common thread that unifies all the activities within our Life Science business is our ongoing commitment to meet the evolving needs of the scientific community and to ensure that we are prepared for whatever the future may hold. The life science industry is highly dynamic. Enabling our customers' continued growth is essential for maintaining our ability to impact life and health with science. We have the resources, expertise, and capabilities to succeed in enabling novel therapeutic modalities—and we have the power to make a positive impact on today's health needs as we support future generations of patients, enabling lifesaving therapies and vaccines that will continue to drive innovation leadership throughout the twenty-first century and beyond.

3 Pioneering Twenty-First-Century Science and Technology Pharmaceutical Solutions Along the Entire "Circle of Life"

For over 350 years, patients have been at the heart of everything we do in healthcare. It drives the discoveries and the technologies we create—and it forms the basis of our future growth. We are proud of our heritage as the world's oldest pharmaceutical and chemical company [27].

Today, we serve patients in around 150 countries, making us one of the largest mid-size companies in the pharmaceutical industry. As of 2022, more than 90 million people per day receive our medicines.

Our Healthcare business follows this guiding principle: Be curious and innovate boldly. Our north star which guides everything we do in healthcare is to help create, improve, and prolong lives—as one for patients.

In the Healthcare sector, our strategy is to grow as a Global Specialty Innovator. Global because we continue to develop and deliver our therapies around the world. Specialty because we solve high unmet medical needs with specialty medicines focused on complex or rare chronic conditions. And Innovator, because we are on the frontlines to develop groundbreaking solutions that offer new hope to patients.

3.1 Our Therapeutic Areas

Within Healthcare, we take a focused leadership approach, which means we are not distracted by high-risk, high-return rewards that are not in our sweet spot. We commit to the key opportunities that address high unmet medical needs and build on our established strong positions. Our work sits within four therapeutic areas.

Oncology, which transforms the standard of care for difficult to treat cancers and unleashes the full potential of some of the most promising approaches in cancer research, specifically around DNA damage. This is focused leadership in action. We leverage our expertise to deliver lifesaving medicines to improve and prolong lives.

Neurology and immunology, which help to improve people's lives with autoimmune and neuroinflammatory conditions. With more than 20 years of experience in multiple sclerosis (MS) care, we remain committed to finding solutions for chronic progressive diseases. With our global footprint and unique position, we are illustrating what focused leadership looks like.

Fertility, which helps to achieve the dream of parenthood through innovative treatment options. As a global leader in fertility, our end-to-end drug portfolio provides treatment solutions for all phases of the in vitro fertilization (IVF) procedure, complemented by innovative, digital solutions to further improve the patient experience along the treatment journey as well as outcomes. Our aspiration to drive innovation from a strong position represents another great example of focused leadership. Today, we are proud to say that over five million babies have been born with the support of our products.

Cardiovascular, metabolism, and endocrinology (CM&E) brings better health through high-quality, affordable medicines. We offer solutions for treatments in diabetes, cardiovascular, thyroid, and endocrinology. This part of our portfolio has a significant impact on patients across the globe and serves as a resilient growth driver complementing our specialty business. We are proud to say that we are a global market leader in several of our therapeutic areas in CM&E and are one of the strongest companies in established portfolios in key growth markets such as China. We provide masterful life cycle management.

3.2 How Our Healthcare Business Strategy Keeps Innovation Alive

Given continuous market uncertainties, increasing competition, and other dynamics, it is increasingly challenging to bring new therapies to market. This, paired with ongoing healthcare reforms, rising inflation, geographical uncertainties, and increased digitalization, shows that our industry is in a constant state of change.

At Merck, we are in an attractive position to take advantage of these macroeconomic shifts. With some of our recent launches, we established our position as a Global Specialty Innovator. Our focused leadership approach will help us grow further and increase stability, resilience, and predictability for our Healthcare business. Likewise, by leveraging our strengths within Life Sciences and Electronics, we show how bioconvergence can deliver new hope to patients and bring the best of Merck Group to market. Two innovations that we've developed cross sector include:

3.2.1 Bioelectronics for Targeted Treatment of Disease

Bioelectronic devices have the potential to not only stimulate but also monitor the disease condition by combining nerve signals with other accessible physiological datasets to create a holistic understanding. Bioelectronic devices show great promise in helping to improve therapeutic outcomes and efficiency for patients with chronic inflammatory diseases. By combining our expertise across electronics, medicines, and drug delivery with the neurostimulation technologies of our external partners, we aim to create novel modalities to enhance the quality of care for patients in several chronic disease areas.

3.2.2 ADCs Driving Innovation in the Fight Against Cancer

ADCs (antibody drug conjugates) unlock immense potential in the treatment of tumor cells by combining the targeting capabilities of monoclonal antibodies with the cancer-killing ability of cytotoxic drug payloads. We have a strategic partnership with Mersana Therapeutics to leverage their proprietary Immunosynthen STING-agonist ADC platform, which partnered with our deep expertise and portfolio of two clinical and nine preclinical assets in the ADC space. It allows us to focus on the discovery of next-generation state-of-the-art ADC drugs. These cross-sector collaborations show how Merck Group is playing to its strengths to create new solutions for patients and the market.

3.3 Our Healthcare Investment in Innovation

Coming off the COVID-19 pandemic, there were many lessons learned to increase speed in study design, setup, and execution. To bring more medicines to more patients faster, we recognized we needed to get very clear on how we manage innovation. Two of our main focus areas relate to how we are rethinking our R&D function to enable faster innovation, and leveraging the latest AI and machine learning solutions to drive new levels of innovation.

3.3.1 Rethinking Our R&D Function to Enable Faster Innovation

Our organic R&D engine is vital to delivering new therapies to the market. However, if we want to further outperform the market in the future, we must increase our rate and speed. We have a goal to double our productivity in R&D and introduce one new product or major indication every 1.5 years on average.

To achieve long-term sustainable growth with this approach, we realize that while our R&D engine is strong, more is needed to sustain our business as a Global Specialty Innovator in the long run.

We will focus our expertise and capabilities and leverage synergies within our existing pipeline to deliver transformative medicines in oncology, neurology, and immunology, augmented by an increased focus on external innovation. This will see us focus on closing pipeline gaps and continuously complement the portfolio via inorganic additions which include, but are not limited to, commercial and late-stage development assets.

The collaboration and commercial license agreement with Mersana Therapeutics to discover novel antibody-drug conjugates (ADCs) marked an important milestone. The option agreement with Nerviano Medical Science for the next generation of PARP inhibitors will be explored to further enhance our portfolio. Our aim is to have more than 50% of new launches result from external co-development partnerships and strategic in-licensing of assets for further in-house development.

With our recent specialty launches, we have made a significant contribution to patients and continue to pursue life-cycle-management opportunities within our established portfolio. We are excited about our pipeline with multiple potential first-in-class or best-in-class assets as well as our highly innovative earlier-stage portfolio.

3.3.2 AI and Machine Learning to Drive New Levels of Innovation

We're making extensive use of AI, machine learning, sensors, and analytics to automate large parts of our supply chain. Our aim is to leverage digital solutions across our entire value chain.

AI in Research

Predictive models enabled by AI are central to our work. In the early prediction of a compound's pharmaceutical properties and in small molecule drug discovery projects, our scientists use AI applications to identify—from millions and millions of drug candidates—the right molecules to target several diseases.

The same applies to navigating the sheer size of the libraries used to screen for new drug candidates. It's now practically impossible for individual researchers to review everything themselves—which is where AI and machine learning can help. Using this technology, we can allow researchers to extract hidden insights from huge datasets. This allows us to predict the properties of a potential compound—meaning that only compounds with desired properties are chosen for synthesis. This saves time and money by preventing work on compounds that are unlikely to be effective.

AI and machine learning in research also support the generation of ideas for entirely novel compounds—where the "invented" molecule is predicted to have all the desired properties required for success. This alone could hugely accelerate the discovery of new effective drugs.

These are just a few of the benefits in the early stages of the drug discovery pipeline, along with countless hours of manual labor that can now be channeled into other innovation activities.

AI in Drug Discovery and Development

In drug discovery and development, the potential of AI applications lies in the optimization of clinical trial designs to make trials faster, smarter, and more ethical.

The Innovative Clinical Trial team is using advanced statistical modeling to find ways to speed up the enrollment process and reduce the amount of trial supplies shipped throughout the world, while also increasing the probability of success. AI can also reduce clinical trial cycle times while improving the costs of productivity and outcomes of clinical development.

To improve the diversity of clinical trials in the healthcare industry, we all need to better reflect the full demographic of patients that will use our drugs. In particular, black, Hispanic, and Asian populations are systematically underrepresented within clinical studies. To address this, the team analyzed the trial data, conducted an in-depth DE&I audit, and placed greater emphasis on clinical trial transparency by providing easier access to clinical trial results and increased utilization of clinical trial data. Our innovative approaches are solving for more than just efficiency and quality—they also improve diversity and inclusion.

AI in Supply

Supply chains in healthcare are incredibly complex, usually involving several independent stakeholders—manufacturers, wholesalers, hospitals, healthcare providers, group purchasing organizations, and many regulatory agencies.

A certain level of automation is already a feature of most supply chain models, with complex algorithms predicting demand and aiding forecasting. However, in recent years, these systems have still required human intervention. As AI and machine learning become increasingly accessible, there is a significant opportunity to create a system that is entirely "self-driven." This move toward greater automation also comes in response to the changing world of healthcare and increased digitization in our daily lives.

The healthcare sector is witnessing a dramatic change in the way goods and services are delivered. Key drivers include a move away from treating shorter episodes of illness toward a greater focus on longer-term wellness and prevention, as well as the changing expectations of patients and consumers.

With the help of machine learning, deep learning, and neuronal networks, we will be able to use our data in an advanced manner to ultimately increase the accuracy of our forecasting and enable real-time decision-making to react to last-minute changes in demands. A digital pharmaceutical supply chain provides real-time data to increase full transparency to visualize and analyze the end-to-end performance along the entire supply value chain, which includes all analytical testing steps, the manufacturing, and packaging of the drug, and its distribution.

3.3.3 The Future of Digital Health and Patient Care

As personalized medicine, online prescribing, more personal health trackers, and sensors come onto the marketplace, the only way to meet changing patient demands is to invest in automation and advanced capabilities in digitization and data. Precision medicine is also becoming widely adopted by patients, as scientists are using highly advanced image processing and omics technologies combined with digital biosensors and mobile fitness and wellness equipment. This enables them to collect physiological and behavioral data on a large scale. This huge data volume can be combined and analyzed using AI and machine learning. In the future, these will help to identify subtle yet measurable indicators of diseases. Below are a few examples of how we are advancing digital health and precision medicine:

Growth Hormone Management with Easypod

Poor adherence to long-term recombinant human growth hormone (r-hGH) treatment can lead to suboptimal clinical outcomes; consequently, supporting and monitoring adherence is a crucial part of patient management. 79% of the specialists involved in growth hormone treatment said that inadequate adherence was one of the main reasons why treatments were not completely successful. But with the help of the injection data stored in the Easypod digital health monitoring device, doctors and their young patients can now observe how the treatment is progressing and, if necessary, find ways in which to ensure the daily injection is never missed. In addition, successful treatments mean cost savings for the healthcare system.

Fertility LifeLines Portal to Support Consumers with Rising Costs

Insurance often won't fully cover fertility treatment, which means people seeking help to start a family can potentially be on the hook for thousands or tens of thousands of dollars out of their pockets. We rolled out a digital tool to help with this issue. Patients can visit the Fertility LifeLines portal, either via a webpage or mobile app, to quickly find out whether they're likely to qualify for financial assistance through the company's medication savings programs. The tool also directs users to other support resources, including the option to connect with a live staff person.

Multiple Sclerosis Leadership and Innovation Network "MS Link" for Better Outcomes

We have been supporting multiple sclerosis (MS) care for more than 20 years. Our investment in the MS Link program (Multiple Sclerosis Leadership and Innovation Network)—a collaborative research network—in which we are partnering with the broader MS community to advance MS research, has a common goal to improve patient outcomes. One of the first collaborations we entered was with UT Southwestern, around 3D MRI technology that is used to look at MS lesions. The applied 3D technology is used to visualize the shape and structure of an MS lesion to better assess and provide an accurate prognosis.

Precision medicine is going to change our healthcare fundamentally. In the future, we see provision and treatment strategies that will be perfectly tailored to each patient. Within Merck, we expect important breakthroughs in personalized patient care, digital health, and advanced laboratory in years to come.

4 Material and Process Solutions to Power Future Waves of Electronics Innovation

Advanced chips are becoming ubiquitous in our lives on many levels. They enable the supercomputers in our pockets known as smartphones, the server farms that return our Internet search results, and artificial intelligence applications such as

AlphaFold, an AI program that can predict protein secondary structures with an accuracy that rivals x-ray structure analyses. The acceleration of digitization across global industries is fueling an exponential growth of data.

Highly impactful technology trends driving demand for data, such as artificial intelligence (AI), 5G networks, big data, and Internet of Things (IoT), all require more powerful chips as well as advanced liquid crystal, OLED, micro-LED display platforms. This, in turn, is driving accelerating demand for more energy-efficient computing, better AI, more data storage, and faster transfer. However, existing semiconductor technologies are struggling to serve newly emerging needs for energy efficiency, processing capacity, and manufacturing scalability that arise from these developments. Merck is working to provide the material solutions needed to meet this challenge.

The dimensional scaling process, often also referred to as Moore's law, has defined the evolution of semiconductors for decades and has led to an exponential increase in performance. This continuous pursuit of higher integration densities, where the transistor numbers on the most advanced chips now exceed 110 billion, has required innovations in specialized materials and processes. The support of integrated supply partners such as Merck in the development of such material and process solutions is an enabling factor for the electronics industry. The materials required are becoming ever more complex and specialized. Today more than half of the non-radioactive elements in the periodic table can be found in semiconductor chips, sharply up from only a scant double handful in the earlier stages of CMOS technology.

Since 2019, the most advanced chips have been made with EUV, which has made it possible to shrink the skyrocketing number of mask layers resulting from multiple patterning to more manageable numbers.

EUV lithography had been eagerly awaited by the industry for many years but was slow to arrive. The main issue was a dearth of EUV photons: the development of suitably bright EUV sources proved to be elusive until about 2015 when engineering breakthroughs raised source powers to levels above 200 mJ/cm^2 at intermediate focus, a level of power that was sufficient to make the use of EUV in production processes economically feasible. Yet EUV lithography has still to reach its full potential: after many dozens of billions of Euros have been spent on the development of the exposure tool hardware, it is being held back by a materials issue that has seen only a small fraction of this investment, maybe only of the order of less than 100 million. The performance of EUV photoresists has become the limiting factor that determines the overall capability of EUV lithography.

In developing EUV resists, photoresist chemists face a perfect storm of limitations that arise from a combination of the very small features that need to be patterned and the high energy of EUV photons:

In EUV lithography, resist features have become so small that the capillary forces between them can easily exceed their mechanical strength, leading to a collapse of the pattern. Here Merck provides a solution in the form of special rinse liquids with lower surface tension that reduce the capillary forces and preserve pattern integrity. Since these rinses are the last liquid to touch the surface, it is imperative that they spin off cleanly and leave no defects or residues. At the same time, achieving the best possible reduction of the capillary forces requires careful tuning of the surfactants in the rinse liquid in order to prevent other degradations of the resist profile. All aqueous development processes in EUV lithography have to use rinse liquids; however, even with their use, resist film thickness is limited to values at or below 40 nm.

The chemically amplified resists (CARs) that have been so successful in all technology nodes since the early 1990s are now running into a number of fundamental limitations related to problems with diffusion control of the acid catalysts and stochastic effects their low EUV absorbance and molecular scale resist non-homogeneity. As a result, CAR performance has stalled at about 12 nm final line/space resolution on the current NA = 0.33 exposure tools. New chemistries are required to push beyond this resolution limit.

The new type of photoresist that has emerged to meet this requirement is the metal-organic resist (MOR). MORs are not chemically amplified but contain metal atoms that have a high intrinsic absorption at EUV wavelength. MORs have EUV absorptions about four times higher than organic polymer-based resists. This higher absorption allows them to have comparable photospeeds to CARs, although there is no amplification of the initial photoevent. The prototypical MOR is based on tin oxido clusters; tin has one of the highest EUV absorption cross sections in the periodic table, and the football-shaped dodeca-tin cluster contains about 60% tin by weight. This new class of resist has the potential to extend EUV lithography below 10 nm linewidths—it is the only type of resist considered to be able to operate at the even lower film thicknesses of 15–16 nm that will be required for the next generation of EUV exposure tools with numerical apertures of NA = 0.55.

Merck is developing two new platforms for EUV resists: we are working on an organic molecular glass resist targeted primarily as a fast resist for pillar patterns, and we are exploring new chemistries for MOR resists aimed at high resolution applications. However, our ambitions in EUV lithography extend to additional new materials beyond rinse liquids and photoresists.

Advanced chips require highly precise placement of successive exposure patterns. The next layer must be aligned to the edges of the previous one to within less than 2 nanometers. Reduction of this edge placement error (EPE) has become a crucial design criterion for new processes, which has led to the adoption of complex self-alignment schemes in which multiple features are derived from a single lithographic exposure step. These processes often require materials with particular "colors," i.e., specific etch resistance profiles, e.g., a dielectric that will be etched by fluorine plasma but not by chlorine, and another one with the opposite behavior. Merck has developed materials and processes for such "multi-color" materials, which are used in advanced chip processes worldwide.

Self-aligned patterning is the most advanced form of classical lithography in which all the pattern information is still directly derived from the information encoded in the masks used for the lithographic exposures. This approach is generally referred to as "top-down lithography": the information transferred to the chip layers flows directly from the circuit layout and mask design (the "top") down to the wafer in the form of a lithographic "pixel" pattern. With increasing transistor density, this information transfer requires higher bandwidths which need to be provided by the exposure tool. EUV lithography is the latest (and possibly last) implementation of this approach. The high cost of EUV lithography, combined with the need to control line edge roughness and edge placement errors more tightly, has given rise to a new paradigm: bottom-up lithography [28].

In bottom-up lithography, part of the information needed to generate the circuit pattern is encoded in the chemical nature of the materials, thus reducing the load on the bandwidth of the exposure.

Top-down lithography will always be required for long range order and to guide connections, whereas bottom-up approaches excel at self-assembly at nanoscale dimensions. Two main areas in which this concept is being implemented are selective deposition and directed self-assembly (DSA).

In area-selective atomic layer deposition (AS-ALD), deposition precursors are designed to react exclusively with one type of surface present on the chip. For example, a layer might contain adjacent lines of a dielectric (e.g., silicon dioxide) and a metal (e.g., copper). The next step in chip manufacture might require the copper lines (and only the copper lines) to be covered by another dielectric. With top-down lithography, this would require the deposition of a layer of the new dielectric on the entire substrate, followed by an exposure step to define a photoresist etch mask, an etch step to transfer the photoresist pattern into the new dielectric, and a photoresist strip step. The bottom-up area-selective ALD approach instead relies on a precursor specifically designed to only deposit on the copper layer. In successive ALD cycles, the precursor builds up the required film thickness atomic layer by atomic layer. The silicon dioxide dielectric right next to the copper line is left untouched. The end result (a dielectric-covered copper line) is the same as that obtained from the top-down process—only that the bottom-up approach requires only a single step and automatically avoids the introduction of new edge placement errors.

These advantages offered by area-selective ALD (AS-ALD) processes have led to their increasing adoption in IC manufacturing. The next frontier for AS-ALD is the move away from 2D to 3D processing, either for selective deposition on 3D device structures such as the Gate-All-Around (GAA) transistor design scheduled to replace FinFET transistors in Logic, topographically selective ALD in which the 3D shape of a feature determines where deposition occurs, or facet-selective ALD, in which deposition occurs on one facet of a crystalline material but not on another, although their chemical composition is identical [29].

At first glance, the selectivity offered by such AS-ALD processes is nothing short of miraculous chemical wizardry. However, the design of the ALD precursors is not left to crystal balls or trial-and-error: the reactivity of a precursor with a surface can be studied effectively with quantum-chemical methods. Merck's quantum chemistry team uses advanced density-functional (DFT) calculations to elucidate the mechanisms of precursor-surface reactions and to design the right precursor for a target surface. Optimized precursor structures are then forwarded to a chemistry team who synthesizes samples for testing and process development in ALD tools.

Our subsidiary Intermolecular in Silicon Valley is specialized to rapidly carry out such testing using advanced combinatorial techniques to hand over processes to customers that meet their specific requirements.

One of the most promising bottom-up patterning technologies is directed self-assembly (DSA). DSA makes use of the fact that in block copolymers (BCPs), the self-interaction of monomer units within a block tends to be more favorable than that between the blocks. Block copolymers, therefore, naturally self-assemble in distinct phases that minimize the area of interaction between blocks. Of these, the hexagonal cylindrical and lamellar phases bear a striking resemblance to the contact hole and line/space structures commonly found in lithographic patterns. With top-down information in the form of guide structures that prescribe the location and orientation of such self-assembled patterns, it is possible to use these polymer phases to define useful circuit patterns. For example, an isolated line/space pattern with a 1:4 size ratio between lines and spaces can be used to guide the assembly of a suitable lamellar block copolymer, resulting in five lines where there previously was only one: the frequency of the original line pattern has increased fivefold. Champion results using block copolymers with specially designed highly phase separating blocks have demonstrated line/space pattern resolution down to 5 nm, a number that is beyond the capability of even the next generation of high NA EUV tools. Similar approaches allow the use of the cylindrical phases for the frequency multiplication of contact hole patterns.

Merck has been a leader in DSA technology since 2010. Building on a partnership with IBM, an early mover in DSA who developed much of the basic technology, we developed the capability to make high-quality material sets for DSA on an industrial scale. Today we can carry out low-temperature anionic polymerizations up to a 1000 L scale to make block copolymers with unprecedented purity and control. Our control of the length of the polymer blocks is precise to plus or minus one monomer unit, and the amount of homopolymer impurities is at very low levels that were previously unavailable commercially. This level of performance is necessary to provide our customers with DSA materials that meet their requirements for reproducibility and low defect levels. Industry leaders have acknowledged that DSA is meeting the requirements for high-end manufacturing: in 2021, researchers from Intel announced at an industry conference that "DSA is a high yielding, low defect technology." [30]

Originally intended as a competitor to advanced lithography techniques such as EUV, DSA has evolved into a complementary technology. The combination of EUV and DSA to achieve high-quality contact hole arrays will serve as an example. Contact hole patterning is a problem area for EUV because the low number of photons per unit volume in EUV lithography can lead to stochastic effects that cause a high variation in contact hole size. In contrast, in DSA the contact hole size is encoded into the properties of the cylindrical BCP—the contact hole literally cannot have a different size. DSA can therefore be used to "rectify" an EUV prepattern, resulting in a highly accurate array of contact holes with very low size variation. Although the original pattern may have size variations and some defects, the DSA process will heal both issues, resulting in a highly uniform and defect-free pattern.

Merck's molecular glass EUV resist can be used to directly form the guide pattern for this kind of process. This approach is particularly attractive since it makes it possible to directly use a pattern made with a high photospeed resistance to guide structure without the need to prepare and etch underlying layers.

The high-performance top-down capabilities of EUV and its high NA extension combined with the magic of bottom-up materials such as DSA and selective deposition will enable the industry to continue dimensional scaling and keep Moore's law alive for more than one, possibly two decades. We therefore have some runway left until chip design needs to fully take off into the third dimension. However, already now there are areas in which even the considerable performance gains from these technologies are woefully inadequate. Computational requirements for the training of AI models have been exploding, growing at rates that greatly exceed the gains provided by Moore's law. For the most advanced class of AI, the Transformer models, computational needs increase by a factor of 275 every 2 years—over 100x faster than provided by Moore's law [31]. These high demands on computer power are already beginning to slow down the development of advanced AI. Obviously, dimensional scaling is not going to provide the answer.

In 1959, Richard Feynman gave a highly influential lecture entitled "There's plenty of room at the bottom," [32] in which he argued that we should make computer circuits with "wires [that] should be 10 or 100 atoms in diameter, and the circuits should be a few thousand angstroms across"—a daring vision at the time when a computer filled a large room, but one that acted as a conceptual blueprint for dimensional scaling. Today the smallest features made by EUV lithography are 8 nm, or about 34 silicon atoms, in size: we have almost realized Feynman's vision.

In 2020, Neil Thompson of MIT gave a counterpoint to Feynman in a paper called "There is plenty of room at the top," in which he showed that for a specific problem, better algorithms had over time led to a performance improvement comparable to that achieved through dimensional scaling [33]. It appears unlikely that training algorithms for advanced AI can be improved to the same extent, but Thompson's observation serves as a reminder that using the circuit performance enabled by dimensional scaling in different ways can lead to large jumps in computational power. A corresponding change on the hardware side is the modification of the basic architecture of computer chips. Traditionally, computer architecture has been based on a concept introduced by John von Neumann in 1945 in which a central processor unit (CPU) is connected to an external memory and moves the results of computations back and forth over a memory bus. Specialized AI processors have already shed the von Neumann architecture to some extent and have moved to more decentralized approaches that provide higher throughput. However, they still have not realized the full potential of neuroscience-driven designs that more closely mimic the way the human brain processes data and lead to massive improvements in computational power and reductions in energy use. Merck is supporting this development through our NeuroCore incubator in Silicon Valley which develops and productizes architectural designs of AI and neuromorphic accelerators enabled by emerging memory technologies. These new device types require new and proprietary material chemistries which we are developing in co-operations with start-ups and other collaborations.

The dimensional scaling progression known as Moore's law has provided the driving force for a transformation of our world to an information-based society that was unimaginable many years ago. However, as Gordon Moore said himself, "no exponential is forever." We still have some time left before we run into the inevitable limit of dimensional scaling, but physics and economics both dictate that there will eventually be such a limit. Electronic materials will be more important than ever as we begin the transition to the next phase of the industry's development in which 3D scaling and heterogeneous integration will be the motors that drive integration density and power improvement in IC performance and costs. Merck is determined to continue to be the company behind the companies that will make this new Moore's law 2.0 a reality.

5 Summary

At the time of the submission of this chapter, the COVID-19 pandemic was still ongoing. However, the rapid development of vaccines, therapeutics, and diagnostics by the global scientific community to fight the pandemic demonstrated the potential of science and technology to achieve significant outcomes for humanity. Scientists, policymakers, and, more broadly, we as a society should ensure the pandemic helps to accelerate scientific momentum.

It must be our collective goal to foster an environment that allows curious scientific minds to outperform in an increasingly complex and connected world. Through initiatives such as the Curious Conference, the Future Insight Prize, and our many other initiatives and collaborations being pursued by Merck across life science, healthcare, and electronics, we look forward to helping unite the scientific community for a better tomorrow.

Part II

Advancing Science and Technology: Examples from Curious 2022 Conference Keynote Speakers

Bringing Photosynthesis 2.0 to Life

Tobias J. Erb

The increasing anthropogenic release of greenhouse gases (GHG), and in particular carbon dioxide (CO_2), has resulted in a dramatic increase in global warming. This has severe consequences for all life and the health of our planet. Understanding and mitigating the effects of climate change has become the key challenge of the twenty-first century. The main underlying cause of human-made climate crisis is an imbalanced cycle of carbon on a global scale: In other words, humans release more CO_2 than can be currently re-captured through natural mechanisms.

While several new concepts and technologies for carbon capture and conversion are currently developed and/or tested, all these technical solutions are still outcompeted by large by biology. Biological CO_2 fixation captures and converts about 400 Gt of CO_2 (equivalent to about 100 Gt C) per year. The incredible capability of nature to sustainably capture and convert atmospheric CO_2 on a global scale serves as both a blueprint and a vision to harness CO_2 as a future carbon source.

Yet, biological systems, such as plants, algae, and bacteria, are (still) limited in their CO_2-fixation capacities (see below), which requires new approaches and methods to overcome these natural boundaries. The emerging field of synthetic biology provides new and exciting opportunities to fundamentally rethink, redesign, and reconstruct photosynthetic CO_2-fixation to create tailor-made and diverse ways for the transformation of CO_2 into useful molecules.

T. J. Erb (✉)
Max Planck Institute for Terrestrial Microbiology, Marburg, Germany
e-mail: toerb@mpi-marburg.mpg.de

U. A. Betz (ed.), *Curious Future Insight*,
https://doi.org/10.1007/978-3-031-41781-8_3

1 Although Evolved over Billions of Years, Natural Photosynthesis Is (Still) Limited

Of the almost 400 Gt CO_2 that are captured and converted naturally per year, more than 90% is fixed through photosynthesis and a single enzyme ribulose-1,5,-bisphosphate carboxylase/oxygenase (Rubisco), which is the key enzyme in the "dark reaction," which is also known as the Calvin-Benson-Bassham (CBB) cycle [1].

Despite its essential role in the global carbon cycle, Rubisco is far from being the optimal solution for CO_2 fixation. The enzyme is actually a relatively slow catalyst that turns over only 5–10 CO_2 molecules per second. Additionally, Rubisco has a strong side reaction with oxygen (O_2) that leads to an oxygenation instead of a carboxylation of ribulose-1,5-bisphosphate. This causes the process of photorespiration, which releases up to almost 30% of the fixed carbon in photosynthesis, strongly limiting photosynthetic yield [2].

Several photosynthetic organisms have evolved different mechanisms to suppress photorespiration, mainly by increasing the local CO_2-concentrations around Rubisco, e.g., through encapsulating the enzyme in dedicated protein compartments (e.g., carboxysomes in cyanobacteria [3] or pyrenoids in green algae [4]), inorganic carbon transporters, or metabolic CO_2 pumps (e.g., in C4- or CAM-plants [5, 6]). Although these work-arounds are efficient measures to increase photosynthetic yield, they rather aim at curing a symptom than the actual root cause. Thus it has been concluded that the CBB cycle is an evolutionary optimized solution, but from different points of view not the optimal solution for biological CO_2-fixation.

2 Beyond Photosynthetic CO_2-Fixation: Discovering New CO_2-Fixing Principles in Different Microbes

Interestingly, Rubisco (and the CBB cycle) is only one of several CO_2-fixing solutions that Nature has invented through the course of evolution. Within the last decade(s) many alternative, so far unknown CO_2-fixing pathways and enzymes have been discovered in different microbes [7, 8]. While they do not reach the global productivity of Rubisco and the CBB cycle, these enzymes and pathways are still responsible for the annual conversion of gigatons of CO_2 and organic carbon on Earth [9–11].

One particular fascinating example is the discovery of enoyl-CoA carboxylases/reductases (ECRs) that catalyze the NADPH-dependent carboxylation of enoyl-CoA thioester into their corresponding alkyl-malonyl-CoA esters [9]. ECRs are particularly interesting, as they fix CO_2 at up to 10-times rates compared to Rubisco (i.e., up to 100 CO_2 molecules s^{-1}), and unlike Rubisco do not show any oxygenase side reaction. These highly favorable parameters make ECRs an interesting alternative to Rubisco-based CO_2-fixation. Yet, the enzyme has not been explored by nature in the context of photosynthetic or autotrophic CO_2-fixation, but primarily in heterotrophic acetate metabolism and natural product biosynthesis.

3 Enoyl-CoA Carboxylases/Reductases: Dissecting a Highly Efficient Principle of CO_2-Fixation

What makes ECRs so much more efficient compared to the key enzyme in photosynthesis and many other CO_2-fixing enzyme? In the last years, great progress was made in respect to understanding the molecular basis of catalysis in this newly discovered class of carboxylases.

Different efforts were undertaken to study the mechanism of CO_2-fixation in ECRs [12–16]. Using NMR, high-resolution MS, and stopped-flow spectroscopy, individual intermediates of the catalytic cycle could be identified [13], and a mechanistic probe was established that allowed to study the reduction and carboxylation steps of ECRs separately from each other (Nature Chem. Biol. 2017). Altogether, this provided a picture of the reaction mechanism, according to which a hydride is transferred from the NADPH cofactor onto the enoyl-CoA substrate to create a reactive enolate. This can either react with CO_2 or collapse into a short-lived intermediate ("C4-ene adduct") that can be stored at the active site and carboxylated into the final product, as soon as CO_2 becomes available.

Notably, and in contrast to Rubisco, ECRs do not react with O_2, which means that they must be able to effectively discriminate between these two gases at the active site. Hence, recent efforts also focused on the question how ECRs bind and control the CO_2 molecule during catalysis. Combining experimental biochemistry, protein crystallography, and advanced computer simulations, it could be shown that only four amino acids are required to create a specific CO_2-binding pocket [15]. Together, these four residues anchor and position the CO_2 molecule for the attack by the reactive enolate created during the catalytic cycle of ECRs. At the same time, these residues also shield the active site from water, which would otherwise interfere with the carboxylation reaction and quench the enolate intermediate into the reduced (side) product [15]. In respect to catalytic rate, it could be shown recently that in ECR from *Kitasatospora setae* high turnover rates are achieved by inter-subunit coupling [16]. Altogether, these studies revealed the molecular details of selective CO_2 binding and C-C-bond formation in a highly efficient carboxylase.

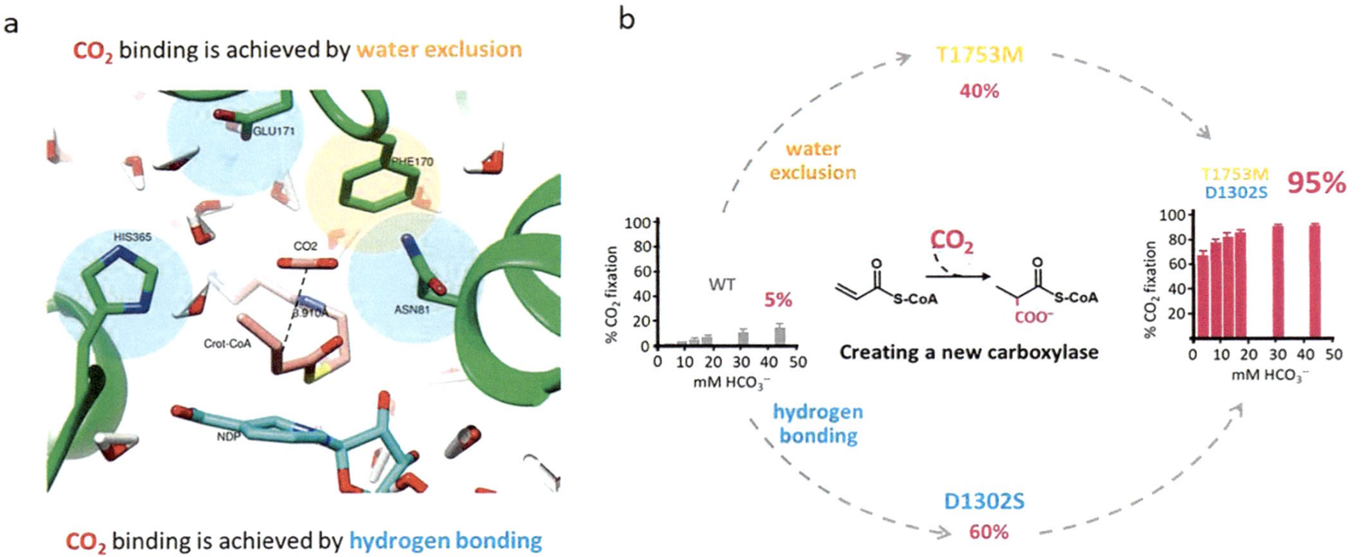

Fig. 1 From understanding CO_2-binding in ECRs to engineering novel CO_2-fixing proteins. (**a**) Four amino acids at the active site of ECRs provide a highly specific CO_2-binding pocket. (**b**) Transferring the molecular basis of selective CO_2-binding into enoyl-CoA reductases converts these enzymes into bona fide CO_2-fixing reductases/carboxylases

4 Insight Must Precede Application: (Re-)Engineering New CO_2-Fixing Enzymes

The mechanistic understanding of ECRs was further used to create new CO_2-fixing enzymes. On the one hand, the substrate spectrum of ECRs could by expanded by active site engineering to create a toolbox of highly active and versatile enzymes that are able to carboxylate more than 20 different substrates [17]. On the other hand, bioinformatics-guided strategies were used to identify scaffolds that might possess a latent carboxylation activity ("sleeping carboxylases"). Notably through these efforts, several enoyl-CoA reductases were identified that in principle feature residues for a (rudimentary) CO_2-binding pocket, and thus have the potential to catalyze a reductive carboxylation (instead of the simple reduction reaction), when external CO_2 is supplied.

Protein engineering was used to convert these reductases into true (reductive) carboxylases by increasing interactions of the proteins with CO_2 and suppressing diffusion of water to the active site. These engineered carboxylases indeed showed improved CO_2-binding and kinetic parameters, which were comparable to naturally existing CO_2-fixing enzymes, demonstrating the successful generation of new CO_2-fixing biocatalysts from first principles [18].

Beyond ECR engineering, another recent example for a designer carboxylase is glycolyl-CoA carboxylase (GCC) a new-to-nature enzyme that was developed recently by combining rational protein engineering with high-throughput microfluidics- and microplate-based screening strategies. During this process, the catalytic efficiency of GCC improved by three orders of magnitude to match the properties of naturally existing CO_2-fixing enzymes [19]. In the following, this enzyme could be successfully interfaced with natural, as well as artificial, photosynthetic metabolism in vitro (see below).

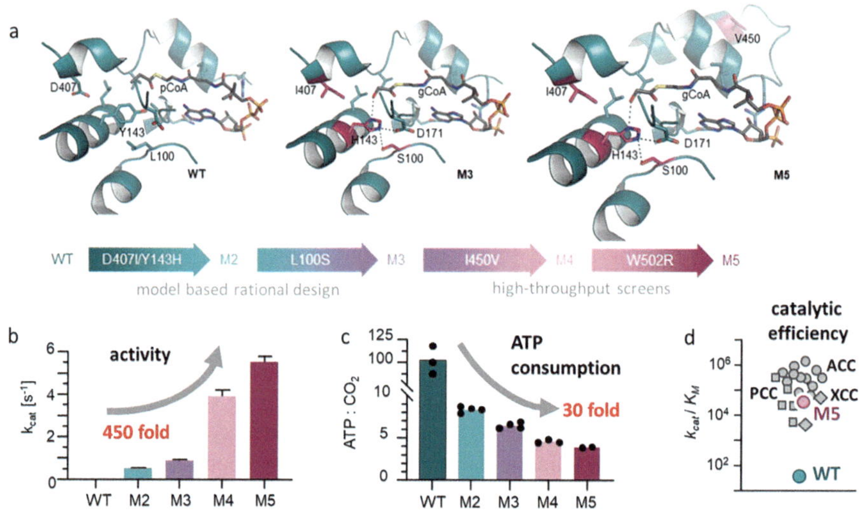

Fig. 2 Developing glycolyl-CoA carboxylase (GCC), a new-to-nature CO_2-fixing enzyme. (**a**) Glycolyl-CoA carboxylase was developed in the scaffold of propionyl-CoA carboxylase (WT) through rational engineering (variants M2 and M3), as well as non-targeted directed evolution (M4 and M5). Final variant M5 contained five mutations. (**b**) Catalytic activity of the five different variants. While the wildtype (WT) showed almost no activity for glycolyl-CoA carboxylation, variant M5 showed catalytic activities comparable to other CoA-carboxylases. (**c**) ATP consumption per CO_2 fixed for the five different variants. While the wildtype showed futile ATP hydrolysis, unfruitful ATP hydrolysis was strongly reduced in variant M5. (**d**) Catalytic efficiency shows that the M5 groups within other, naturally existing CoA-carboxylases

5 Metabolic Retrosynthesis: Exploring Those CO_2-Fixation Pathways Nature Did Not Invent

As mentioned earlier, besides the CBB cycle of photosynthesis, nature has evolved eight alternative CO_2-fixation pathways [7, 8]. While these eight different solutions demonstrate the impressive creativity of evolution, they only reflect a small fraction of the theoretically possible pathways that nature could have explored [20]. Is it actually possible to explore this *terra incognita* and realize completely novel CO_2-fixation pathways that nature has not invented (yet)?

Through the emerging field of synthetic biology, it has become possible to build novel biological solutions from first principle. In contrast to metabolic engineering efforts that mainly focus on optimizing existing pathways, such synthetic biology approaches carry a more disruptive potential by expanding the naturally existing metabolic space through radically new solutions [21]. Notably, these solutions could be of improved thermodynamic and kinetic efficiency, thus allowing for higher

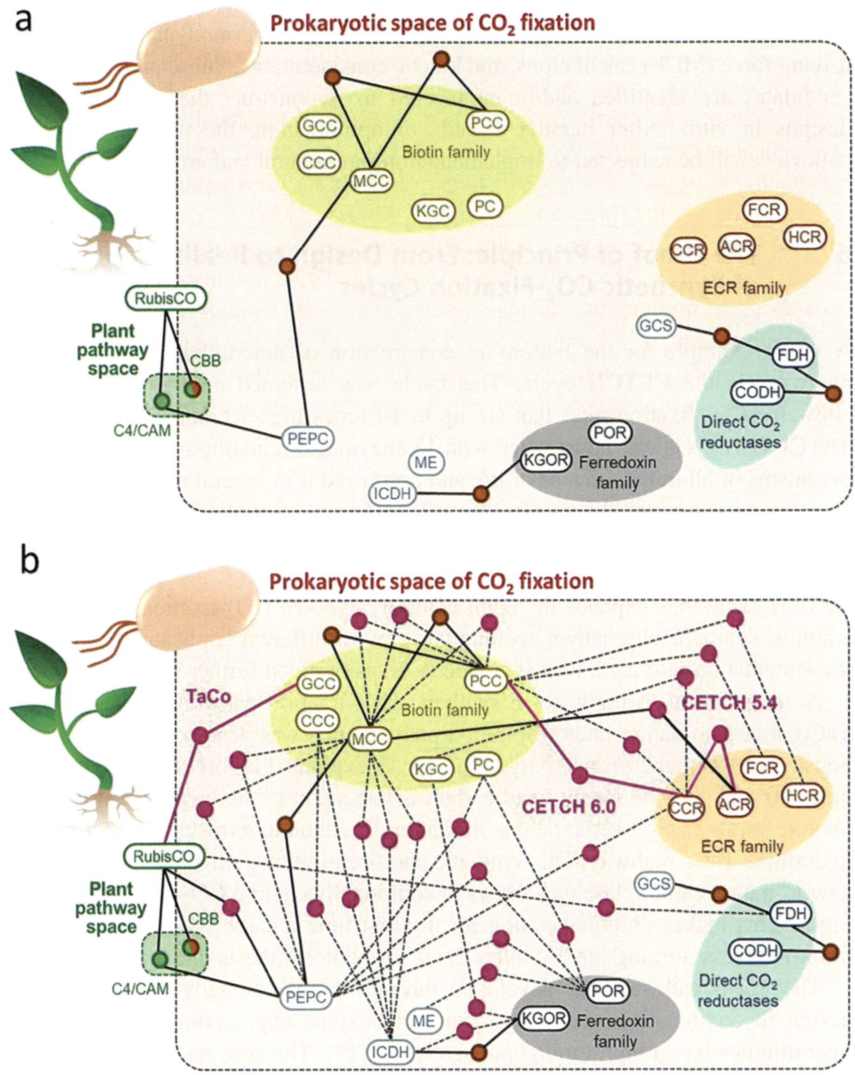

Fig. 3 The metabolic space of CO_2-fixation. (**a**) Natural space: Nature has evolved seven different CO_2-fixation pathways (nodes) that are built around different carboxylases (round boxes). (**b**) Natural and synthetic space: Shown are 28 additional new-to-nature pathways that have been designed, which are again covering only a fraction of the theoretical possible space

fluxes (i.e., CO_2-capturing rates) at lower energetic demand (e.g., ATP consumption) compared to their natural counterparts.

To realize these new-to-nature solutions, the concept of "metabolic retrosynthesis" has been pioneered, according to which novel biosynthetic routes are drafted in a first, theoretical phase. These different solutions are evaluated in a second phase using physical-chemical principles, such as thermodynamics, max-min driving force (MDF) calculations, and kinetic considerations. Subsequently, enzyme candidates are identified and/or engineered to re-construct the most promising designs in vitro. After iterative rounds of optimization, the improved in vitro pathways will be subjected to implementation into natural and artificial cells.

6 The Proof of Principle: From Design to Realization of Synthetic CO_2-Fixation Cycles

A prime example for the bottom-up construction of new-to-nature CO_2-fixation pathways is the CETCH cycle. This cycle was designed around ECRs, which allow for CO_2-fixation rates that are up to 10 times higher compared to Rubisco. The CETCH cycle was constructed with 17 enzymes originating from nine different organisms of all three domains of life and optimized it in several rounds by enzyme engineering and metabolic proofreading [22]. The optimized CETCH cycle (version 5.4) converts CO_2 into organic molecules faster than the natural CO_2-fixation pathway of photosynthesis and notably at 20% less energy per CO_2 fixed. The CETCH cycle thus expands the eight naturally evolved CO_2-fixation pathways by a ninths, synthetic alternative, opening the way for different applications, including implementation into natural or synthetic cells, as outlined further below.

Another recent example for a synthetic CO_2-fixation pathway is the so-called TaCo, a new-to-nature photorespiration pathway that was designed and realized to boost photosynthetic productivity [19, 23]. As explained before, in photosynthesis up to 30% of the previously fixed carbon is lost again to the environment through photorespiration. To overcome this limitation, "metabolic retrosynthesis" was used to draft the TaCo pathway. This synthetic photorespiration pathway features GCC, a new-to-nature carboxylase that allows to additionally capture CO_2 instead of releasing it. This makes photorespiration for the first time a carbon-positive (i.e., CO_2-fixing) process, turning the "Achilles' heel" of photosynthesis into an asset [23].

The TaCo pathway was developed through computationally guided pathway design in combination with state-of-the art enzyme engineering combined with microfluidics-based high-throughput screening [19]. The core reaction sequence of the TaCo pathway consists of three new-to-nature enzymes, including the completely novel CO_2-fixing enzyme GCC (see above). Modeling shows that the TaCo pathway dramatically reduces energetic demands of photorespiration by approx. 30% (ATP) and 20% (NAD(P)H), respectively, while at the same time increasing carbon efficiency between 40 and 150%, independent of Rubisco's oxygenation activity.

7 Mimicking Evolution: Optimizing Synthetic CO_2-Fixation Pathways Through Machine Learning Approaches

Despite the successful development of several new-to-nature networks, such as the CETCH cycle or the TaCo pathway, the optimization of these systems still provides a challenge. This is because the individual parts (i.e., enzymes) used to reconstruct these systems are derived from very different biological backgrounds, which makes their interactions hard to predict. These unknown interactions include allosteric interactions and inhibitions, stability and reactivity of metabolites, as well as their undesired degradation or turnover through enzyme promiscuity. Overall, this causes a nonlinear complexity that is virtually impossible to improve through purely rational approaches.

Moreover, while in the context of living cells, optimization of biological networks happens through natural selection; this principle is not simply applicable to complex in vitro networks, which are not directly genetically encoded. Thus, the in vitro optimization of synthetic pathways like the CETCH cycle asks for novel approaches to improve these new-to-nature systems in a fast and efficient manner.

To alleviate the rapid improvement of biological networks outside of a cellular context, machine learning-guided workflows have been developed. These strategies use iterative design-build-learn cycles to enable the systematic optimization of a given biological target function. One example is METIS (named after the ancient goddess of wisdom and crafts Μῆτις, lit. "wise counsel"), a workflow that is able to search the combinatorial space of a complex in vitro system over several rounds for a (local) optimum. To find such optima, METIS relies on automated experimentation for prototyping different combinations, which is followed by subsequent analysis and machine learning-guided prediction of an improved set of combinations [24].

The METIS workflow was recently used to improve productivity of the CETCH cycle, which features 26 components encompassing 17 enzymes, and different cofactors and salts in vitro, spanning a theoretical space of approx. 10^{25} combinations. Exploring this theoretical space over only eight rounds of active learning with approx. 1000 different variants improved CO_2-fixation productivity tenfold compared to the initial described CETCH cycle version 5.4 [24]. These efforts resulted in the most efficient CO_2-fixation system described to date, demonstrating how machine learning can be combined with lab automation to systematically optimize synthetic biological networks in a data-driven fashion, paralleling the mechanisms of natural evolution.

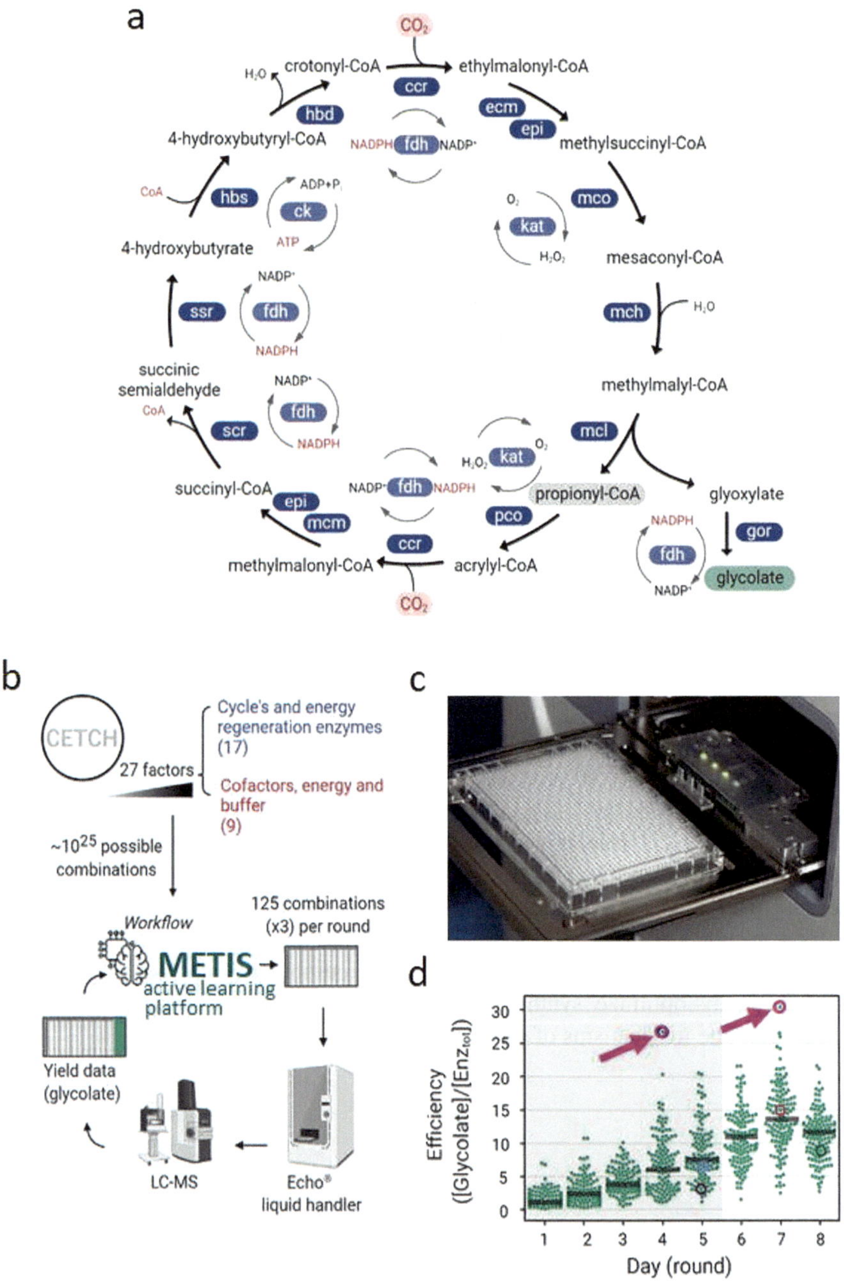

Fig. 4 Optimizing the CETCH cycle by active learning and lab automation. (**a**) Outline of the CETCH cycle. (**b**) Setup and workflow of the design-build-test cycle, including the METIS active learning platform. (**c**) Picture of lab automation with liquid handling robots. (**d**) Improvement of

8 Building Complex Molecules: Expanding Synthetic CO_2-Fixation by Biosynthetic Modules

In its basic version, the CETCH cycle and other in vitro pathways capture and convert CO_2 into relatively simple chemical building blocks. However, to make such pathways truly versatile synthesis platforms, it is essential to connect them with different biosynthetic modules downstream to create new routes into multi-carbon compounds.

To demonstrate the versatility of the CETCH cycle, the cycle was recently coupled with different biosynthetic modules to establish direct routes from CO_2 into acetyl- and malonyl-CoA. These intermediates were subsequently diversified through the action of different terpene and polyketide synthases into a multitude of different compounds [25, 26].

Combining more than 30 enzymes in one reaction pot allowed the synthesis of different mono- and sesquiterpenes, such as limonene, sabinene, α-pinene, α-bisabolene, and β-farnesene from CO_2 [25]. In another example, CO_2 could be directly converted into 1,3,5,7,9,11,13-pentadecaheptaene, an all-trans polyene, which upon chemical hydrogenation leads to pentadecane, a prime component of diesel fuel [25]. Finally, the production of the polyketide 6-DEBS, the precursor of the antibiotic erythromycin, directly from CO_2 was also achieved recently through the coupling of more than 50 different reactions in one condition [26].

These experiments show how synthetic CO_2-fixation pathways can be extended to produce different value-added compounds, which paves the way to develop customized routes for the synthesis of any given molecule from CO_2. One important challenge that remains to be solved is the stability and robustness of such in vitro systems, and the constant supply or regeneration of cofactors, such as ATP and NAD(P)H, over long time periods.

9 Building Molecular Complexity: Realizing Artificial Chloroplasts

Photosynthesis takes place in the chloroplast, which is a highly integrated molecular machine that converts light into chemical energy, which in turn is used to drive CO_2-fixation. A defining goal for bottom-up synthetic biology is to develop integrated systems that approach or even rival the light-driven CO_2-fixation capabilities of natural chloroplasts.

In a multidisciplinary effort, synthetic biology and soft-matter engineering were recently combined to create an "artificial chloroplast" for the light-driven capture

productivity of the CETCH cycle over eight active learning cycles. Individual points represent different pathway variants. Pink arrows highlight highly favorable CETCH variants that are tenfold improved in CO_2-fixation efficiency compared to the initial cycle and conversion of CO_2 [27]. To realize this project, in a first step, a microfluidic platform to encapsulate and operate photosynthetic membranes in cell-sized micro-droplets was developed. These micro-droplets could be energized by light to power single enzymes as well as more complex enzyme cascades. Furthermore, multiplexing methods were established to follow the activity of hundreds of droplets in parallel and real-time [27].

In a second step, this microfluidics platform was used to encapsulate the complete CETCH cycle together with photosynthetic membranes. In the final setup, 18 enzymes worked together to continuously convert CO_2 into glycolate. These results demonstrated the successful coupling of natural light-conversion modules and synthetic carbon-fixation pathways into one integrated system that is functionally equivalent to a chloroplast and even exceeds its natural counterpart in CO_2-fixation efficiency. Although currently limited to 2 hours stability, these "artificial chloroplasts" are a big leap forward toward the construction of complex catalytic systems that show the intricacies of natural existing cells.

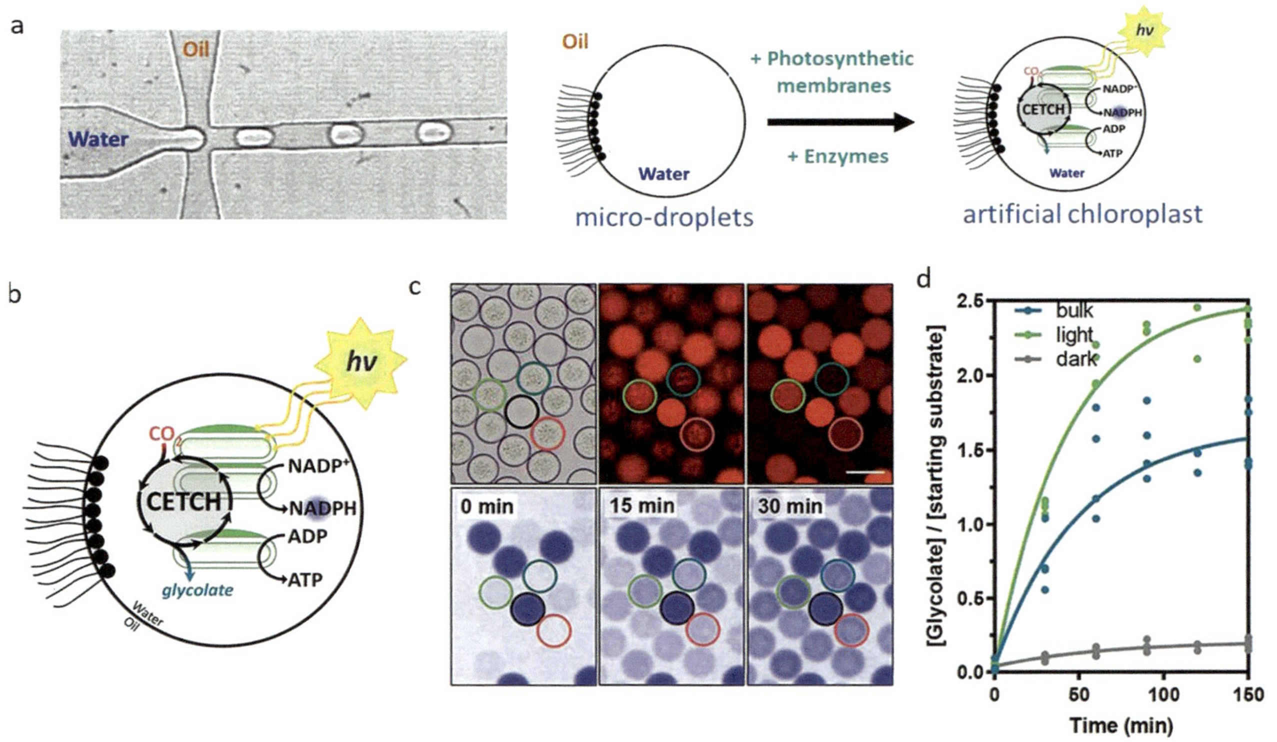

Fig. 5 The "artificial chloroplast": A light-powered micro-compartment for CO_2 capture (**a**) Microfluidics as enabling technology to create cell-sized micro-compartments that can be equipped with photosynthetic membranes and enzymes. (**b**) Construction plan of the artificial chloroplast. (**c**) Microscopic picture of artificial chloroplasts in action. (**d**) Light-powered production of glycolate from CO_2 in the artificial chloroplast compared to bulk assays and the dark control

10 Implement and Conquer: Bringing Synthetic CO_2 Fixation into Living Cells

To demonstrate the proof-of-principle and to harness their full potential for biotechnology and agriculture, the CETCH cycle and other synthetic CO_2-fixation pathways will need to be successfully implemented into living systems. However, realizing complex synthetic pathways in living cells poses several technical and biological challenges. This includes methods to encode, assemble, and transfer the necessary genetic information into a suitable host cell, as well as the integration of the new set of reactions into the native genetic and metabolic networks of the respective host.

A promising strategy to achieve this goal stepwise is growth-coupled dependencies, which have successfully used to implement "simpler" new-to-nature pathways in the recent past [28, 29]. According to this strategy, one (or several) essential metabolites in the native metabolic network are completely isolated, creating an auxotrophy. This auxotroph can be rescued through implementation of the new-to-nature pathway that is able to synthesize the missing metabolite. Once the new-to-nature pathway is active and enables growth, adaptive laboratory evolution can be used to optimize further integration and/or coordination of the new-to-nature pathway with the endogenous metabolism. These dependencies can be further extended, until not only one essential metabolite, but all the biomass is formed through the synthetic pathway.

While there are still many challenging steps ahead before the successful implementation of complex synthetic CO_2-fixation pathways, such as the CETCH cycle, in living cells, these efforts will probe the plasticity and adoptability of modern cells toward radically new "designer metabolism," and at the same time also have very practical applications for realizing a sustainable world of tomorrow.

Acknowledgments I would like to thank the many friends, colleagues, and mentors that encouraged, supported, and inspired me, in particular Georg Fuchs, Birgit Alber, John A. Gerlt, Ivan Berg, and Julia Vorholt. I have been blessed to work in and with a team of outstanding scientists and wonderful human beings, and have received generous support through many funding agencies in different countries. I am very grateful for the long-term trust and support by the Max Planck Society. Finally, I would like to express my very personal thanks to my parents, Heidi and Hansjörg Erb, who have always supported me along the way, as well as my family, Annette, Florian, and Jule, my joy and my personal source of energy and inspiration. This text is based on thoughts and elements of several publications and presentations from our lab over the years.

Literature

1. Bassham JA, Benson AA, Kay LD, Harris AZ, Wilson AT, Calvin M. The path of carbon in photosynthesis. XXI. The cyclic regeneration of carbon dioxide acceptor. J Am Chem Soc. 1954;76(7):1760–70.
2. Walker BJ, VanLoocke A, Bernacchi CJ, Ort DR. The costs of photorespiration to food production now and in the future. Annu Rev. Plant Biol. 2016;67:107–29.
3. Turmo A, Gonzalez-Esquer CR, Kerfeld CA. Carboxysomes: metabolic modules for CO2 fixation. FEMS Microb Lett. 2017;364(18):fnx176.

4. Wunder T, Mueller-Cajar O. Biomolecular condensates in photosynthesis and metabolism. Curr Op Plant Biol. 2020;58:1–7.
5. Hatch MD, Slack CR. Photosynthesis by sugar-cane leaves. A new carboxylation reaction and the pathway of sugar formation. Biochem J. 1966;101:103–11.
6. Osmond CB. Crassulacean acid metabolism: a curiosity in context. Annu Rev Plant Physiol. 1978;29:379–414.
7. Berg IA. Ecological aspects of the distribution of different autotrophic CO2 fixation pathways. App Env Microbiol. 2011;77(6):1925–36.
8. Bierbaumer S, Nattermann M, Schulz L, et al. Enzymatic conversion of CO2: From natural to artificial utilization. Chem Rev. 2023;123(9):5702–54.
9. Erb TJ, Berg IA, Brecht V, Müller M, Fuchs G, Alber BE. Synthesis of C5-dicarboxylic acids from C2-units involving crotonyl-CoA carboxylase/reductase: the ethylmalonyl-CoA pathway. Proc Natl Acad Sci. 2007;104(25):10631–6.
10. Könneke M, Schubert DM, Brown PC, et al. Ammonia-oxidizing archaea use the most energy-efficient aerobic pathway for CO2 fixation. Proc Natl Acad Sci. 2014;111(22):8239–44.
11. Schada von Borzyskowski L, Severi F, Krüger K, et al. Marine Proteobacteria metabolize glycolate via the β-hydroxyaspartate cycle. Nature. 2019;575(7783):500–4.
12. Erb TJ, Brecht V, Fuchs G, Müller M, Alber BE. Carboxylation mechanism and stereochemistry of crotonyl-CoA carboxylase/reductase, a carboxylating enoyl-thioester reductase. Proc Natl Acad Sci. 2009;106(22):8871–6.
13. Rosenthal RG, Ebert MO, Kiefer P, Peter DM, Vorholt JA, Erb TJ. Direct evidence for a covalent ene adduct intermediate in NAD(P)H-dependent enzymes. Nat Chem Biol. 2014;10 (1):50–5.
14. Rosenthal RG, Vögeli B, Wagner T, Shima S, Erb TJ. A conserved threonine prevents self-intoxication of enoyl-thioester reductases. Nat Chem Biol. 2017;13(7):745–9.
15. Stoffel GMM, Saez DA, DeMirci H, et al. Four amino acids define the CO2 binding pocket of enoyl-CoA carboxylases/reductases. Proc Natl Acad Sci. 2019;116(28):13964–9.
16. DeMirci H, Rao Y, Stoffel GM, et al. Intersubunit coupling enables fast CO2-fixation by reductive carboxylases. ACS Centr Sci. 2022;8(8):1091–101.
17. Peter DM, Schada von Borzyskowski L, Kiefer P, Christen P, Vorholt JA, Erb TJ. Screening and engineering the synthetic potential of carboxylating reductases from central metabolism and polyketide biosynthesis. Angew Chem Int Ed. 2015;54(45):13457–61.
18. Bernhardsgrütter I, Schell K, Peter DM, et al. Awakening the sleeping carboxylase function of enzymes: engineering the natural CO2-binding potential of reductases. J Am Chem Soc. 2019;141(25):9778–82.
19. Scheffen M, Marchal DG, Beneyton T, et al. A new-to-nature carboxylation module to improve natural and synthetic CO2 fixation. Nat Catal. 2021;4(2):105–15.
20. Wurtzel ET, Vickers CE, Hanson AD, et al. Revolutionizing agriculture with synthetic biology. Nat Plants. 2019;5(12):1207–10.
21. Erb TJ, Jones PR, Bar-Even A. Synthetic metabolism: metabolic engineering meets enzyme design. Curr Op Chem Biol. 2017;37:56–62.
22. Schwander T, Schada von Borzyskowski L, Burgener S, et al. A synthetic pathway for the fixation of carbon dioxide in vitro. Science. 2016;354(6314):900–4.
23. Trudeau DL, Edlich-Muth C, Zarzycki J, et al. Design and in vitro realization of carbon-conserving photorespiration. Proc Natl Acad Sci. 2019;115(49):E11455–64.
24. Pandi A, Diehl C, Yazdizadeh Kharrazi A, Scholz SA, et al. A versatile active learning workflow for optimization of genetic and metabolic networks. Nat Comm. 2022;13(1):1–15.
25. Sundaram S, Diehl C, Cortina NS, Bamberger J, Paczia N, Erb TJ. A modular in vitro platform for the production of terpenes and polyketides from CO2. Angew Chem Int Ed. 2020;60(30): 16420–5.
26. Diehl C, Gerlinger PD, Paczia N, Erb TJ. Synthetic anaplerotic modules for the direct synthesis of complex molecules from CO2. Nat Chem Biol. 2023;19(2):168–75.

27. Miller TE, Beneyton T, Schwander T, et al. Bottom-up construction of a chloroplast mimic capable of light-driven synthetic CO2 fixation. Science. 2020;368(6491):649–54.
28. Satanowski A, Dronsella B, Noor E, et al. Awakening a latent carbon fixation cycle in Escherichia coli. Nat Comm. 2020;11(1):5812.
29. Kim S, Lindner SN, Aslan S, Yishai O, Wenk S, Schann K, Bar-Even A. Growth of E. coli on formate and methanol via the reductive glycine pathway. Nat Chem Biol. 2020;16(5):538–45.

Medical Implications of Functional and Destructive Cellular Motions: Curiosity-Driven Open Issues

Andre Rivalta, Disha-Gajanan Hiregange, Tanaya Bose, Gil Fridkin, K. Shanmugha Rajan, Ada Yonath, Ella Zimmerman, Anat Bashan, and Hagith Yonath

1 Introduction

Albert Einstein alleged "I have no special talent. I am only passionately curious," and Albert Szent-Gyorgyi added: "Discovery consists of looking at the same thing as everyone else and thinking something different." These Nobel Prize laureates were right. Curiosity has long been the driving force of the biological and biomedical sciences, and curiosity-driven research has contributed significantly to our understanding of both simple and complex life processes. Consequently, sophisticated lessons originated from seemingly simple questions, pushed by the natural and insatiable human thirst for knowledge. In fact, curiosity yielded huge dividends in both the short and the long run: we wouldn't enjoy the benefits of the GPS (or Global Positioning System), if not for the very theory of relativity Einstein proposed, to name an example. Indeed, in research, as a fundamental human impulse, curiosity is a never-ending source of new, head-scratching questions. Thus, several basic issues, some of which with medical implications, are still puzzling us.

Here, we focus on very few selected biomedical non-resolved questions, by attempting to enlighten the broad topic of the relation between structural flexibility and functional appropriateness. Specifically, in the context of our efforts to engage the results of the emerging scientific efforts toward solutions for human health and wellbeing, we are relating to several open questions and focus specifically on a few of them, all associated with protein biosynthesis in healthy as well as in sick cells.

A. Rivalta (✉) · D.-G. Hiregange · T. Bose · K. S. Rajan · A. Yonath · E. Zimmerman · A. Bashan
Department of Chemical and Structural Biology, Weizmann Inst of Science, Rehovot, Israel
e-mail: Ada.Yonath@weizmann.ac.il

G. Fridkin
Department of Organic Chemistry, Israel Institute for Biological Research, Ness Ziona, Israel

H. Yonath
Sheba Medical Center, Ramat-Gan and Tel-Aviv University, Tel-Aviv, Israel

U. A. Betz (ed.), *Curious Future Insight*,
https://doi.org/10.1007/978-3-031-41781-8_4

2 Linking Structural Flexibility to Functional Suitability

2.1 The Main Protein Biosynthetic Machine, the Ribosome

Ribosomes are the universal multi-RNA-protein cellular assemblies that translate the genetic code into proteins (Fig. 1). They are present in massive numbers in all living organisms, including proliferating human cells, which may contain over 3.3 million ribosomes each. Mammalian ribosomes consist of about 81 r-proteins (ribosomal proteins) and 4 rRNA (ribosomal RNA) chains containing over 6000 rRNAs, many of which participate directly in the protein production process. Their primary functions, namely, efficient genetic code decoding, peptide bond formation, protein elongation, and tRNA release, are performed mostly by the rRNA while being assisted or controlled by ribosomal proteins.

All ribosomes consist of two riboprotein subunits of unequal size, which associate upon the initiation of protein biosynthesis (Figs. 1 and 2) and dissociate once it is terminated. The protein biosynthesis process is performed cooperatively by the two ribosomal subunits and requires signaling between the various functional sites, which are located within the ribosome rather far from each other. The small ribosomal subunit plays a key role in facilitating the initiation of the translation process and in the accurate decoding of the genetic message of the mRNA by controlling the fidelity of codon-anticodon interactions. The large ribosomal subunit catalyzes peptide bond formation and guarantees the elongation of nascent proteins by channeling them into their exit tunnel. This mode of operation is valid even in the presence of several types of RNA modifications, for example, in the highly modified *Leishmania* ribosome [1, 2].

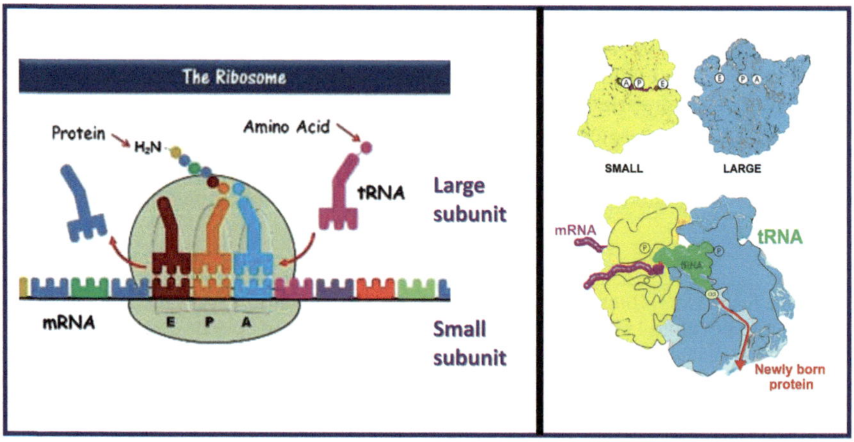

Fig. 1 An overall description of the ribosome's actions. Left – as a cartoon. Right relates to the ribosome's structure

During each cycle of the elongation event, a new peptide bond is formed by a multi-component cooperative event. mRNA carries the genetic instructions for the amino acid sequence of the produced proteins to the ribosome, and amino-acylated tRNA molecules, each specific to a natural amino acid, are delivering the amino acids to the ribosome. All tRNA molecules share a similar L-shape structure, and although they are built mainly of double helices, their functional sites are located within their single-stranded regions. These include the tRNA anticodon stem-loop that participates in the decoding by base-pairing with the mRNA, and the 5′ end of the universal CCA sequence, which carries the amino acids at its other end.

The ribosome possesses three tRNA binding sites, called A, P, and E, each located on both ribosomal subunits (Figs. 1 and 2). The A-site hosts the aminoacylated tRNA, the P-site is the peptidyl tRNA location, and the E-site designates the exiting deacylated tRNA path. The elongation of the polypeptide chain is associated with A- > P- > E translocation of the mRNA chain, together with the tRNA molecules associated with it. Once a peptide bond is created, the peptidyl chain is detached from its tRNA, and the deacylated tRNA molecule exits the ribosome through the E-site, while the A-site tRNA is translocated, presumably by a rotatory motion [3, 4] to the P-site. The so-obtained nascent proteins exit from the ribosome through the protein exit tunnel, which is actively involved in the process of co-translational protein folding (e.g., [5]) as it possesses discriminating properties, and hence can participate in regulating the intracellular co-translational processes. Interestingly, this tunnel could be biochemically and structurally detected even before the structure of ribosome was determined [6–9], and despite initial hints indicating post-translational folding [10, 11], until recently it was assumed to be a rather passive protected path for the nascent peptides.

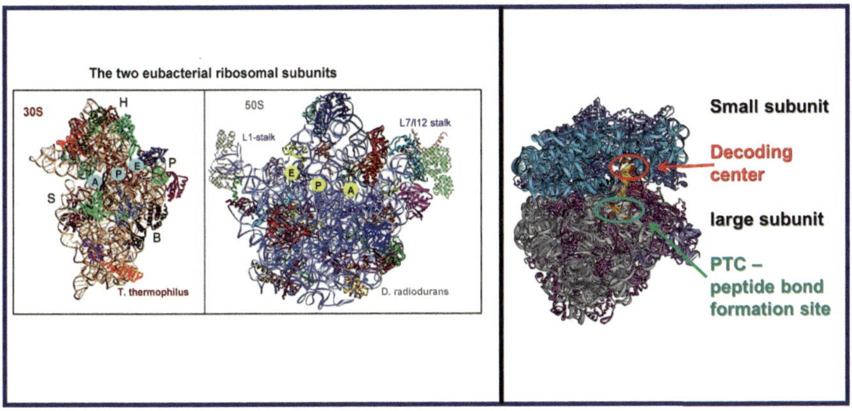

Fig. 2 The structures of bacterial ribosomal subunits with designations of the tRNA binding sites. In both parts – the ribosomal RNA is shown by lines, and the proteins as coils. Left – the two subunits, Right – the assembled ribosome

Currently, thanks to thoughtful studies, it is clear that the quality control of ribosomes creation and function is rather complicated. Also, we understand better the initial framework of origin of life and have the ability to focus on the contribution of selected structural features to the creation of several ribosome's sub-regions [12]. Still, the complete picture of the ribosome's evolution is not fully clear.

2.1.1 Open Major Issues

- While the overall ribosomes' function is rather well understood for all kingdoms of life, our current understanding of all factors controlling and regulating the processes involved in cellular protein production, including the dynamics of protein turnover within cells and tissues, is still not complete. Thus, only a part of the complex regulatory network that controls the overall process has been toughly studied. Consequently, our understanding of the signaling pathways that initiate and terminate protein biosynthesis is only partially uncovered [13].
- Currently, thanks to thoughtful studies, it is clear that the quality control of ribosomes creation and function is rather complicated. Also, we understand better the initial framework of origin of life (see below) and have the ability to focus on the contribution of selected structural features to the creation of several ribosome's sub-regions [12]. Still, the complete picture of the ribosome's evolution is not fully clear.
- As cell vitality requires fast and smooth processing of protein formation, the ribosome must possess features participating in processes allowing response to cellular signals. Do we expect to completely discover and understand these features?
- Efficient processivity of the ribosome catalytic activities depends on accurate positioning of the ribosomal substrates, and it has been suggested that disorder of the PTC may have a functional role in this process. Is this a result of a natural strategy to minimize cell function under hostile conditions?
- RNA modifications are common in biology and are particularly prevalent in rRNA. The specific function of each of these modifications is not fully understood. It is thought that they may play a role in the structure and stability of ribosomes as well as in the regulation of protein synthesis. In addition, they may be involved in ribosome maturation [14], or in the evolution and adaptation to different environments as well as to "enemies," such as antibiotics, by acquiring resistance, e.g., A2058G in bacteria ([15]; [16]; [17]; [18]), but most of them are still elusive.

2.2 Bacterial Growth Under Stress

Bacteria devised various mechanisms for responding to hostile growth environments. For example, they can maintain their existence under stressful conditions by temporarily stopping their normal life, thanks to a specific "hibernation mode" which temporarily stops ribosomal activity [19, 20]. Other survival mechanisms showing adaption to harsh surrounding environments have also been uncovered. For example, *Deinococcus radiodurans* (*D. radiodurans*) is an extremely robust Gram-positive mesophilic eubacterium that nevertheless shares extensive similarities with *Escherichia coli* and *T. thermophilus*. It was originally identified as a contaminant of irradiated canned meat. Currently it is isolated from environments that are either very rich or extremely poor in organic nutrients. As this bacterium lives under stress, its ribosomes are more stable than those of other bacteria, and therefore their large subunits (called D50S) could be among the first ribosomal particles to crystallize, and yielded high resolution structural details, especially of the PTC [21–23] which facilitated further sophisticated studies that shed light on the origin of life [24–28].

D. radiodurans contains a few stress-related features, among them the general stress protein CTC, which undergoes conformational changes upon binding the substrate analog ASM (tRNA acceptor stem mimic) to the ribosome, although it does not interact directly with the bound ASM. CTC is a ribosomal protein that regulates tRNA binding to ribosomes [29]. Among the known CTC proteins, the *D. radiodurans* CTC is the longest. In D50S it has three domains. One is located on the ribosome's solvent side and is similar to the single domain *E. coli* protein L25. The combination of this and the second *D. radiodurans* CTC domain resembles the *T. thermophilus* homologue, TL5. The third, which seems to control tightly the A-site tRNA binding, is unique to *D. radiodurans* (Fig. 3). In fact, in D50S each of the domains of protein CTC has a defined task. The N-terminal domain stabilizes the intersubunit-bridge confining the A-site-tRNA entrance. The middle domain protects the intersubunit B1a bridge even at harsh conditions, like elevated temperatures, and the C-terminal domain that can undergo substantial conformational rearrangements upon substrate binding indicates that CTC participates in biosynthesis-control under stressful conditions. Thus, the interactions of CTC with the solvent side of the large subunit central protuberance, its ability to enhance the stability of the B1b intersubunit bridge, and its involvement in controlling the A-site tRNA binding by space exclusion seem to indicate a part of the mechanisms that *D. radiodurans* developed for its survival under stress.

2.2.1 Open Major Issues

- What were the driving forces that created the *D. radiodurans* robust bacteria? When was it created?
- How was the natural biosynthesis of the general stress protein CTC controlled by various alternative factors?
- Do the structures of the *D. radiodurans* CTC domains indicate the *D. radiodurans* development path? Dose the structure of CTC of *D. radiodurans* have any bearing on its development path?
- Is protein CTC part of the stress response of *D. radiodurans*?
- Are there other stress-related features in the biology of *D. radiodurans?*

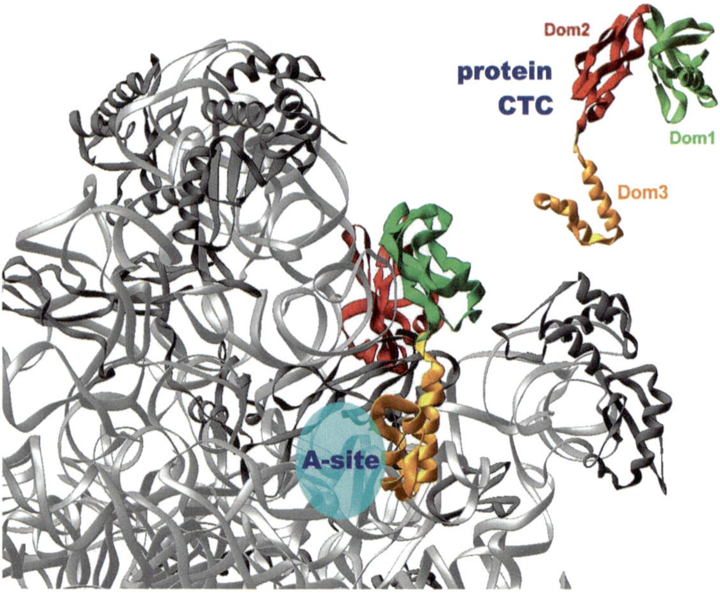

Fig. 3 (a) A zoom into part of D50S, in which the proteins and the RNA backbones are shown in gray, except for protein CTC, which undergoes substantial conformational changes upon ASM binding, although it does not interact with it. CTC domain-1 resembles protein E. coli L25h is shown in red; domain-2, which together with domain 1 resembles protein TL5 of T. thermophilus is shown in green; and domain-3, which is unique to D. radiodurans, is shown in gold. The position of PTC is shown in cyan. The top-right insert shows the structure of protein CTC in D. radiodurans

2.3 Functional Motions and Protective Flexibility

2.3.1 Post-Peptide-Bond Formation: Gating and Discrimination

The nascent proteins exit from the ribosome was assumed to be an automatic process after their termination. Actually, it was also assumed to be a fast process, required for efficient overall cellular function. However, conversely it was found that not only the cells include features facilitating control of this event; even the ribosomes are capable of controlling it. Thus, residing on the exit tunnel walls near its exit, and stretching around its opening, ribosomal protein L22 seems to mediate ribosome response to cellular regulatory signals, since it can swing across the tunnel, and cause gating as well as elongation arrest [21, 23]. Protein L22 consists of a single globular domain and a highly conserved hairpin that has a unique twisted conformation [30, 31]. Within the ribosome, it is positioned with its globular domain on the surface of the large subunit, where it can sense the conditions on the ribosome's periphery, whereas its hairpin lines the protein exit tunnel wall and extends approximately 30 Å away from the protein core (Fig. 4). The rather flexible tip of protein L22 can flip across the tunnel and interact with its both sides,

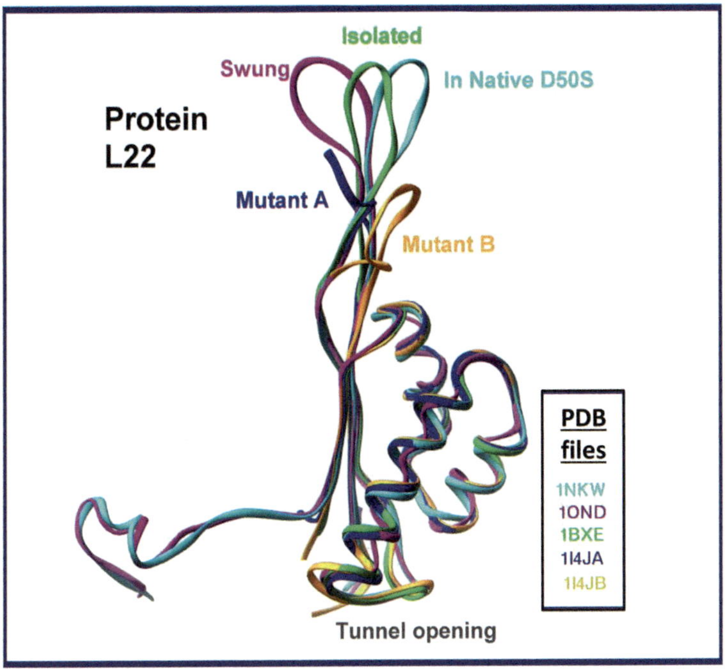

Fig. 4 Protein L22 and its flexible tip, as found in various bacterial ribosomes. The PDB file numbers are given in the inserted box

suggesting that a similar swing is involved in the regulatory role assigned to the tunnel.

2.3.2 Open Major Issues

- Is a single protein sufficient for maintaining and performing gating of the nascent proteins exit tunnel? Logically, it seems that a single protein is not sufficient for such a complicated concerted role.
- The nascent protein exit tunnel is rather long in molecular terms. Why is such a long path needed to protect the nascent protein while being synthetized?
- Is the discrimination at the tunnel exit obtained by the flexibility of the long arm of protein L22, the only mean for controlling the pace of protein biosynthesis?

2.4 Origin of Life: The Proto Ribosome—Symmetry at the Active Site of the Ribosome: Structural and Functional Implications

The PTC, namely, the peptidyl transferase center, is the rRNA region residing within the ribosome large subunits where peptide bonds are being formed. This universal semi-symmetrical region, comprising of about 200 rRNA nucleotides, has been

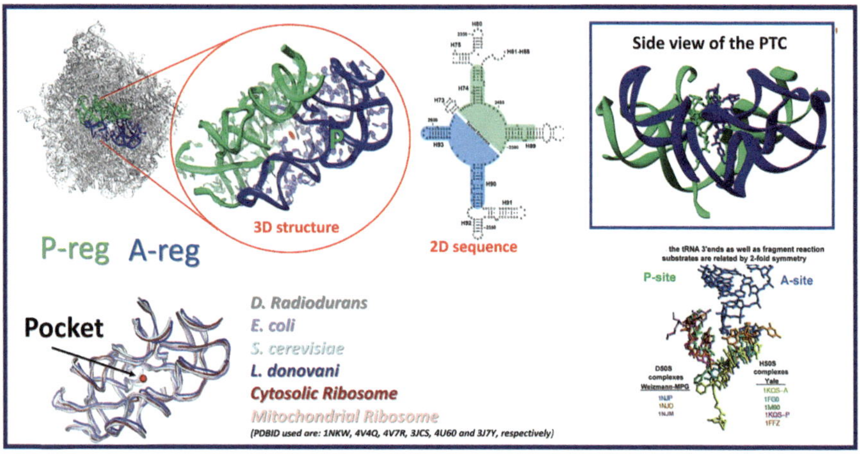

Fig. 5 The protoribosome concept. The upper part shows the top shows the structure and the sequence of the contemporary PTC with its two subregions A and P. the lower part indicates its universality

identified in and around the PTC in the large ribosomal subunits of ribosome from all kingdoms of life. Apart from confirming that the ribosome is an RNA machine, namely, a ribozyme, this finding evoked the suggestion that this region was originally a site useful for RNA world interactions, in which amino acids could also dimerize. It later evolved to become the ribosome active center, where the ASMs of A and P tRNAs meet at the entrance to the nascent proteins exit tunnel.

Thus, the ribosomal active site, where the peptide bonds are being formed, is situated within a universal semi-symmetrical region that is embedded in the otherwise asymmetric ribosome structure. This highly conserved region may be the remnant of the protoribosome, which seems to be a dimeric prebiotic machine that initially catalyzed prebiotic reactions, including the rather late formation of chemical bonds (Fig. 5).

In laboratory experiments aimed at imitating the formation of such pockets, namely, constructions meant to mimic the protoribosome, a marked preference of pockets composed of two identical chains, resembling the current PTC P-site, was observed. These pockets were obtained by dimerization of RNA chains of sequences resembling mainly of the P-region of the contemporary PTC, and thus may indicate that the protoribosome was originally a symmetrical homodimer, namely, a pocket made of RNA chains of the same sequence in each of its parts [32]. This concept is in line with the assumption that originally the protoribosome provided almost equal RNA interactions to both its substrates, located at each side of the pocket and created the peptide bonds [26, 27].

Later, with the evolving preference of initial oligopeptides, like those that seem to be able to stabilize the protoribosome, alongside the evolving optimization of the protoribosome into an RNA machine, each of its two parts was independently adjusted to fulfill its role in peptide bond formation, namely, the entrance to the amino acid charged ASM to A-site and the pushing out the uncharged ASM from the P-site. In this way, the protoribosome adapted to the specific polypeptides' formation requirements, a process through which the protoribosome matured into the PTC contemporary form. Thus, the transition from homo- to hetero-dimers occurred by optimizing the functionality of the protoribosome toward its development to a molecular machine, and the protoribosome conception indicated how the RNA world could be linked to modern life [32].

Furthermore, seemingly crucial features to the polymerase activity of the early ribosome were also identified [33], and models of the PTC activity with initial peptides as larger substrates have indicated which of them could be the initial amino acids [34].

2.4.1 Open Major Issues

- Did the protoribosome appear spontaneously in the prebiotic world? Why did it appear?
- Was the protoribosome a symmetrical pocket? How was it produced?
- How did the transition from the initial homo- to the later hetero-dimers occur?
- How was the functionality of the protoribosome optimized?
- What are the evolutionary roots of the ribosome?

3 Protein Biosynthesis in Medicine

3.1 Ribosomal Antibiotics: Contemporary Challenges in Medical Usage of Antibiotics

Owing to the vital role played by the ribosomes, many antibiotics target them and successfully obstruct the key ribosomal functions [35–43]. However, the extremely fast increase in antibiotic resistance of many pathogenic bacteria alongside the slow progress (actually negligible) in developing new antibiotics by pharma companies worldwide causes serious medical issues [44–47], including inability to use the antibiotics for treating infections. This leads to severe consequences, including prolonged illness, disability, and even death. Furthermore, most antibiotics do not distinguish between pathogens and nonpathogenic bacteria [48] which may, in turn, affect overall health.

About half of the currently used antibiotics target the process of protein biosynthesis, as it is a key process of life, by blocking the internal ribosome's active sites, such as the mRNA decoding path, the PTC, or the protein exit tunnel. In contrast, instead of targeting the internal ribosomal active for reducing and/or controlling antibiotics resistance, pathogens peripheral species-specific sites, are (Fig. 6), selected [39, 49]. These peripheral sites are identified in various steps, the first of which is performed by structural comparisons to non-pathogenic bacteria.

In principle, as each pathogen tends to contain more than a single such site, several matching molecules may be designed. Hence, for each pathogen several lead

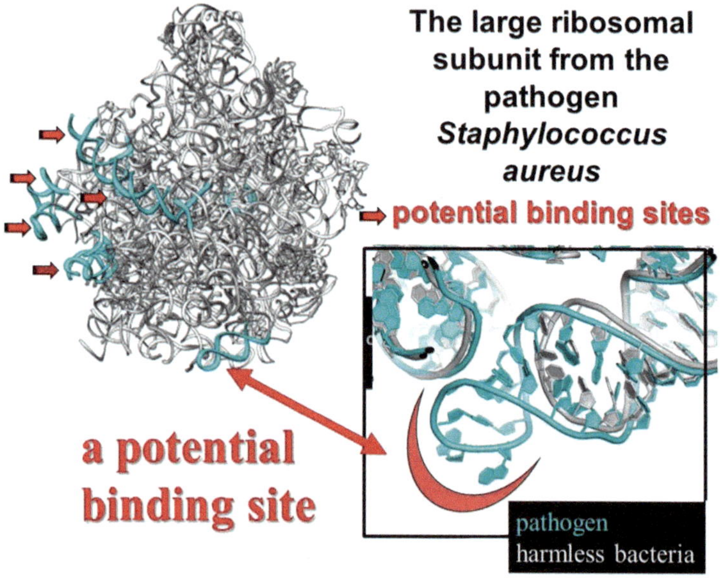

Fig. 6 Pathogens unique ribosomes peripheral species-specific structural sites, which may be selected to attach environmentally friendly compounds for the development of novel anti-pathogens drugs

compounds that have the potential to be further developed into a next-generation antibiotic drug are being identified. Thus, although not understanding the reasons for the existence of the pathogen-specific peripheral exposed regions, selective binding of antisense oligonucleotides (ASOs) or their chemical imitations [50] could be achieved.

It was shown that lipid nanoparticles (LNPs) are suitable for broad-spectrum nucleic acid delivery [51], as well as means for human vaccination, for example, in the COVID-19 Pfizer-BioNTech vaccine [52]. Combining ASO approaches with the nanoparticles-based nucleic acid delivery can potentially lead to the design of "pathogen-specific antibiotics," in contrast to the current preference for broad-spectrum antimicrobial drugs that target the highly conserved functional sites of many different species, including helpful bacteria.

In short, the inhibition of protein biosynthesis by the newly identified potential sites for binding of molecules composed of nucleic acids, oligopeptides, etc., which can be optimized in terms of their antibiotics action, chemical properties, poisonous level, modes of penetration, delivery style, and biodegradability.

3.1.1 Open Major Issues

- What is the origin of the differences between the ribosomes surfaces of pathogenic and benign bacteria?

- Will the alternative approach, namely, using for antibiotic design pathogen-specific ribosomal peripheral features instead of the ribosomal active regions, lead to a slower pace of resistance appearance?

3.2 The Eukaryotic Ribosome as a Medicinal Target

Insight into the structural mechanisms of the pathways of disease of protein biosynthesis should provide the basis for innovative structure-based drug discovery, as well as for the development of novel therapeutic approaches for several human diseases, such as cancers, genetic disorders, and infectious diseases. Thus, owing to the immense importance of protein biosynthesis, intense studies on the ribosome's structural elements and the mechanisms and on the dynamics of biological and disease-linked cellular pathways involved in protein biosynthesis are being performed. For example, it was shown that most natural rRNA modifications cluster around functionally important regions of the ribosome, including the decoding center and the peptidyl transferase center, where they are thought to stabilize folding and the tertiary structure of RNA at these functionally important sites [53–57]. Nevertheless, despite the intensive scientific activity, the natural creation of ribosomes is still considered a difficult puzzle [14, 58, 59].

The investigations on non-coding RNAs that are implicated in human disease yielded some less expected results, including mRNA usage for vaccination by the design of nano-chemically stabilized chains that allows for sufficient tissue and cellular penetration. Among the unforeseen results, it was found that even a single RNA modification can play a critical role in the assembly of the ribosome. For example, a single methylation of ribosomal RNA was found to gate the correct assembly of functional ribosomes [14]. An additional example concerns ribosomes involvement in normal and "ill" expression of proteins in a rather complicated and still not fully investigated manner. Thus, tissue-specific regulation of protein expression pattern in mammals was discovered even in highly complicated events, connected to male fertility [60]. Additionally, since rRNA does not act only as the central scaffold for ribosomal subunits but also, in fact mainly, serves as the center for catalytic activity in ribosome biogenesis, abnormal pre-rRNA processing may cause defects in ribosomal functions, such as unusually early aging, disrupted cardiac protein balance, and induced cardiac hypertrophy, blood and neurodegenerative diseases.

3.3 Ribosomopathies and Somatic Mutations

Ribosomopathies are congenital ribosome-malfunction-related diseases, connected to malfunctioning or reduced function of translating ribosomes, which are characterized by defects in RPs, rRNA processing, or in the assembly of the ribosomes [13, 61, 62]. They include blood diseases like Diamond-Blackfan anemia (DBA) and the Shwachman-Diamond syndrome (SDS), X-linked dyskeratosis

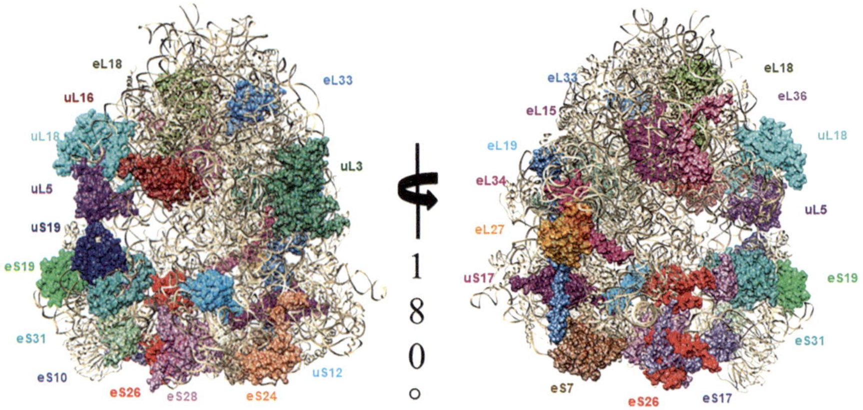

Fig. 7 Shows all DBA genetically mutated ribosomal proteins. Note their location on or close to the ribosomes surface

congenita (DC), cartilage hair hypoplasia (CHH), and Treacher Collins syndrome (TCS) [61]. Among them, *RPS19* is the first ribosomal gene that was implicated in human disease [63]. It is the most frequently mutated gene in DBA with a total of over 77 mutations that are mostly either whole gene deletion, translocation, or truncation [64].

The molecular pathogenesis studies of DBA patients showed that approximately half of all DBA cases are attributed to mutations in the genes of the ribosomal proteins *RPL35A, RPS36, RPS2, RPS7, RPS10, RPS15, RPS17, RPS19, RPS26, RPS24, RPS27A, RPS28 RPL2, RPL5, RPL7, RPL9, RPL14, RPL19, RPL23A, RPL26, RPL35, RPL36, RPS7, RPS8, RPS10, RPL11* [62, 65], all located on or close to the ribosomes surface (Fig. 7). Among those, in ~50% of DBA patients, the genes coding for proteins RPS19 (called also eS19), RPL5 (uL18), RPL11 (uL5), and RPS10 (eS10) are the most frequently mutated [66].

Many unanswered questions are associated with ribosomopathies, for example, so far, the mutations of the DBA encoding genes have been identified only at the genetic level. Also, a search for common denominators among the various mutations, performed at the genes as well as the proteins level, did not reveal any clear linkage between clinical syndromes and the types of their associated mutations or their locations. Furthermore, still there is no answer for the fascinating query: providing DBA disease originates from the ribosomal mutations, how can the mutations triggered by these mutations cause such significant medical problems only in selected parts of the body, although the ribosomes are necessary in all tissues?

These findings corroborate the view that the actual basis for the DBA and other ribosomopathies is caused by the existence of fewer ribosomes. In fact, currently it is not known if the gene-products are indeed expressed and incorporated in the ribosomes of the patients, or if the many mutated genes of the ribosomal proteins exist only in the genes of their ribosomes. It is commonly suggested, but not proven, that the structural alterations caused by the mutations prevent their incorporation into partially pre-formed ribosomes, which consequently hinder ribosome assembly, thus resulting in a severe reduction in the number of functioning ribosomes.

However, conflicts between this general view and the observed clinical diversity, namely, the observed connections between the patients' medical symptoms and the r-proteins mutations, point at mutated r-proteins involvement in protein biosynthesis, and challenge the ribosome-deficiency common view. In fact, this opinion oversights several unambiguous indications of explicit connections between specific mutations and specific types/appearances of the diseases (e.g., the previously noted links between particular ribosome's mutations and several types of cancer, as well as of the various DBA's individual tissue-specific symptoms, and their associated unique physical personal abnormalities, like short height, or cleft palate, that are connected to mutated *RPL5* and *RPL11*, or the specific genetic connections to defected heart, or renal anomalies, or bone marrow failure syndromes).

Moreover, this common view opposes findings confirming the existence of ribosome-incorporated mutated r-proteins, like (a) those related to the connection between ribosomes regulation and the under expression of ribosomal protein RPL22 [67]; (b) those regulating the expression of stress-response regulating genes, like *RPL3* [68]; (c) the uS12-mutant ribosomes that were shown to promote dysmorphism in a recently described ribosomopathies diseases, via increased levels of amino acid disincorporation [69]; (d) those connected to the finding that ribosomal ambiguity mutation Rps2-A226Y that is involved in mice early aging [70]; (e) the connection between modified rRNA base 1248 that is related to common cancer-connected genes, such as p53 [71]; (f) those containing laboratory engineered selected mutation R116D in *RPS3*, which led, as originally planned, to faulty mRNA translation [72]; (g) significant changes that were identified by ribosome footprinting or polysomal RNA sequencing [73].

Other members of this group include blood diseases like 5q- myelodysplastic syndrome. These diseases are associated with genetic mutations of the biosynthetic machinery, including mutations residing in ribosomal components, which may cause mistakes in the process of genetic-code translation, or lead to problems in ribosomes biogenesis, namely, the creation of partially assembled ribosomes owing to mutations in their components.

It was also found that many somatic mutations occur in different ribosomal parts in about 25% of all cancers. Examples include the recurrent R98S mutation of ribosomal protein *RPL10*, which normally facilitates the IRES-dependent translation [74], or mutations in *RPS15* in chronic lymphocytic leukemia (CLL) patients [75] and deletion of *RPS14* in 5q-myelodysplastic syndrome [76, 77].

3.3.1 Open Major Issues

- Despite the impressive recent advances in understanding ribosomopathies, several basic intriguing questions still exist and require further conceptual and/or practical studies. Among them, those relating to the possible generalization vs. ribosome specialization need further attention.

- Are all the genetic mutations expressed? In other words, do the mutations dysregulate the ribosome or just lead to a smaller number of correctly functioning ribosome? Or, do both suggested mechanisms exist?
- How can self-renewing tissues happen in cells that are assumed to have a combination of malfunctioning and well-functioning ribosomes?
- What are the bases of the tissue-specificity of DBA and the other ribosomopathies?
- Are there any common denominators among the various cancer mutations?
- What is the linkage between clinical syndromes and the types of the mutations or their locations?
- A highly intriguing point: although the mutations in DBA are expected to create very small fluctuations in the ribosomes structure, they were shown to exceedingly influence its assembly or perform its functions. How does this happen? How can such small alterations cause such significant medical problems? How are these problems expressed only in a part of the body, although the ribosomes are necessary and function in all tissues?

4 Conclusions

This manuscript describes studies that were mainly driven by curiosity, which opened various paths, including medical research, which were not reachable otherwise. It also highlights the need of more research in almost all points discussed here, including origin of life on earth, the roles played by rRNA modifications, the relationship between disorder and ribosome function, and the origins of ribosomopathies.

References

1. Shalev-Benami M, Zhang Y, Matzov D, et al. 2.8-Å Cryo-EM structure of the large ribosomal subunit from the eukaryotic parasite Leishmania. Cell Rep. 2016;16:288–94. https://doi.org/10.1016/j.celrep.2016.06.014.
2. Shalev-Benami M, Zhang Y, Rozenberg H, et al. Atomic resolution snapshot of Leishmania ribosome inhibition by the aminoglycoside paromomycin. Nat Commun. 2017;8:1589. https://doi.org/10.1038/s41467-017-01664-4.
3. Gindulyte A, Bashan A, Agmon I, et al. The transition state for formation of the peptide bond in the ribosome. Proc Natl Acad Sci U S A. 2006;103:13327–32. https://doi.org/10.1073/pnas.0606027103.
4. Huang L, Krupkin M, Bashan A, et al. Protoribosome by quantum kernel energy method. Proc Natl Acad Sci U S A. 2013;110:14900–5. https://doi.org/10.1073/pnas.1314112110.
5. Javed A, Christodoulou J, Cabrita LD, Orlova EV. The ribosome and its role in protein folding: looking through a magnifying glass. Acta Crystallogr D Struct Biol. 2017;73:509–21. https://doi.org/10.1107/S2059798317007446.
6. Blobel G, Sabatini DD. Controlled proteolysis of nascent polypeptides in rat liver cell fractions. J Cell Biol. 1970;45:130–45. https://doi.org/10.1083/jcb.45.1.130.

7. Malkin LI, Rich A. Partial resistance of nascent polypeptide chains to proteolytic digestion due to ribosomal shielding. J Mol Biol. 1967;26:329–46. https://doi.org/10.1016/0022-2836(67)90301-4.
8. Milligan RA, Unwin PN. Location of exit channel for nascent protein in 80S ribosome. Nature. 1986;319:693–5. https://doi.org/10.1038/319693a0.
9. Yonath A, Leonard KR, Wittmann HG. A tunnel in the large ribosomal subunit revealed by three-dimensional image reconstruction. Science. 1987;236:813–6. https://doi.org/10.1126/science.3576200.
10. Amit M, Berisio R, Baram D, et al. A crevice adjoining the ribosome tunnel: hints for cotranslational folding. FEBS Lett. 2005;579:3207–13. https://doi.org/10.1016/j.febslet.2005.03.023.
11. Bashan A, Yonath A. Ribosome crystallography: catalysis and evolution of peptide-bond formation, nascent chain elongation and its co-translational folding. Biochem Soc Trans. 2005;33:488–92. https://doi.org/10.1042/BST0330488.
12. Bowman JC, Petrov AS, Frenkel-Pinter M, et al. Root of the tree: the significance, evolution, and origins of the ribosome. Chem Rev. 2020;120:4848–78. https://doi.org/10.1021/acs.chemrev.9b00742.
13. Jiao L, Liu Y, Yu X-Y, et al. Ribosome biogenesis in disease: new players and therapeutic targets. Sig Transduct Target Ther. 2023;8:15. https://doi.org/10.1038/s41392-022-01285-4.
14. Yelland JN, Bravo JPK, Black JJ, et al. A single 2′-O-methylation of ribosomal RNA gates assembly of a functional ribosome. Nat Struct Mol Biol. 2022;30:91. https://doi.org/10.1038/s41594-022-00891-8.
15. Cundliffe E. How antibiotic-producing organisms avoid suicide. Annu Rev Microbiol. 1989;43:207–33. https://doi.org/10.1146/annurev.mi.43.100189.001231.
16. Weisblum B. Erythromycin resistance by ribosome modification. Antimicrob Agents Chemother. 1995;39:577–85. https://doi.org/10.1128/AAC.39.3.577.
17. Pfister P, Corti N, Hobbie S, et al. 23S rRNA base pair 2057–2611 determines ketolide susceptibility and fitness cost of the macrolide resistance mutation 2058A→G. Proc Natl Acad Sci U S A. 2005;102:5180–5. https://doi.org/10.1073/pnas.0501598102.
18. Berisio R, Corti N, Pfister P, et al. 23S rRNA 2058A→G alteration mediates Ketolide resistance in combination with deletion in L22. Antimicrob Agents Chemother. 2006;50:3816–23. https://doi.org/10.1128/AAC.00767-06.
19. Matzov D, Aibara S, Basu A, et al. The cryo-EM structure of hibernating 100S ribosome dimer from pathogenic Staphylococcus aureus. Nat Commun. 2017a;8:723. https://doi.org/10.1038/s41467-017-00753-8.
20. Matzov D, Bashan A, Yap MF, Yonath A. Stress response as implemented by hibernating ribosomes: a structural overview. FEBS J. 2019;286:3558–65. https://doi.org/10.1111/febs.14968.
21. Bashan A, Agmon I, Zarivach R, et al. Structural basis of the ribosomal machinery for peptide bond formation, translocation, and nascent chain progression. Mol Cell. 2003;11:91–102.
22. Harms J, Schluenzen F, Zarivach R, et al. High resolution structure of the large ribosomal subunit from a mesophilic eubacterium. Cell. 2001;107:679–88. https://doi.org/10.1016/S0092-8674(01)00546-3.
23. Zarivach R, Bashan A, Berisio R, et al. Functional aspects of ribosomal architecture: symmetry, chirality and regulation. J Phys Org Chem. 2004;17:901–12. https://doi.org/10.1002/poc.831.
24. Agmon I, Auerbach T, Baram D, et al. On peptide bond formation, translocation, nascent protein progression and the regulatory properties of ribosomes. Delivered on 20 October 2002 at the 28th FEBS meeting in Istanbul. Eur J Biochem. 2003;270:2543–56. https://doi.org/10.1046/j.1432-1033.2003.03634.x.
25. Agmon I, Bashan A, Yonath A. On ribosome conservation and evolution. Israel J Ecol Evol. 2006;52:359–74. https://doi.org/10.1560/IJEE_52_3-4_359.

26. Bose T, Fridkin G, Bashan A, Yonath A. Origin of life: chiral short RNA chains capable of non-enzymatic peptide bond formation. Isr J Chem. 2021;61:863–72. https://doi.org/10.1002/ijch.202100054.
27. Bose T, Fridkin G, Davidovich C, et al. Origin of life: protoribosome forms peptide bonds and links RNA and protein dominated worlds. Nucleic Acids Res. 2022;50:1815–28. https://doi.org/10.1093/nar/gkac052.
28. Davidovich C, Belousoff M, Wekselman I, et al. The proto-ribosome: an ancient Nano-machine for peptide bond formation. Isr J Chem. 2010;50:29–35. https://doi.org/10.1002/ijch.201000012.
29. Krupkin M, Wekselman I, Matzov D, et al. Avilamycin and evernimicin induce structural changes in rProteins uL16 and CTC that enhance the inhibition of A-site tRNA binding. Proc Natl Acad Sci U S A. 2016;113:E6796. https://doi.org/10.1073/pnas.1614297113.
30. Halfon Y, Matzov D, Eyal Z, et al. Exit tunnel modulation as resistance mechanism of S. aureus erythromycin resistant mutant. Sci Rep. 2019;9:11460. https://doi.org/10.1038/s41598-019-48019-1.
31. Wekselman I, Zimmerman E, Davidovich C, et al. The ribosomal protein uL22 modulates the shape of the protein exit tunnel. Structure. 2017;25:1233–1241.e3. https://doi.org/10.1016/j.str.2017.06.004.
32. Bashan A, Belousoff M, Davidovich C, Yonath A. Origins of life and evolution of the biosphere: the journal of the International Society for the Study of the origin of life. Orig Life Evol Biosph. 2010;40:347–497. https://doi.org/10.1007/s11084-010-9213-2.
33. Fox GE, Tran Q, Yonath A. An exit cavity was crucial to the polymerase activity of the early ribosome. Astrobiology. 2012;12:57–60. https://doi.org/10.1089/ast.2011.0692.
34. Agmon I. Prebiotic assembly of cloverleaf tRNA, its Aminoacylation and the origin of coding, inferred from acceptor stem coding-triplets. Int J Mol Sci. 2022;23:15756. https://doi.org/10.3390/ijms232415756.
35. Auerbach T, Bashan A, Yonath A. Ribosomal antibiotics: structural basis for resistance, synergism and selectivity. Trends Biotechnol. 2004;22:570–6. https://doi.org/10.1016/j.tibtech.2004.09.006.
36. Belousoff MJ, Shapira T, Bashan A, et al. Crystal structure of the synergistic antibiotic pair, lankamycin and lankacidin, in complex with the large ribosomal subunit. Proc Natl Acad Sci U S A. 2011;108:2717–22. https://doi.org/10.1073/pnas.1019406108.
37. Berisio R, Schluenzen F, Harms J, et al. Structural insight into the role of the ribosomal tunnel in cellular regulation. Nat Struct Biol. 2003;10:366–70. https://doi.org/10.1038/nsb915.
38. Harms JM, Bartels H, Schlünzen F, Yonath A. Antibiotics acting on the translational machinery. J Cell Sci. 2003;116:1391–3. https://doi.org/10.1242/jcs.00365.
39. Matzov D, Bashan A, Yonath A. A bright future for antibiotics? Annu Rev Biochem. 2017b;86:567–83. https://doi.org/10.1146/annurev-biochem-061516-044617.
40. Schlünzen F, Harms JM, Franceschi F, et al. Structural basis for the antibiotic activity of ketolides and azalides. Structure. 2003;11:329–38. https://doi.org/10.1016/s0969-2126(03)00022-4.
41. Schlünzen F, Zarivach R, Harms J, et al. Structural basis for the interaction of antibiotics with the peptidyl transferase Centre in eubacteria. Nature. 2001;413:814–21. https://doi.org/10.1038/35101544.
42. Tenson T, Mankin A. Antibiotics and the ribosome. Mol Microbiol. 2006;59:1664–77. https://doi.org/10.1111/j.1365-2958.2006.05063.x.
43. Yonath A. Antibiotics targeting ribosomes: resistance, selectivity, synergism and cellular regulation. Annu Rev Biochem. 2005;74:649–79. https://doi.org/10.1146/annurev.biochem.74.082803.133130.
44. Kurosu M, Siricilla S, Mitachi K. Advances in MRSA drug discovery: where are we and where do we need to be? Expert Opin Drug Discov. 2013;8:1095–116. https://doi.org/10.1517/17460441.2013.807246.
45. Ventola CL. The antibiotic resistance crisis: part 1: causes and threats. P T. 2015;40:277–83.

46. WHO. Antimicrobial resistance: global report on surveillance. Geneva: WHO; 2014.
47. WHO. Global antimicrobial resistance and use surveillance system (GLASS) report 2022. Geneva: WHO; 2022.
48. Willing BP, Russell SL, Finlay BB. Shifting the balance: antibiotic effects on host-microbiota mutualism. Nat Rev Microbiol. 2011;9:233–43. https://doi.org/10.1038/nrmicro2536.
49. Cimicata G, Bose T, Fridkin G, et al (2018) Contemporary challenges in medical usage of antibiotics. In: Transformative roles of science in society: from emerging basic science toward solutions for People's wellbeing. Vatican City, pp 97–101.
50. Gagliardi M, Ashizawa AT. The challenges and strategies of antisense oligonucleotide drug delivery. Biomedicine. 2021;9:433. https://doi.org/10.3390/biomedicines9040433.
51. Hejdankova Z, Vanek V, Sedlak F, et al. Lipid nanoparticles for broad- Spectrum nucleic acid delivery. Adv Funct Materials. 2021;31:2101391. https://doi.org/10.1002/adfm.202101391.
52. Badiani AA, Patel JA, Ziolkowski K, Nielsen FBH. Pfizer: the miracle vaccine for COVID-19? Public Health Pract (Oxf). 2020;1:100061. https://doi.org/10.1016/j.puhip.2020.100061.
53. Decatur WA, Fournier MJ. rRNA modifications and ribosome function. Trends Biochem Sci. 2002;27:344–51. https://doi.org/10.1016/S0968-0004(02)02109-6.
54. Kim DF, Green R. Base-pairing between 23S rRNA and tRNA in the ribosomal a site. Mol Cell. 1999;4:859–64. https://doi.org/10.1016/s1097-2765(00)80395-0.
55. Panse VG, Johnson AW. Maturation of eukaryotic ribosomes: acquisition of functionality. Trends Biochem Sci. 2010;35:260–6. https://doi.org/10.1016/j.tibs.2010.01.001.
56. Polikanov YS, Melnikov SV, Söll D, Steitz TA. Structural insights into the role of rRNA modifications in protein synthesis and ribosome assembly. Nat Struct Mol Biol. 2015;22:342–4. https://doi.org/10.1038/nsmb.2992.
57. Sloan KE, Warda AS, Sharma S, et al. Tuning the ribosome: the influence of rRNA modification on eukaryotic ribosome biogenesis and function. RNA Biol. 2017;14:1138–52. https://doi.org/10.1080/15476286.2016.1259781.
58. Baßler J, Hurt E. Eukaryotic ribosome assembly. Annu Rev Biochem. 2019;88:281–306. https://doi.org/10.1146/annurev-biochem-013118-110817.
59. Kressler D, Hurt E, Baßler J. A puzzle of life: crafting ribosomal subunits. Trends Biochem Sci. 2017;42:640–54. https://doi.org/10.1016/j.tibs.2017.05.005.
60. Li H, Huo Y, He X, et al. A male germ-cell-specific ribosome controls male fertility. Nature. 2022;612:725–31. https://doi.org/10.1038/s41586-022-05508-0.
61. Mills EW, Green R. Ribosomopathies: There's strength in numbers. Science. 2017;358: eaan2755. https://doi.org/10.1126/science.aan2755.
62. Narla A, Ebert BL. Ribosomopathies: human disorders of ribosome dysfunction. Blood. 2010;115:3196–205. https://doi.org/10.1182/blood-2009-10-178129.
63. Lipton JM, Ellis SR. Diamond-Blackfan anemia: diagnosis, treatment, and molecular pathogenesis. Hematol Oncol Clin North Am. 2009;23:261–82. https://doi.org/10.1016/j.hoc.2009.01.004.
64. Campagnoli MF, Ramenghi U, Armiraglio M, et al. RPS19 mutations in patients with diamond-Blackfan anemia. Hum Mutat. 2008;29:911–20. https://doi.org/10.1002/humu.20752.
65. Da Costa L, Moniz H, Simansour M, et al. Diamond-Blackfan anemia, ribosome and erythropoiesis. Transfus Clin Biol. 2010;17:112–9. https://doi.org/10.1016/j.tracli.2010.06.001.
66. Kampen KR, Sulima SO, Vereecke S, De Keersmaecker K. Hallmarks of ribosomopathies. Nucleic Acids Res. 2020;48:1013–28. https://doi.org/10.1093/nar/gkz637.
67. Yang M, Sun H, Wang H, et al. Down-regulation of ribosomal protein L22 in non-small cell lung cancer. Med Oncol. 2013;30:646. https://doi.org/10.1007/s12032-013-0646-0.
68. Pecoraro A, Carotenuto P, Russo G, Russo A. Ribosomal protein uL3 targets E2F1 and cyclin D1 in cancer cell response to nucleolar stress. Sci Rep. 2019;9:15431. https://doi.org/10.1038/s41598-019-51723-7.
69. Sulima SO, Kampen KR, De Keersmaecker K. Cancer biogenesis in Ribosomopathies. Cell. 2019;8(3):229. https://doi.org/10.3390/cells8030229.

70. Moore J, Akbergenov R, Nigri M, et al. Random errors in protein synthesis activate an age-dependent program of muscle atrophy in mice. Commun Biol. 2021;4:703. https://doi.org/10.1038/s42003-021-02204-z.
71. Babaian A, Rothe K, Girodat D, et al. Loss of m1acp3Ψ ribosomal RNA modification is a major feature of cancer. Cell Rep. 2020;31:107611. https://doi.org/10.1016/j.celrep.2020.107611.
72. Simms CL, Kim KQ, Yan LL, et al. Interactions between the mRNA and Rps3/uS3 at the entry tunnel of the ribosomal small subunit are important for no-go decay. PLoS Genet. 2018;14: e1007818. https://doi.org/10.1371/journal.pgen.1007818.
73. Kampen KR, Fancello L, Girardi T, et al. Translatome analysis reveals altered serine and glycine metabolism in T-cell acute lymphoblastic leukemia cells. Nat Commun. 2019a;10: 2542. https://doi.org/10.1038/s41467-019-10508-2.
74. Kampen KR, Sulima SO, Verbelen B, et al. The ribosomal RPL10 R98S mutation drives IRES-dependent BCL-2 translation in T-ALL. Leukemia. 2019b;33:319–32. https://doi.org/10.1038/s41375-018-0176-z.
75. Ljungström V, Rosenquist R. Not so lost in translation: RPS15 mutations in CLL. Blood. 2018;132:2317–9. https://doi.org/10.1182/blood-2018-09-875179.
76. Dutt S, Narla A, Lin K, et al. Haploinsufficiency for ribosomal protein genes causes selective activation of p53 in human erythroid progenitor cells. Blood. 2011;117:2567–76. https://doi.org/10.1182/blood-2010-07-295238.
77. Vlachos A. Acquired ribosomopathies in leukemia and solid tumors. Hematology. 2017;2017: 716–9. https://doi.org/10.1182/asheducation-2017.1.716.

Cellular Responses to DNA Damage: Translating Mechanistic Insights Towards New Therapies

Stephen P. Jackson

Abstract

All organisms possess DNA repair and associated DNA damage response (DDR) mechanisms to detect, signal and repair diverse forms of DNA damage. In addition to their fundamental importance for myriad aspects of cell physiology, DDR processes also have medical importance, as underlined by their deregulation or inactivation causing developmental defects, cancer predisposition, stem-cell exhaustion, infertility, immune-deficiencies, inflammation, neurodegeneration and/or premature ageing. New DDR proteins and regulators undoubtedly await discovery. Moreover, we still only have a rudimentary understanding of how human DDR components functionally connect as a network and are regulated by factors including cell-cycle stage, transcription, replication, chromatin, drugs and physiological stress. We also currently lack accurate and predictive models for DDR processes and how their functions connect to health and disease. In this chapter, I briefly survey the various forms of DNA damage and explain the general principles and processes by which cells respond to and repair them. I then go on to explain how our increasing knowledge of such processes is providing insights into human disease mechanisms, and is also paving the way for the development of new drugs, including some that are already extending the lives of cancer patients worldwide.

S. P. Jackson (✉)
Cancer Research UK Cambridge Institute, Li Ka Shing Building, University of Cambridge, Cambridge, UK

The Gurdon Institute, and Department of Biochemistry, University of Cambridge, Cambridge, UK
e-mail: steve.jackson@cruk.cam.ac.uk

U. A. Betz (ed.), *Curious Future Insight*,
https://doi.org/10.1007/978-3-031-41781-8_5

1 Introduction to DNA Damage and Its Repair

The DNA in every cell in the human body—and indeed the DNA in all cells in every organism on Earth—is continually being damaged [1]. This damage in part arises from actions of various external sources, including physical stresses, ultraviolet light and ionising radiations, components of burnt tobacco products and both natural and man-made environmental chemicals. Crucially, DNA damage also arises through the effects of myriad endogenously generated chemicals that arise as a part of normal physiology, including reactive oxygen species generated as by-products of oxidative metabolism. Other lesions in DNA can arise because of misincorporations of the wrong base, incorporations of corrupted DNA bases or ribonucleotides during DNA replication. Strikingly, even taking aside the actions of external factors, just about every one of the ~50 trillion cells in the human body sustains thousands of DNA lesions per day.

To deal with this onslaught, our cells, and those of all organisms, have evolved sophisticated molecular mechanisms to detect, signal the presence of and repair DNA damage in its myriad forms. As described further below, there are various DNA repair and associated pathways, each largely dedicated to a certain type of DNA damage. The efficiency and accuracy of such mechanisms allows sufficient genomic integrity and stability for organisms to carry out their normal lifespans, and usually enables the accurate passage of the genetic blueprint through the germline from generation to generation. Nevertheless, errors in DNA repair processes inevitably arise from time to time, leading to mutations, some with pathological consequences. Mutations range from single base changes and small insertions and deletions in the DNA sequence to larger deletions, chromosomal rearrangements and loss or amplifications of genomic regions. By mutating gene products or changing their expression, or by leading to defects in cell proliferation or chromosome segregation, mutations can impair key cellular functions, thereby contributing to a range of human pathologies, including neurodegenerative diseases, infertility, immune dysfunction, type-2 diabetes, premature aging and cancer [2, 3].

2 A Brief Survey of DNA Lesions and Their Dedicated Repair Systems

DNA damage ranges from the relatively simple cases where the wrong base, or chemically modified/corrupted base, is misincorporated into the DNA or when DNA bases become damaged via modifications such as alkylation or oxidation (Fig. 1). Other lesions include inter-strand DNA cross-links generated by agents such as platinum-based anticancer drugs, DNA-protein cross-links induced by radiations and by certain environmental and endogenously arising chemicals, and 6-4 photoproducts and pyrimidine dimers induced by the actions of ultraviolet light. In addition, the phosphodiester backbone of DNA can be broken via unrestrained actions of various cellular nucleases, by physical forces, via radiation, and certain reactive chemicals. DNA breaks are classed as single-strand breaks (SSBs) where

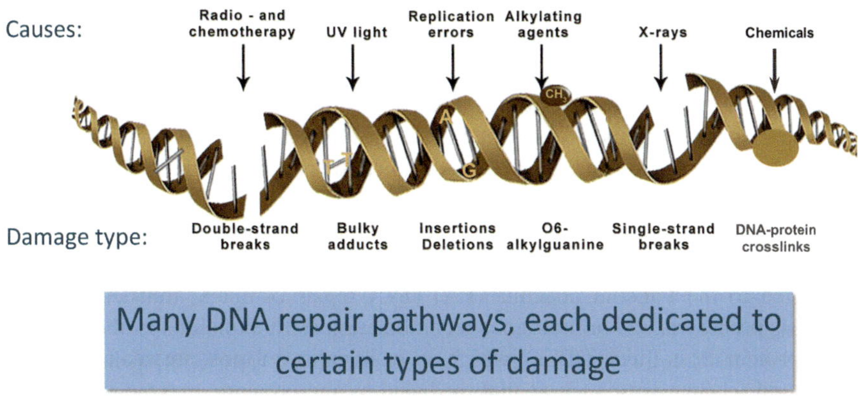

Fig. 1 DNA damage: many types and many cellular responses

one strand of the DNA duplex is broken, or double-strand breaks (DSBs) where two SSBs occur on opposite strands of the DNA duplex in relatively close proximity. DSBs are widely regarded as the most toxic form of DNA damage. Indeed, while agents such as ionising radiation produce relatively few DSBs compared to other types of DNA lesions, the toxicity of such agents is largely driven by DSB formation and the intrinsic difficulty in effectively repairing this type of DNA lesion. Notably, as I discuss further below in the context of cancer therapy, DSBs can also arise through the processing of other DNA lesions, by the actions of nucleases and when certain lesions such as SSBs are encountered by the DNA replication apparatus during S phase of the cell proliferation cycle.

Over the past few decades, we have learned that there exists a set of DNA repair systems, each largely dedicated to a certain type of DNA damage ([2–4]; Fig. 1). Consequently, defects in such pathways generally yield hypersensitivity to a certain type of DNA damaging agent and/or defects in repair of certain types of DNA damage, but not others. Whereas some alkylated base lesions can be repaired by direct protein-mediated reversal by proteins called DNA alkyl-guanine alkyl transferases (AGTs), many inappropriate or damaged bases are recognised by various specialised DNA glycosylase enzymes that promote base removal, leaving abasic sites that are targeted by apurinic/apyrimidinic endonuclease (APE) enzymes to yield SSBs that are then repaired by SSB repair, a system that occurs by either so-called long-patch or short-patch mechanisms involving DNA polymerase and DNA ligase activities, and sometimes endonuclease activities. SSB repair is promoted by the DNA damage binding protein, poly (ADP-ribose) polymerase 1 (PARP1), and to a lesser extent the related protein, PARP2. Upon binding to DNA SSBs (or DSBs), PARP1 becomes activated enzymatically and then carries out auto poly-ADP-ribosylation (PARylation) and also generates poly-(ADP-ribose) (PAR) chains on histones and certain other proteins. These PARylation processes help to regulate chromatin structure to promote BER/SSB repair, help recruitment of other proteins such as XRCC1, and cause dissociation of PARP from the SSB to allow downstream repair processes to ensue. As described in later sections of this chapter, PARP1/2 enzymatic inhibitors have been shown to have exciting utility in treating various cancers.

DNA base mismatches as well as small insertions and deletions on one of the DNA strands are recognised and dealt with by the DNA mismatch repair (MMR) system, while bulky base adducts such as those caused by ultraviolet light or certain chemicals are dealt with by nucleotide excision repair (NER), which operates via both transcription-coupled and global genome-tailored mechanisms. Inter-strand DNA cross-links are repaired usually in the context of replication stalling by factors connected to the Fanconi anaemia (FA) DNA repair complex, translesion DNA synthesis processes, and homologous recombination (HR; explained further below). DNA-protein cross-links (DPCs), which occur in various forms, are dealt with both by general mechanisms, such as that mediated by the protease enzyme SPRTN, as well as more dedicated systems for specific DPCs, such as those generated through abortive actions of DNA topoisomerase I or topoisomerase II enzymes (TOP1 and TOP2, respectively). Once the attached protein is cleared away, ensuing DNA repair then generally funnels into SSB repair, NER or DSB repair pathways that are described below.

DSBs are repaired by several processes: homologous recombination (HR), non-homologous end-joining (NHEJ), micro-homology-mediated end-joining (MMEJ), and a variant of this called theta-mediated end-joining (TMEJ). In NHEJ, DSB ends are recognised by the Ku/DNA-PKcs protein kinase complex [5], and then, after any needed "cleaning up" of DNA ends via nuclease enzymes such as Artemis and the actions of polynucleotide kinase-phosphatase (PNKP), the two ends are aligned and ligated via the activity of DNA ligase IV (LIG4) in association with its partner protein XRCC4 as well as accessory proteins XLF (NHEJ1) and PAXX. Notable features of NHEJ are that it is generally highly efficient and can operate across all stages of the cell cycle, and in non-proliferating cells. Nevertheless, the mechanism of NHEJ makes it intrinsically prone to inducing insertions and deletions (albeit usually relatively short range) at the site of repair. This mutagenesis is largely inconsequential to somatic cells given that they are diploid and the fact that most of the genome does not encode protein sequence, but nevertheless can sometimes give rise to mutations with pathologic potential. Notably, this feature of NHEJ is harnessed in genetic engineering via clustered regularly interspaced short palindromic repeat (CRISPR) and related technologies, wherein, for example, a site-specific genomic DSB generated by the CRISPR-associated nuclease Cas9 is repaired by inaccurate NHEJ, leading to mutations and often inactivation of the targeted locus. In cases where NHEJ does not occur, alternative pathways such as MMEJ and TMEJ can operate, the latter employing the DNA polymerase and helicase enzyme POL Theta, and in both cases this normally generates sequence deletions through DNA-end joining taking place via regions of short micro-homologies.

The alternative type of process that can repair DSBs is termed homologous recombination (HR). While HR can occur through various mechanisms, a salient, common feature of HR is that the DSB-bearing DNA damaged molecule is repaired by it entering into synapses with an undamaged partner DNA molecule, usually the sister chromatid. A key early step in HR is the 5' to 3' nucleolytic removal of one of the two strands of the DNA double helix, a process called resection. Resection is mediated by the protein CtIP in association with the MRE11-RAD50-NBS1 (MRN) complex and also promoted by enzymes including exonuclease 1 (EXO1), DNA2, and the BLM helicase, and is controlled by the activities of BRCA1 and associated factors, cyclin dependent kinase activity and the DNA-PKcs-related protein kinase ATM. Resection leads to the generation of 3' single-stranded DNA tails. These are initially bound and stabilised by replication protein A (RPA), which is then replaced by the recombinase protein, RAD51, in the process facilitated and controlled by BRCA2 and associated factors. The RAD51 nucleoprotein filament then mediates strand invasion into the partner DNA molecule, with the 3' end of the DNA tail serving as a primer for template-dependent DNA synthesis. Ensuing processes lead to recombination intermediates that, when resolved, generally result in accurate repair of the DSB. While this accuracy is a key feature of HR, the requirement in higher eukaryotes for a sister chromatid means that it is restricted to cells in late S phase or G2 phase of the cell cycle. Consequently, HR is not generally used outside of these cell cycle stages or in non-proliferating cells, as its utilisation in such contexts could lead to the generation of chromosomal rearrangements, loss of heterozygosity or other potentially pathologic products.

3 The Broader DNA Damage Response Network

Over the past two decades, it has become increasingly clear that, rather than working as a set of separate pathways, DNA repair processes are subject to broader, more collective control in which there is considerable communication and crosstalk between the different DNA repair pathways and between these and a range of other cellular processes [4]. This coordination and integration, into what is often termed the DNA damage response (DDR; Fig. 2), allows for cells to use repair processes judiciously, in the right order, time and space and in the most appropriate contexts such as cell cycle stage, replication or transcription contexts, and chromatin status. This network also leads to situations where certain DNA repair pathways can serve as backups for one another. Such relationships likely evolved to provide failsafe systems to maximise genome stability, but as described further below, these relationships also provide opportunities for therapeutic applications.

Many of the abovementioned mechanisms are tightly controlled by reversible post-translational protein modifications including but not limited to phosphorylation, acetylation, ubiquitylation, sumoylation and poly(ADP-ribosyl)ation [6]. As such, they are underpinned by the actions of a plethora of enzymes that catalyse the generation of, or bring about the reversal of, these modifications. Studying these enzymes has provided us with great insights into the control of DNA repair and associated DDR processes, and has also provided opportunities for developing drugs and drug-like compounds for use as experimental tools and for potential exploitation as therapeutic agents.

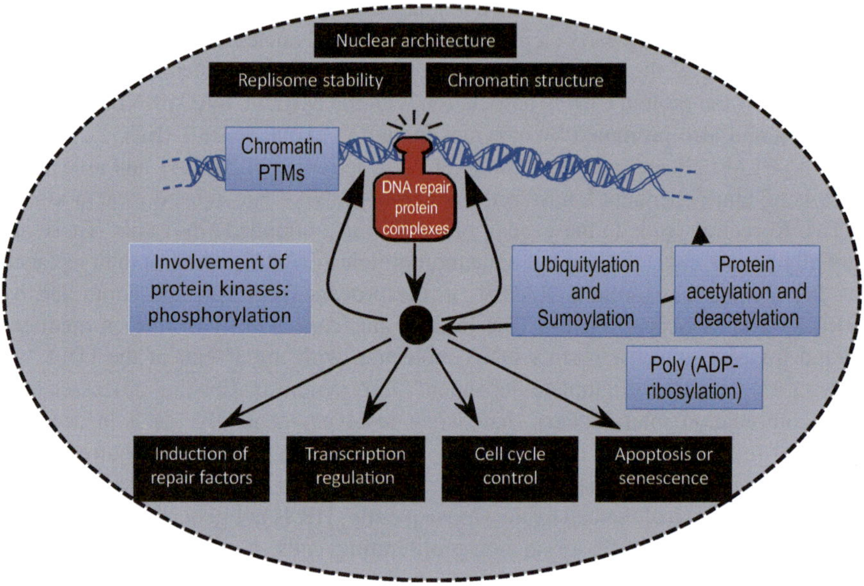

Fig. 2 Schematic of cellular DNA damage responses (DDR)

In the face of high levels of and/or persistent DNA damage, proliferating cells usually slow or stop progression through the cell cycle—mediating so-called cell cycle checkpoints. These mechanisms are presumed to have evolved to maximise genome stability by allowing time for repair before proceeding with key and vulnerable processes such as DNA replication, chromosomal segregation and mitosis. Furthermore, in certain settings, unrepaired DNA damage or ongoing DNA damage can trigger cells to enter apoptosis or other forms of programmed cell death, or long-term proliferation arrest, sometimes called cell senescence. These may have evolved as anticancer mechanisms through them preventing the proliferation of damaged cells potentially on a path towards being cancerous.

The DDR network has been particularly extensively studied in the context of DNA DSBs, which sometimes take a considerable time, ranging from minutes to hours, to be repaired. During this timeframe, cells launch a range of broader responses, often including the transcriptional and post-transcriptional induction of factors that directly or indirectly promote DNA repair and cellular integrity, and post-translational activation and recruitment of DNA repair factors to the vicinity of DNA lesions. In addition to promoting the accuracy and efficiency of DNA repair processes, such mechanisms can also influence and are influenced by factors such as chromatin structure, cell cycle status and progression, metabolic factors and cellular identity.

A quintessential example of this broader DDR is provided by cells forming, at sites of unrepaired DSBs, so-called DNA damage foci that can be readily detected by indirect immunofluorescence microscopy studies [5, 6]. Fundamental to these foci is phosphorylation of the histone H2A variant, H2AX, by the protein kinases ATM, ATR and/or DNA-PK, with this phosphorylation often being propagated for considerable distances (up to several mega-bases) along chromatin flanking a DSB site. Phosphorylated H2AX, termed γH2AX (gamma-H2AX), is then recognised by certain DDR components, most notably MDC1, which then serves as a scaffold for the regulated recruitment of various other factors, including ubiquitin and SUMO E3 ligases such as RNF8, RNF168, RNF138 and RNF4, whose concerted activities in turn control the recruitment, activities and interactions of other factors in DNA-damage foci. These additional factors include the highly studied protein 53BP1 and its partner proteins such as the so-called shieldin complex that, with 53BP1, help control whether a DSB is repaired by end-joining or HR mechanisms.

4 DNA Damage Repair and Its Connections to Cancer

DNA damage and cellular responses to it are of crucial relevance to cancer and cancer research. First, inaccurate DNA damage repair gives rise to mutations or DNA copy-number changes, which when occurring in tumour suppressor genes or oncogenes, are crucial for fuelling carcinogenesis. Second, inherited or acquired DDR defects, such as those caused by mutations in genes for factors such as BRCA1, BRCA2, ATM or TP53, give rise to heightened cancer predisposition, partly through enhanced mutagenesis and in some cases also due to them allowing less constrained cell proliferation in the face of ongoing DNA damage, telomere attrition and/or replication stress. Third, aside from surgery, radiotherapy and chemotherapies—which remain the most frequently used and impactful approaches to cancer therapy—work primarily by inflicting DNA damage, with efficacy being achieved through production of damage in cancer cells and side-effects largely been caused by damage to normal cells of the patient. Fourth, we know that DNA repair and associated mechanisms interface with many other cancer relevant processes, ranging from metabolism, inflammation and both innate and adaptive immunity. Fifth, it is becoming increasingly clear that DDR biomarkers have diagnostic and prognostic potential in cancer, in part because genome instability fuels carcinogenesis and also because it is often an ongoing process in malignant cells, manifested by spontaneous DNA damage formation and replication stress. Finally, as discussed in greater depth below, we have learned over the past two decades that various DDR enzymes represent attractive targets for anticancer drug discovery.

5 My Forays into DNA Repair: Driven by Curiosity

One could say that my whole career in the DNA repair arena arose from a chance occurrence: through me identifying and then studying a DNA-damage activated protein kinase, named DNA-PK (DNA-dependent protein kinase). In a weekend experiment during my post-doctoral studies, driven not by a specific hypothesis but curiosity, I wondered whether there might exist in human cells, a protein kinase that was stimulated by DNA? By carrying out a relatively simple biochemical experiment with extracts of human HeLa cells incubated with ATP and with or without restriction-enzyme linearised plasmid DNA (I had this at hand because I was carrying out some cloning experiments at that time), I found that the answer was a definite yes! That is, by Western blotting, I observed that my favourite transcription factor, Sp1, became phosphorylated in these reactions but only when DNA was present [7]. Curiously, my immediate attempts to reproduce this result failed. Driven by a wish to understand this apparent lack of reproducibility, I came to realise that these experiments would only produce Sp1 phosphorylation if the plasmid DNA was linearised—this did not occur with the same DNA molecule in supercoiled, circular form. While I initially hypothesised that it was lack of supercoiling that allowed the kinase to be active, it was a while later that in my own group with my first PhD student, Tanya Gottlieb, we realised that this was because DNA-PK is activated by DNA DSBs [8].

Our work, along with parallel studies from William Dynan and colleagues [8, 9], established that DNA-PK comprises three polypeptides: a large catalytic subunit DNA-PKcs and a dimeric DNA binding component called Ku (whose subunits are generally referred to as Ku70 and Ku80). Fuelled by the idea that the kinase was activated by DNA breaks, ensuing work by my group, our collaborators and others demonstrated that cells deficient in Ku70, Ku80 or DNA-PKcs are hypersensitive to ionising radiation as a consequence of being defective in DNA DSB repair by NHEJ. As I mention later, one could regard my whole career in DNA repair has been based on what the protein, Ku, does.

6 The Concept of and Applications for Drugging DNA Repair

With these things in mind, it was notable in the late 1990s that work in my lab and elsewhere demonstrated that certain DNA repair enzymes, including PARP1, DNA-PK and the DNA-PK-related kinases ATM and ATR, could be inhibited by chemical molecules with drug-like properties. Furthermore, our experiments provided indications that such compounds had more profound effects on sensitising certain cancer cells to DNA damaging agents than some other cell types. Indeed, in 1997, with help from the University of Cambridge and cancer research organisations funding my lab at that time, I conceived and founded the company KuDOS Pharmaceuticals (so named based on my career in DNA repair being based on what "Ku does"). The concept behind KuDOS was to use biochemical assays for DDR enzymes as the basis for developing high throughput chemical screening approaches to identify hits that could then be developed into potent and selective small molecule DNA repair inhibitor drugs.

It took some time, however, to pull in substantial funding for the company because, at that time, the concept of inhibiting DNA repair enzymes was counter-intuitive, and indeed many considered an unworthwhile if not a stupid undertaking. "Why would anyone want to inhibit DNA repair, when DNA repair mechanisms protect against various human diseases?"

The key narrative that in the end was persuasive to funders, underpinned by some initial data that my group had generated, was that certain cancer cells are likely to be much more reliant on certain DNA repair pathways than normal cells in the body. As I explained previously in this chapter, this could first reflect the heightened proliferation status of cancer cells and the fact that they are often proliferating under conditions that generate chronic DNA damage and other things such as DNA replication stress. Second, even in the mid-1990s, there was a growing realisation that loss of or deregulation of DNA repair processes is usually associated with—and often core to—the genesis of many cancers. Indeed, DNA repair/DDR pathway dysfunction has now come to be regarded as one of the fundamental drivers of carcinogenesis [10], as well as a source of evolved resistance to various anticancer agents.

7 From Ku to KuDOS

The above factors, coupled with the idea that certain DNA repair pathways can act as backup pathways for one another, led me to a key hypothesis as described in the KuDOS business plan—that in specific cases, DNA repair inhibition would have strong selective effects on cancer cells compared to normal cells. Highlighting an extreme paradigm for such an effect in the business plan and my presentations, was the concept of synthetic lethality, initially identified in model organisms but now widely recognised as operating across essentially all of biology, including in human cells. In this scenario, if there are two pathways, controlled by genes depicted as A and B in Fig. 3, that work as alternative pathways carrying out an essential cellular process, the cell or organism can lose one or the other process but cannot lose both and still remain viable. While this concept was initially generated through genetic considerations, the concept extolled in the KuDOS business plan was to translate this into a pharmacological application: if one could generate drugs inhibiting the

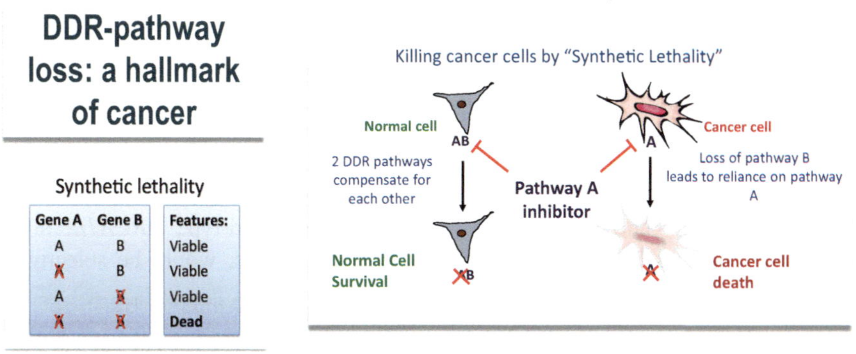

Fig. 3 DNA repair defects as potential Achilles heels of cancer cells

product of gene B, then these drugs would selectively kill cancer cells lacking pathway A but not normal cells that retain this pathway (Fig. 3).

Over the course of a few years, KuDOS developed small molecule chemical inhibitors of various DDR enzymes, including DNA-PK, ATM, ATR, and PARP1/2. In addition to us obtaining encouraging data showing that some such compounds sensitised cancer cells to radiation and certain anticancer chemotherapeutics, we explored potential standalone activities. The most notable of these studies emanated from a collaboration between KuDOS, my academic lab at the University of Cambridge and the academic group of Professor Alan Ashworth and Dr. Chris Lord at the Institute of Cancer Research in London. Alongside other researchers, Ashworth and colleagues had worked for several years on BRCA1 and BRCA2, helping to establish their connections to DNA repair processes, in particular HR. Consequently, the Ashworth group had various isogenic mammalian cell lines proficient or deficient in BRCA1 or BRCA2. Our collaboration explored the potential for standalone activity of KuDOS drugs in such settings. Strikingly, in a set of eureka-moment studies, this work demonstrated that small molecule inhibitors of PARP are extremely toxic to cells lacking BRCA1 or BRCA2 function but have much, much less impact on cells homozygous or heterozygous for functional BRCA1 or BRCA2. Indeed, in the initial cell lines we analysed, this differential sensitivity was over 1000-fold! These findings, which were published alongside a related study from Professor Thomas Helleday, Nicola Curtin and their colleagues [11, 12], had various mechanistic ramifications as well as potential clinical application.

Based on the above findings, a model was proposed, which has now been expanded and refined, wherein inhibition of PARP leads to delayed DNA SSB repair and also persistence of inhibited PARP on SSB-containing lesions—so-called PARP trapping (being unable to mediate auto-PARylation, the enzyme is unable to promote its own dissociation from the damaged DNA). While impaired SSB repair and PARP trapping are not a very toxic under most settings, they can become much more problematic during S phase. Thus, when encountered by replication forks, these structures can generate replication stalling and replication-fork collapse, associated with the formation of one-ended DNA DSBs. Notably, such pathological structures are not overtly toxic to most normal cells because they can be readily repaired by HR processes. However, in cells deficient in HR, such as those defective in BRCA1/2, this recombination does not take place effectively, leading to cell death. Our work thus suggested that PARP inhibitors would be selectively toxic to cancers arising in *BRCA1/2*-mutant carrier patients. In this setting, while the normal cells of the patient retain one copy of the *BRCA1/2* gene and would therefore be predicted to be relatively resistant to PARP inhibition, the fact that cancer cells in these individuals generally lose the function of the wild-type *BRCA1/2* allele suggested that these cancer cells, or even their precursors, would be selectively killed by PARP inhibition.

8 The PARP Inhibitor Olaparib Heralds a New Arena for Cancer Therapy

Having developed the potent and selective PARP1/2 inhibitor, olaparib, and having formulated it in a manner that allowed its dosing into the first human subjects, the way was paved for KuDOS to initiate phase 1 clinical trials [13]. However, the onerous costs associated with clinical development and the lack of sufficient capital for this at the time in the UK venture capital (VC) sector led the company to explore partnership opportunities, leading to KuDOS being acquired by AstraZeneca in 2005/2006. Swiftly ensuing clinical development indicated lack of overt toxicity and, moreover, identified various patients whose tumours displayed growth inhibition or marked atrophy upon treatment. Strikingly, various such patients were known *BRCA1/2*-mutation carriers and/or had a strong family history of relevant cancers. Further clinical development demonstrated the true utility of this opportunity and, in December 2014, olaparib (brand name Lynparza) was approved by the US Food and Drug Administration (FDA) and the European Medicines Agency (EMA) to treat ovarian cancer patients with BRCA mutations and/or whose tumours had initially responded to DNA cross-linking platinum-based chemotherapy. This made olaparib the world's first approved DNA-repair-enzyme inhibitor drug, the first marketed PARP inhibitor, and the first drug targeting inherited cancer predisposition. It is also notable that olaparib and other subsequently approved PARP inhibitors fall within the arena of "precision medicines", being tailored to certain patients who could be medically diagnosed, in this case via *BRCA1/2* mutations or associated genomic features indicative of homologous recombination deficiency (HRD; an acronym that I coined at a KuDOS meeting).

Since that time, olaparib (now co-developed by AstraZeneca and Merck & Co.) and other PARP inhibitors have been approved for clinical applications, particularly for cancers with inherited or somatically acquired *BRCA1/2* mutations, in broader HRD settings, as well as in certain apparently non-HRD contexts, with FDA approval now applying in ovarian, breast, pancreatic and prostate cancer settings.

It is exciting to see that various companies worldwide now have programs and clinical trials utilising inhibitors of other DNA repair enzymes including ATR, ATM, DNA Polymerase Theta (Pol Theta) and others. Of particular note, Pol Theta inhibitors have potential utility in BRCA1/2 deficient cancers, including those that have evolved certain types of resistance mechanism to PARP inhibitors. Another intriguing synthetic lethality opportunity that is being exploited relates to ATR inhibitors in the context of cancer cells experiencing high levels of replication stress. Another is based on the realisation that the DNA repair helicase, WRN, displays synthetic lethality with genomic mutational landscapes generated in cancers with deficiencies in DNA mismatch repair. Hopefully, such drugs will show clear clinical benefits and thereby make it to registration over the coming few years.

As might be expected, the earlier use of PARP inhibitors in clinical trials is leading to greater benefits for the patient, as exemplified by recent trials. This presumably reflects the fact that earlier treatments can take place when there is less heterogeneity in cancer cell populations within the patient, thereby reducing the potential for the existence and evolution of cell-clones with drug resistance.

Unfortunately, as is the case with most anticancer therapies, evolution of resistance is evident with DDR inhibitors, including PARP inhibitors, even in BRCA1/2 deficient settings [14]. Fuelled by this and curiosity for understanding underlying molecular mechanisms, there has been considerable research effort in trying to define mechanisms for resistance. So far, work has established that this can arise in various ways. First, DDR inhibitor drugs, like other drugs, are sometimes subject to attenuation via drug metabolism, drug efflux or other pharmacological processes. In addition, it has become clear that certain BRCA mutations can be subject to circumvention via generation of compensatory mutations, such as frameshifts, elsewhere in the *BRCA1/2* gene coding sequence, leading to re-expression of functional or partially functional BRCA1/2 proteins. Alternatively—and intriguing from a mechanistic point of view—various laboratories, including my own, have shown that deregulation of certain other DDR components can "rewire" DNA repair processes such that PARP inhibitor resistance can arise even when BRCA1/2 functionality remains compromised. Most notable in this regard are mutations in, or down-regulation of, 53BP1 and associated factors such as shieldin components, which allow some HR restoration in BRCA1 deficient settings [14].

But could this information on resistance mechanisms be of any use to cancer patients? Evidence suggests that this may indeed be the case in certain situations. For example, we and others have found that while loss of shieldin components promotes resistance to PARP inhibitors, BRCA1 deficient cells with 53BP1 or shieldin dysfunction are nevertheless still hypersensitive to Pol Theta inhibitors, and may also be particularly sensitive to other therapeutic agents, including DNA cross-linking agents and ionising radiation. Optimistically then, this leads to the hypothesis that if we understand resistance mechanisms in molecular detail and can identify potential collateral vulnerabilities that such resistance mechanisms sometimes provide, then these may be exploitable in clinical settings. Coupled together with earlier detection and molecularly targeted patient monitoring, there is a prospect of clinicians anticipating and even exploiting resistance mechanisms, either through drug combinations, or via changing therapies during the course of the patient's journey, based on molecular features collected from biopsies, liquid biopsies or other sources.

Finally, it seems that there might even be opportunities for PARP and other DDR enzyme inhibitors in prophylactic settings. It seems that functional loss of BRCA1/2 and other drivers of HRD often takes place relatively early during the evolution of cancer, potentially even well before the clone of cells has established full malignant capacity. If and when this occurs, then it may be possible to wipe out those early cancer cells or cancer-precursor cells via periodic treatment with PARP inhibitors or other DDR enzyme inhibitors, thereby preventing cancer formation and giving patients an alternative to drastic surgical procedures that are currently sought by *BRCA1/2* mutation carrier individuals. Together with developments in other areas of cancer research and clinical developments, we can optimistically consider the prospect of more holistic and comprehensive approaches that will in due course allow patients to survive and live fulfilling lives much longer, even if their cancers are not fully curable.

9 Conclusions and Perspectives

Through intensive worldwide efforts, we now know that there are well over 200 human proteins with strong connections to DNA repair and DDR processes and, while much is known about many of these, even those best characterised are not yet fully understood. Moreover, we are only just now really beginning to understand the complexity and sophistication of the functional interaction landscape between these various DDR proteins and their connections to myriad other cellular processes. It is in these domains that my laboratory and many others are now actively engaged, with such work as always being spearheaded by curiosity, clear hypotheses, and the applications of new technologies and approaches. While we already know about certain opportunities for translating such new knowledge towards better understanding and treating cancer, I am confident that many more opportunities await identification and exploitation.

It is also worthwhile recognising that outside of cancer, DNA damage and its repair are also associated with a wide range of other human pathologies, and therefore that modulators of DNA repair/DDR could find utility in various other settings. For example, a company that I co-founded, Mission Therapeutics, is exploring various clinical opportunities for inhibitors of deubiquitylating enzymes (DUBs) that play key roles in DNA repair and also other cellular process. Another arena—currently being explored by my laboratory and a company that I founded, Adrestia Therapeutics—is to exploit a concept termed synthetic rescue. In this scenario, an anomaly or pathology caused by a defect in one gene/process can be rescued by changing another gene, or by a drug targeting the product of that other gene. Synthetic rescue and synthetic lethality can therefore be considered as opposite sides of the same coin. In addition to identifying therapeutic opportunities for human genetic diseases and potentially related sporadic diseases, exploration of synthetic rescue is providing exciting mechanistic insights into fundamental aspects of cell and disease biology. Finally, it is interesting to speculate that if we could find drugs or other therapeutic modalities that could regulate—particularly increase—the efficiency and/or accuracy of DNA repair processes, these could provide opportunities for delaying or preventing a whole range of conditions caused by defective DNA repair and accrued mutations. If they could be developed, DNA repair/DDR enhancer drugs might even be able to stave off the ravages of cellular decline which undoubtedly represent a fundamental cornerstone of what we regard as natural ageing.

If we remain ambitious and driven by curiosity, the future seems bright!

Acknowledgements For their advice, encouragement wisdom and humour, I thank all members of my laboratory, past and present as well as the innumerable colleagues I have had the pleasure to interact with in academia, industry and business over the years. As this chapter is a personal perspective and is intended for a generalist audience rather than specialists, I have limited the number of references cited and apologise to others whose fine research and publications I have not cited. Work in my research laboratory is currently supported by grants from Cancer Research UK, the European Research Council (ERC), GSK and the Mark Foundation for Cancer Research.

References

1. Lindahl T. Instability and decay of the primary structure of DNA. Nature. 1993;362:709–15.
2. Jackson SP, Bartek J. The DNA-damage response in human biology and disease. Nature. 2009;461:1071–8.
3. Hoeijmakers JHJ. DNA damage, aging, and cancer. N Engl J Med. 2009;361:1475–85.
4. Ciccia A, Elledge SJ. The DNA damage response: making it safe to play with knives. Mol Cell. 2010;40:179–204.
5. Blackford AN, Jackson SP. ATM, ATR, and DNA-PK: the trinity at the heart of the DNA damage response. Mol Cell. 2017;66:801–17.
6. Polo SE, Jackson SP. Dynamics of DNA damage response proteins at DNA breaks: a focus on protein modifications. Genes Dev. 2011;25:409–33.
7. Jackson SP, Macdonald JJ, Lees-Miller S, Tjian R. GC box binding induces phosphorylation of Sp1 by a DNA-dependent protein kinase. Cell. 1990;63:155–65.
8. Gottlieb TM, Jackson SP. The DNA-dependent protein kinase: requirement for DNA ends and association with Ku antigen. Cell. 1993;72:131–42.
9. Dvir A, Peterson SR, Knuth MW, Lu H, Dynan WS. Proc Natl Acad Sci U S A. 1992 Dec 15;89 (24):11920–4.
10. Hanahan D. Hallmarks of cancer: new dimensions. Cancer Discov. 2022;12:31–46.
11. Farmer H, et al. Targeting the DNA repair defect in BRCA mutant cells as a therapeutic strategy. Nature. 2005;434:917–21.
12. Bryant HE, et al. Specific killing of BRCA2-deficient tumours with inhibitors of poly(ADP-ribose) polymerase. Nature. 2005;434:913–7.
13. https://www.icr.ac.uk/news-archive/kudos-pharmaceuticals-first-patient-treated-with-new-anti-cancer-agent
14. Dias MP, Moser SC, Ganesan S, Jonkers J. Understanding and overcoming resistance to PARP inhibitors in cancer therapy. Nat Rev Clin Oncol. 2021;18:773–91.

From Waste to Food: Toward the Creation of a Sustainable Food Generator

Ting Lu and Stephen Techtmann

1 Introduction

Food is the most fundamental human need. It provides us energy and nutrition for all kinds of activities. Ever since the origin of life, food has been the most basic security required by all living creatures, including human beings and will continue to be. However, despite our advances in science and technology, food insecurity remains as a global challenge. According to a 2020 report from the United Nations [1], there are 690 million people suffering from hunger and, each year, the world's hungry population grows by 12 million. Notably, over 21% of children under age 5 are stunted, and around 33% of women at reproductive age suffer from anemia [1, 2]. Additionally, food insecurity can become escalated in the presence of different factors such as conflicts, economic shocks, public health challenges, and weather extremes. For instance, the COVID-19 pandemic has led to an increase of 83–132 million in the hungry population [1]. Making the issue more concerning is that arable land continues to be lost due to climate change and economic development while the world population will reach 10 billion by 2050 [3].

Food insecurity also leads to malnutrition, which is associated with prohibitive health and social costs. With the current dietary pattern, the total health and social cost in 2030 will reach 1.3 and 1.7 trillion US dollars, respectively [1]. These costs can be reduced vastly by healthy diets, as reflected by reports showing that healthy diets may reduce up to 97% of the health cost and 41–74% of greenhouse gas emission [1]. However, healthy diets are five times more expensive than starchy

T. Lu (✉)
Department of Bioengineering, University of Illinois Urbana-Champaign, Urbana, IL, USA
e-mail: luting@illinois.edu

S. Techtmann
Department of Biological Sciences, Michigan Technological University, Houghton, MI, USA
e-mail: smtechtm@mtu.edu

U. A. Betz (ed.), *Curious Future Insight*,
https://doi.org/10.1007/978-3-031-41781-8_6

staples; as a result, over 3 billion people around the world do not have regular access to healthy diets [4].

Concurrently, our planet faces another pressing threat, plastic pollution [5]. Plastics are polymers with remarkable physiochemical properties that are used everywhere from food packaging to construction, textiles, electronics, and others. Every year, the world produces 380 million tons of new plastics [6]. As of 2015, 6.3 billion tons of plastics have been generated and, by 2050, the total plastics are estimated to reach 34 billion tons [7, 8]. As they are inexpensive, plastics are often disposed after a single use. Studies show that only 9% of end-of-life plastics are recycled, 12% of them are incinerated, and the vast majority (79%) are accumulated in landfills and other natural environments [8].

The mismanagement of used plastics causes adverse impacts on the environment, wildlife, and human health. It contributes to climate change, with an impact equivalent to the annual emission of 850 million metric tons of greenhouse gases [9]. In some regions, up to 60% of plastic waste is mismanaged [10], which disperses via waterways and is transported by wind. Strikingly, 83% of tap water samples are shown to contain microplastics [11]. Plastics enter the oceans eventually. Currently, there are 150 million metric tons of plastics in the oceans and, each year, eight million metric tons of plastics are added [12]. Plastics also kill millions of animals and, through entanglement and ingestion, are crippling and depleting more than 1300 species [13]. Additionally, plastic waste is harmful to human health. For instance, Bisphenol A, a building block of plastic used for beverage containers and plastic dinnerware, is shown to reduce the levels of sex and thyroid hormones [14], and over 95% of adults in the USA have had detectable Bisphenol A in their urine [15].

Tackling any of these two challenges—food insecurity and plastic pollution—will be outstanding. Addressing the two together? That will be incredible.

2 Vision and Strategy

Inspired by the old adage of "one man's trash is another man's treasure," we envision a dream generator of future food (Fig. 1): A sustainable, efficient, and versatile technology that converts end-of-life plastics and other waste streams into edible food [16]. By doing so, we will reduce plastic waste while increasing food supply, thereby solving the two grand challenges simultaneously. Additionally, we hope the food produced from this generator will not only provide sufficient energy but also bear no toxicity, contain full nutrition, offer health-promoting benefits, and allow personalization.

How to realize such a vision? We turn to microorganisms, tiny organisms that colonize environments on our planet from water to the soil and to the animal body and are known for their diverse functionalities. In natural settings, they often exist in the form of communities where cellular cooperation, competition, and other interactions shape ecosystem composition and function. Here, we propose to harness designer communities of microbes to convert waste to food and further augment the

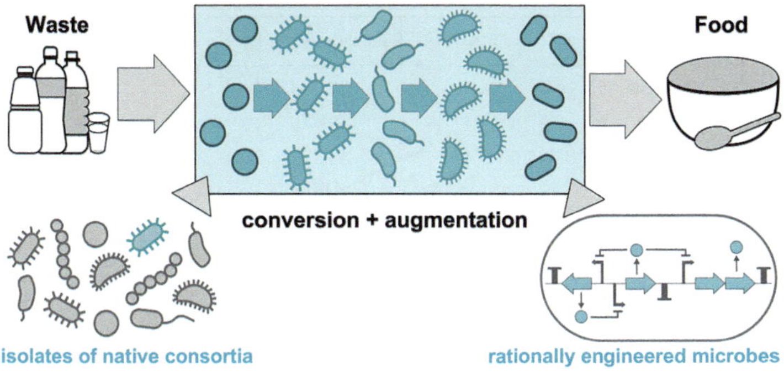

Fig. 1 Conceptual illustration of a technology that turns plastic waste into edible food. Centering around the technology are designer microbial communities which are assembled with isolates of native consortia or rationally engineered strains and capable of converting end-of-life plastics to edible food with tailored nutritional and health-benefiting ingredients

nutrient contents and health benefits of the food. To develop such communities, one way is to isolate functional species from naturally occurring microbiomes in various habitats. Alternatively, they can be tailor-made. Using a synthetic biology approach, we can genetically and metabolically reprogram microbes to produce desired functions by engineering gene circuits and introducing them into target species. These two complementary approaches allow us to achieve the waste-to-food conversion and augmentation.

3 Designer Microbes to Degrade Plastics

The increasing concern of plastic pollution has fostered the development of different mitigation strategies. One class of strategies is physiochemical approaches, including thermal, mechanical, and chemical recycling [17]. Another is biological degradation, which harnesses enzymes and live microorganisms to break down polymers into monomeric substances and even further metabolize them [18]. Despite their unique strengths, each of the strategies has limitations. Mechanical recycling is hampered by complicated sorting and inferior properties of recycled products, thermal recycling results in incomplete combustion, and release of toxic compounds while chemical recycling is hindered by strict catalyst and processing requirements [19–22]. The biological route eliminates complex or energy-intensive processing needs; additionally, it yields cellular biomass as a potential food source that is unavailable with physiochemical approaches. However, plastic biodegradation is typically slow in terms of the rate of degradation.

To generate potentially edible biomass while efficiently degrading plastics, we developed an integrated workflow that combines chemical pre-processing with microbial metabolism to enable the conversion from plastic waste to edible food (Fig. 2). Specifically, the workflow involves three steps. First, end-of-life plastics are depolymerized with chemical reagents or thermochemical approaches into

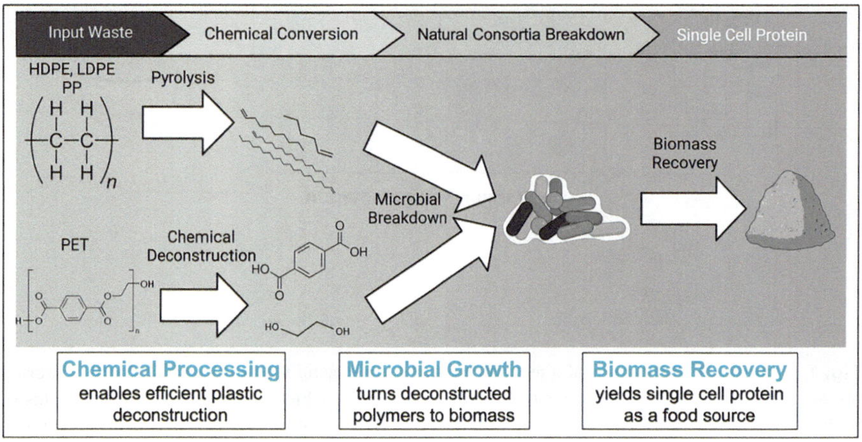

Fig. 2 An integrated chemical and biological workflow that converts used plastics into edible biomass as a food source. The workflow involves three steps, including chemical deconstruction, microbial conversation, and biomass recovery and processing

monomeric substances or more biodegradable compounds. For example, polyethylene terephthalate (PET) can be deconstructed through hydrolysis while polyethylene may be depolymerized with pyrolysis. Second, deconstructed plastics are mixed with designer microorganisms to fulfill fermentation, through which plastics are consumed as carbon sources by microbes to produce cellular biomass. Our previous work along with others has supported the combined chemical and biological processing of plastic wastes to produce biomass and other valuable chemicals [23–25]. Third, biomass is recovered and further processed to yield single cell protein (SCP) as a food source.

One essential step here is to develop designer microbes that metabolize depolymerized plastics. To that end, rationally engineered microbial strains and their communities provide a compelling option. By introducing gene circuits into target species, microbes can be tailored in terms of its metabolic capacity with a synthetic biology approach. Using PET as an example, its hydrolysis results in two monomeric chemicals, terephthalic acid and ethylene glycol. To degrade these components, microbial species such as *Pseudomonas*, *Rhodococcus*, and *Bacillus* can be selected as cellular chasses because of their degradation potential and amenability for engineering. Heterologous pathways are subsequently introduced to enable and enhance the metabolism of the chemicals. Here, enzyme optimization through directed evolution and rational engineering is critical. Additionally, to increase the overall degradation efficiency, an important strategy is the construction of synthetic communities of microbes. As ecosystems allow to fulfill complex tasks and enhance function robustness [26–28], communities-based engineering is particularly compelling given the facts that plastics with distinct physiochemical properties are typically mixed in nature and, even for a single plastic, it is often composed of mixed monomeric chemicals. Along this rational engineering direction, there have been exciting advances and tremendous interest [29–34]; nevertheless, it needs significant further investigations.

Another strategy to obtain plastic-degrading microbes is to screen for isolates from native microbiomes. Natural microbial communities are vast stores of biotechnologically relevant enzymes. Native plastic-degrading microorganisms have been isolated from diverse contaminated sites [33]. The pathway for the complete degradation of PET has been extensively studied in isolates such as *Ideonella sakaiensis* [35]. This biochemical pathway involves esterase enzymes known as PETases that depolymerize PET into MHET or BHET, which is terephthalate with either a single or two ethylene glycol groups. Further enzymatic degradation removes the ethylene glycol groups. Further metabolism of the PET monomers is then catalyzed through biochemical pathways for terephthalate and ethylene glycol degradation. Similarly, isolates have been identified with capabilities for polyethylene and polystyrene degradation [36, 37]. These isolates and the pathways for biochemical degradation have been characterized, and enzymes of interest have been identified. These natural isolates have shown promise for plastic upcycling and could be used for waste upcycling for food production. One of the rate-limiting steps for plastic and other waste upcycling is the depolymerization of polymeric waste streams and the understanding of the biochemical pathways for deconstruction and metabolism. Natural isolates are a vast store of biotechnological potential and may help to speed up the upcycling of increasingly diverse waste streams.

Importantly, while the discussion here uses PET as our primary example to illustrate the concept, the integrated workflow and the creation of designer microbes are potentially applicable to different types of plastic waste.

4 Microbial Biomass as a Food Source

The United Nations' Sustainable Development Goal 2 aims to reach zero global hunger. Novel solutions are required to meet this goal. The world population is estimated to reach over 10 billion people by the year 2050 [3]. As the world's population grows, climate change further exacerbates some of the challenges to accomplishing the goal. For example, climate change is leading to the loss of arable land and changes in water availability. Traditional agricultural practices are land and water intensive. Therefore, there is a need to develop creative solutions to reach the nutritional needs of a growing population while making more efficient use of water and land resources. One creative solution is cellular agriculture, particularly the production of single-cell protein (SCP)—an alternative protein source from microbial cells [38, 39]. One of the appeals of SCP is that microbial biomass can be produced from diverse feedstocks including low-quality feedstocks such as waste streams and inedible plant materials.

The use of microbial cells as food and nutritional supplements has been common for many years. In addition to yeast, which have been used in numerous fermentation products and algae such as *Chlorella* and *Spirulina* that have been used as a nutritional supplement, the use of whole microbial cells as complete food products is growing in popularity [40]. For example, mycoproteins such as those produced from *Fusarium* spp. have obtained the Generally Recognized as Safe (GRAS) status. Extensive safety testing was performed to demonstrate the efficacy of using fungi such as *Fusarium* spp. as a food product. Bacterial cells have also been used as a SCP source [41]. Most of the SCP products that are considered GRAS by the US FDA are fungal-based mycoproteins. However, some bacterial products have obtained GRAS status for agricultural applications as a product for animal feed products. For example, methylotrophic bacteria have been grown on waste methane and used as feed for animal feed [42, 43]. This product additionally has undergone extensive testing for the safety of these foods.

One of the challenges of using SCP, especially bacterial SCP, as a food for humans is the high nucleic acid content in microbial biomass, which can lead to the production of uric acid during metabolism and thus contribute to the development of gout [44]. Therefore, to use bacteria as SCP, the nucleic acid content must be tightly controlled. Methods have been developed to decrease the nucleic acid content of foods through processes such as heat treatment [45].

While there is precedent for the use of microbial biomass as a food product, there is still a need for improved methods for assessing the safety and nutrition of microbial food products. Precision fermentation is a booming industry that uses microbes to generate food products as whole foods with specific compounds. In these cases, product safety must be thoroughly evaluated due to the alternative route of food production. While animal testing is commonplace to ensure the safety of food products, alternative methods that enable rapid prototyping of microbial-based food would be a great advance. One such approach is based on the metabolomics of food to identify putative toxins and nutritional content. Untargeted metabolomics has been an increasingly common method for studying the suite of metabolites found in an organism. The same techniques can be used in a food screening protocol. Methods developed for high-resolution mass spectrometry and untargeted metabolomics to identify putative toxins in plant biomass; similar approaches could be used for screening SCP products [46, 47].

5 Tailored Microbes to Enhance Safety, Nutrition, and Health Benefits

An ideal future food shall provide sufficient energy, bear no toxicity, contain required nutrition, offer health-promoting benefits, and further allow personalization. To transform plastic-derived biomass into such a food, one strategy is to tailor microbes to produce different beneficial ingredients. Supporting the feasibility of the concept, food microbes have been metabolically engineered and utilized in precision fermentation. For example, we have adopted lactic acid bacteria (LAB) as platform chasses for function enhancement, as LAB are commonly used in food fermentation and confer health benefits naturally. To facilitate the utilization of LAB, we developed a pathway engineering platform, involving a shuttle for hosting target networks and associated strategies for gene editing of large and small DNA parts, for efficient manipulation of LAB gene networks [48]. As a demonstration, *Lactococcus lactis* was engineered to overproduce nisin, a broad-spectrum, FDA-approved antimicrobial peptide widely used as a natural food preservative. The platform allowed to optimize an eleven-gene nisin biosynthesis pathway, resulting in an almost six-time increase in nisin production [49]. Additionally, by altering the amino acid sequence, we created nisin variants with an enhanced stability that allow Queso Fresco to be stored without detectable the pathogen *Listeria monocytogenes* for over 28 days when used together with other compounds [50]. Such designer strains serve as candidate species to be integrated with plastic-degrading microbes to enhance food safety.

In addition to synthesizing food preservatives, microbes can be engineered to enhance nutrient and health-promoting contents. For example, one possible product is amino acids including glutamate—which serves as a neurotransmitter and helps learning and memory—and arginine which supports heart health and muscle building. Another possibility is polyunsaturated fatty acids that are found in cold-water fish and shown to reduce heart disease, depression, and attention-deficit hyperactivity disorder. Through the division of labor among strains, engineered microbial consortia can be created by establishing one strain that degrades deconstructed plastics to secrete metabolites and another consuming the metabolites to produce desired ingredients. For instance, *R. jostii* can be designed to break down PET hydrolysate to produce intermediates that are fed to microbes that produce amino acids or polyunsaturated fatty acids. Similarly, microbes can be metabolically rewired to produce flavor molecules, such as sweeteners, and then integrated with plastic-degrading strains to increase the palatability of the food.

6 Conclusions, Future Directions, and Implications

Sustainable development necessitates the ability to produce nutritious food to feed a growing population and meet the needs of a changing world. Here we described recent work that points the potential for waste streams such as plastic waste to be upcycled into nutritious products including SCP. This work leverages advances in plastic degradation and alternative protein production to develop a food generator that is fueled with waste streams. One of the core components of this generator is designer microbial communities. Microorganisms have an incredible ability to metabolize diverse feedstocks and are engineerable to produce various commodity chemicals. So far, studies have focused on developing engineered organisms for the conversion of waste streams to valuable products such as food, fuels, and nutritional products. However, with the increase in the diversity of waste streams and products, a deeper integration of biological and chemical methods will be valuable. Additionally, microbial communities have been shown to exhibit specialization and division of labor for tailoring functions to accommodate diverse feedstocks and to produce different end products. With the advances in engineering technologies, a food generator that accommodates increasingly diverse feedstocks and desired end products will be realized.

Future research is needed to realize the potential and versatility of the waste-to-food generation system. As mentioned earlier, this technology must ensure that the food product is safe and provides all needed nutrition. The safety of food products is often assessed through animal testing. The use of untargeted metabolomics and genomic methods can help to speed the prototyping and adoption of alternative SCP targets; metabolomic strategies may help to identify putative toxins as well as ensure that none of the products or residual components of the feedstock are present in the recovered microbial biomass. This type of information will facilitate the identification of candidate strains and inform appropriate bioprocessing conditions. Additionally, advances in bioinformatic identification of putative allergens and toxins could aid in strain screening for SCP production. Several genomic databases of toxins and allergens have been developed for food product screening. While these databases are helpful first screens, more research must be performed to clarify the importance of finding a positive match in one of these databases. Previous work has shown that many potential allergens identified in these databases have homologs in organisms that are non-allergenic. One such example is aldehyde dehydrogenase, a common enzyme across the tree of life but in some cases acts as an orphan allergen [51]. Through biochemical and clinical tests, it was shown that the vast majority of proteins annotated as allergenic forms of acetaldehyde dehydrogenase that had IgE reactivity did not elicit a response in the skin prick test, which suggested that while bioinformatic methods may be a first screening tool biochemical and clinical tests are needed.

Although our focus here is plastic-to-food conversion, other waste streams can be exploited as potential feedstocks for food production. One attractive candidate is inedible plant biomass. As inedible biomass often takes up a significant fraction of total crop biomass and many non-edible plants can grow in poor soil on marginal lands, converting inedible biomass into nutritious food would greatly advance our ability of food generation. Plant biomass is composed of polymers, including lignin and cellulose, which store a large amount of energy that could be potentially accessible via microbial metabolism. In fact, a diverse set of bacteria and fungi have evolved pathways for degrading plant polymers; we envision that designer microbial communities can be developed to consume the inedible material to produce nutritious compounds. Another appealing possibility is agricultural and food waste. While these waste streams are often used for composting or disposal, they can be harnessed as feedstocks for microbial biomass production. Together, diverse low-cost and low-quality feedstocks can be utilized for SCP production.

Our society faces the growing challenges of food insecurity and excess waste. The microbial-based conversion of waste to food provides a promising avenue to address these two challenges. Looking forward, continued and collective efforts are needed to support a healthy and sustainable development of the planet.

Acknowledgments This work was supported by the Future Insight Prize and the Defense Advanced Research Projects Agency ReSource program via the cooperative agreement HR00112020033. The views, opinions, and/or findings expressed are those of the author and should not be interpreted as representing the official views or policies of the Department of Defense or the US Government.

References

1. World Health Organization. The state of food security and nutrition in the world 2020: transforming food systems for affordable healthy diets (Vol. 2020). Food & Agriculture Org; 2020.
2. World Health Organization. The global health observatory; 2020. https://www.who.int/data/gho/data/indicators/indicator-details/GHO/prevalence-of-anaemia-in-women-of-reproductive-age-(-).
3. Food and Agriculture Organization, Food. The future of food and agriculture: alternative pathways to 2050. Food and Agriculture Organization of the United Nations Rome; 2018.
4. Carducci B, Keats EC, Ruel M, Haddad L, Osendarp SJM, Bhutta ZA. Food systems, diets and nutrition in the wake of COVID-19. Nature Food. 2021;2(2):68–70.
5. Borrelle SB, Ringma J, Law KL, Monnahan CC, Lebreton L, McGivern A, et al. Predicted growth in plastic waste exceeds efforts to mitigate plastic pollution. Science. 2020;369(6510): 1515–8.
6. Ritchie H, Roser M. Plastic pollution. Our world in data; 2018
7. Geyer R. Production, use, and fate of synthetic polymers. In: Plastic waste and recycling. Academic Press; 2020. p. 13–32.
8. Geyer R, Jambeck JR, Law KL. Production, use, and fate of all plastics ever made. Sci Adv. 2017;3(7):e1700782.
9. Dai L, Ruan R, You S, Lei H. Paths to sustainable plastic waste recycling. Science. 2022;377 (6609):934.
10. Organisation for Economic Co-operation and Development. Plastic pollution is growing relentlessly as waste management and recycling fall short, says OECD; 2022.
11. Uhl W, Eftekhardadkhah M, Svendsen C. Mapping microplastic in Norwegian drinking water. Atlantica. 2018;185:491–7.
12. Agenda I. The new plastics economy rethinking the future of plastics. In World Economic Forum (Vol. 36); 2016.
13. Bergmann M, Tekman MB, Gutow L. Sea change for plastic pollution. Nature. 2017;544 (7650):297.
14. Rochester JR. Bisphenol a and human health: a review of the literature. Reprod Toxicol. 2013;42:132–55.
15. Calafat AM, Kuklenyik Z, Reidy JA, Caudill SP, Ekong J, Needham LL. Urinary concentrations of bisphenol a and 4-nonylphenol in a human reference population. Environ Health Perspect. 2005;113(4):391–5.
16. Schaerer LG, Wu R, Putman LI, Pearce JM, Lu T, Shonnard DR, et al. Killing two birds with one stone: chemical and biological upcycling of polyethylene terephthalate plastics into food. Trends Biotechnol. 2022;41(2):184–96.
17. Singh B, Sharma N. Mechanistic implications of plastic degradation. Polym Degrad Stab. 2008;93(3):561–84.
18. Wei R, Tiso T, Bertling J, O'Connor K, Blank LM, Bornscheuer UT. Possibilities and limitations of biotechnological plastic degradation and recycling. Nature Catalysis. 2020;3 (11):867–71.
19. Ragaert K, Delva L, Van Geem K. Mechanical and chemical recycling of solid plastic waste. Waste Manag. 2017;69:24–58.
20. Villain F, Coudane J, Vert M. Thermal degradation of polyethylene terephthalate: study of polymer stabilization. Polym Degrad Stab. 1995;49(3):393–7.
21. Sinha V, Patel MR, Patel JV. PET waste management by chemical recycling: a review. J Polym Environ. 2010;18(1):8–25.

22. Kawai F, Kawabata T, Oda M. Current state and perspectives related to the polyethylene terephthalate hydrolases available for biorecycling. ACS Sustain Chem Eng. 2020;8(24): 8894–908.
23. Byrne E, Schaerer LG, Kulas DG, Ankathi SK, Putman LI, Codere KR, et al. Pyrolysis-aided microbial biodegradation of high-density polyethylene plastic by environmental inocula enrichment cultures. ACS Sustain Chem Eng. 2022;10(6):2022–33.
24. Werner AZ, Clare R, Mand TD, Pardo I, Ramirez KJ, Haugen SJ, et al. Tandem chemical deconstruction and biological upcycling of poly (ethylene terephthalate) to β-ketoadipic acid by pseudomonas putida KT2440. Metab Eng. 2021;67:250–61.
25. Guzik MW, Kenny ST, Duane GF, Casey E, Woods T, Babu RP, et al. Conversion of post consumer polyethylene to the biodegradable polymer polyhydroxyalkanoate. Appl Microbiol Biotechnol. 2014;98:4223–32.
26. Deter HS, Lu T. Engineering microbial consortia with rationally designed cellular interactions. Curr Opin Biotechnol. 2022;76:102730.
27. Bittihn P, Din MO, Tsimring LS, Hasty J. Rational engineering of synthetic microbial systems: from single cells to consortia. Curr Opin Microbiol. 2018;45:92–9.
28. McCarty NS, Ledesma-Amaro R. Synthetic biology tools to engineer microbial communities for biotechnology. Trends Biotechnol. 2019;37(2):181–97.
29. Meyer-Cifuentes IE, Werner J, Jehmlich N, Will SE, Neumann-Schaal M, Öztürk B. Synergistic biodegradation of aromatic-aliphatic copolyester plastic by a marine microbial consortium. Nat Commun. 2020;11(1):5790.
30. Skariyachan S, Taskeen N, Kishore AP, Krishna BV, Naidu G. Novel consortia of *enterobacter* and *pseudomonas* formulated from cow dung exhibited enhanced biodegradation of polyethylene and polypropylene. J Environ Manag. 2021;284:112030.
31. Qi X, Ma Y, Chang H, Li B, Ding M, Yuan Y. Evaluation of PET degradation using artificial microbial consortia. Front Microbiol. 2021;12:778828.
32. Skariyachan S, Taskeen N, Kishore AP, Krishna BV. Recent advances in plastic degradation-from microbial consortia-based methods to data sciences and computational biology driven approaches. J Hazard Mater. 2021;426:128086.
33. Ali SS, Elsamahy T, Al-Tohamy R, Zhu D, Mahmoud YAG, Koutra E, et al. Plastic wastes biodegradation: mechanisms, challenges and future prospects. Sci Total Environ. 2021;780: 146590.
34. Nguyen PQ, Huang X, Collins DS, Collins JJ, Lu T. Harnessing synthetic biology to enhance ocean health. Trends Biotechnol. 2023;41:860–874.
35. Yoshida S, Hiraga K, Takehana T, Taniguchi I, Yamaji H, Maeda Y, et al. A bacterium that degrades and assimilates poly (ethylene terephthalate). Science. 2016;351(6278):1196–9.
36. Restrepo-Flórez JM, Bassi A, Thompson MR. Microbial degradation and deterioration of polyethylene–a review. Int Biodeterior Biodegradation. 2014;88:83–90.
37. Ho BT, Roberts TK, Lucas S. An overview on biodegradation of polystyrene and modified polystyrene: the microbial approach. Crit Rev Biotechnol. 2018;38(2):308–20.
38. Bratosin BC, Darjan S, Vodnar DC. Single cell protein: a potential substitute in human and animal nutrition. Sustainability. 2021;13(16):9284.
39. Choi KR, Yu HE, Lee SY. Microbial food: microorganisms repurposed for our food. Microb Biotechnol. 2022;15(1):18–25.
40. Ravindra P. Value-added food: single cell protein. Biotechnol Adv. 2000;18(6):459–79.
41. Nyyssölä A, Suhonen A, Ritala A, Oksman-Caldentey KM. The role of single cell protein in cellular agriculture. Curr Opin Biotechnol. 2022;75:102686.
42. Braude R, Hosking ZD, Mitchell KG, Plonka S, Sambrook IE. Pruteen, a new source of protein for growing pigs. I Metabolic experiment: utilization of nitrogen. Livest Prod Sci. 1977;4(1): 79–89.
43. Hardy RW, Patro B, Pujol-Baxley C, Marx CJ, Feinberg L. Partial replacement of soybean meal with Methylobacterium extorquens single-cell protein in feeds for rainbow trout (Oncorhynchus mykiss Walbaum). Aquac Res. 2018;49(6):2218–24.

44. Wiebe M. Myco-protein from fusarium venenatum: a well-established product for human consumption. Appl Microbiol Biotechnol. 2002;58:421–7.
45. Ritala A, Häkkinen ST, Toivari M, Wiebe MG. Single cell protein—state-of-the-art, industrial landscape and patents 2001–2016. Front Microbiol. 2017;8:2009.
46. Pearce JM, Khaksari M, Denkenberger D. Preliminary automated determination of edibility of alternative foods: non-targeted screening for toxins in red maple leaf concentrate. Plan Theory. 2019;8(5):110.
47. Breuer SW, Toppen L, Schum SK, Pearce JM. Open source software toolchain for automated non-targeted screening for toxins in alternative foods. MethodsX. 2021;8:101551.
48. Kong W, Kapuganti VS, Lu T. A gene network engineering platform for lactic acid bacteria. Nucleic Acids Res. 2016;44(4):e37.
49. Kong W, Lu T. Cloning and optimization of a nisin biosynthesis pathway for bacteriocin harvest. ACS Synth Biol. 2014;3(7):439–45.
50. Ibarra-Sánchez LA, Kong W, Lu T, Miller MJ. Efficacy of nisin derivatives with improved biochemical characteristics, alone and in combination with endolysin PlyP100 to control *Listeria monocytogenes* in laboratory-scale Queso Fresco. Food Microbiol. 2021;94:103668.
51. Glenn KC, Silvanovich A, Lee SG, Allen A, Park S, Dunn SE, et al. Biochemical and clinical studies of putative allergens to assess what distinguishes them from other non-allergenic proteins in the same family. Transgenic Res. 2022;31(4–5):507–24.

are not well defined, and they do not present an exciting vision that can attract excellent young researchers to the field. Next, I would like to mention that the belief that we know "everything" that should be known about molecules is far from correct, and I would like to propose what I consider is the next vision for chemistry, which can be summarized as "From Matter to Life." This goal covers all fields of chemistry and beyond, and it relates to two important issues: (1) what we should know about life in order to attempt to produce synthetic life in the laboratory and (2) what should be developed in terms of chemical processes in order to synthesize life.

The process of "synthesizing life" is a far-reaching goal, and it will definitely not be achieved in the coming decade and even beyond. However, by studying what is involved in the chemistry of life and how we can start to explore those processes in the laboratories, these issues can lead to enormous advancements in many fields and in developing "life-improving" materials and methods.

Researchers are already aiming to produce life in the lab, and several topics that are essential for the transition from matter to life have already been investigated very extensively. However, there is no general consensus among chemists that this is an important goal, on which the field should focus, and that chemists from all sub-disciplines can contribute to it.

Next, I will present several examples of subjects that should be explored on our way from Matter to Life. I will also show that these subjects may contribute now not only to the main goal—they may also introduce new technologies and methods. First, I will briefly list some of the subjects that have already been extensively researched; then I will describe in more detail subjects, which, in my opinion, are underappreciated or are less explored.

1. *Self-assembly.* It is well recognized that the self-assembly of molecular systems to supra-molecular systems is essential for the existence of any form of life. Indeed, in the last decades much scientific effort focused on this subject [2]. Self-assembled systems define the boundaries of organisms as well as the compartments within the organism. However, self-organization also defines the molecules within compartments [3]. In most of the self-assembled structure studies, there is only one or very few units that are organized in pseudo-periodic supramolecular structures. There are still many challenges in this field of research; among them is to imbed functionalities within structural elements like membranes, and assemble different molecules into predefined non-periodic structures. Hence, despite that this field has already been recognized as an essential research topic for approaching "life-like" structures, there is still much to explore on the way to achieving fully controlled supramolecular structures.
2. *Energizing.* In living organisms, energy is most commonly stored in adenosine triphosphate (ATP). While ATP stores energy, the sources of energy are different, among them glucose, oxygen, amino acids, and lipids. Living organisms typically consume energy all the time, and as a result, their chemical systems are never in equilibrium. At present, when one attempts to introduce a functioning device into a living organism, one must equip it with a battery that will provide the necessary energy. The concept of using molecules, which already exist in the organism, as a

source of energy for electronic and other devices is very attractive. Indeed, extensive research in this direction has already been performed [4], but there is still a need for optimization and development [5]. Energy is commonly provided by the binding of ATP to various systems, however it is possible that other molecules, beside ATP, can be used.

3. *Very long-range information transport.* The information transfer in living organisms can be divided into a relatively intermediate range, up to hundreds of nanometers, or a very long range. This subject has been the focus of extensive research. The electron transfer for many nanometers and its efficiency are not fully understood, and we find it surprising that in many cases very efficient electron transfer processes can exceed hundreds of nanometers [6, 7]. There is another regime, in the nervous system, for example, that information is transferred in the axon for very long distances, up to meters, very efficiently. It is known that in this case typically ions transfer the information. The velocity of the information transfer is up to 100 meter/sec. Interestingly, the "action potential" that transfers the information does not decay as a function of length. However, despite the vast information existing on information transfer in biology, very few artificial systems that use similar mechanisms have been developed [8, 9]. Exploring this field will advance our basic understanding of charge transport, and there may be important applications in medicine as well as in various technologies.
4. *Why are proteins "so big"?* This is a well-recognized question, why proteins are much bigger than their reactive site. Many researchers have addressed this issue, and usually the explanation given was that the structural properties affect, for example, the entropy involved in their reactivity [10]. Recently it was demonstrated that charge polarization might play an important role in the protein-protein association reaction [11]. When two proteins interact by an electrostatic interaction, the polarizability of the protein defines how much charge can be accumulated at the bond. As more charge is accumulated, the bond becomes stronger, and the reaction becomes faster. Hence, the domains of the protein, which are relatively far from the reaction site, contribute to the protein's polarizability, and therefore define the reaction rates. This concept of increasing the polarizability of molecules by attaching groups that are not directly involved in the reaction, but can contribute to the polarizability, may be used in other cases, besides the association of proteins. However, it has been little explored.

There are certainly many more subjects that can fall under the title "from matter to life" that are currently explored. The list above is only a small sample that demonstrates that even in subjects that are studied extensively, there are still very important open questions that should be addressed.

Many phenomena in biology present a challenge for chemists. Some of them are rarely addressed. Next, I will provide several examples of these phenomena, with no attempt to cover them all, but some examples of the way one can develop fresh thinking in chemistry based on bio-processes:

1. *Synthesizing without separation.* Commonly in chemistry, a synthetic process involves several stages. After each stage, the desired product is separated from the mixture and is used as a reagent for the next step. Interestingly, this is not the way molecules are synthesized in biology. In biology, typically reactions occur in a "soup" containing many other molecules. However, somehow a specific product can be formed with almost no "by-products." It is true that many of the reactions in biology are catalyzed by enzymes, but it is still quite surprising that other molecules do not interfere with enzymatic processes. The high specificity of the reactions in biology may result from the molecules reacting in a highly crowded environment, i.e., the cell, in which molecules that react could be placed in the vicinity of each other and perhaps even in the favorable orientation. However, in crowded environment, the motions of the molecules and their conformers are limited; therefore, the entropy is low, and the processes are mainly governed by enthalpy[12, 13]. The number of studies of chemical reactivity in a confined environment is very limited [14–16], and confinement is not commonly used to govern chemical processes or to avoid by-products. It is clear that the field of "chemistry without separation" can teach us on how process occur in life, but it may also be of a high economic value, if multiple-step processes can be reduced to chemistry with no separation.
2. *Chirality – not only structure.*
   Pasteur recognized that "Chirality is a signature of life." [17] Indeed, important molecules in living organisms are chiral; these molecules include amino acids, proteins, DNA, and sugars. One may wonder why chirality is so well preserved throughout evolution. Chemists put much effort into producing chiral molecules with well-defined handedness, since they are required for pharmaceutical applications and for agro-related products. For a long time, it has been assumed that chirality serves only as structural motif and that chiral molecules do not have any special electronic properties. However, in the last two decades new properties of chiral molecules have been found, which at the first glance, seem not to be related to bio-functionality, but with time, they have been found to be very relevant [18]. It became apparent that chiral molecules are spin filters, namely, they transfer electrons with one specific spin. Which spin is transferred depends on the handedness of the molecule and the electron's direction of motion. This effect is called the chiral-induced spin polarization (CISS) effect [19]. Moreover, it was found that when chiral molecules are charge polarized, due to an electric field, the charge polarization is accompanied by spin polarization. This property was found to enhance redox reactions in which oxygen molecules such as respiration and photosynthesis are involved. It also enhances the oxygen evolution reaction, used to produce hydrogen, and the efficiency of fuel cells. The CISS effect is also involved in long-range electron transfer and in enantiospecific reactions [20]. It seems that oxygen-based life could not be developed without chiral molecules [18]. Hence, Pasteur's statement actually has deep foundations, and indeed life, as we know it, cannot exist without chirality.

   In relation to the "Matter to Life" effort, it is important to understand how in the early stage of life formation, chirality could be introduced from achiral molecules with no chiral molecule as a catalyst [21]. In recent studies, there are indications that magnetic surfaces can induce the breaking of symmetry and the formation of chiral products when all the reagents are achiral. This subject of how chiral molecules can be synthesized efficiently from achiral reagents without chiral catalysts is of major importance in chemistry and may open new directions in breaking symmetry in chemical processes.

3. *The role of surfaces. Heterogeneous catalysts and magnetic surfaces*
   Heterogeneous catalysis is a well-established technique and is in use in numerous industrial applications. However, there are some elements of it that have been much less explored, for example, catalysis by organic molecules that are adsorbed on surfaces. In biology, surfaces and interfaces are of major importance, and they define compartments, control input, and output from these compartments and produce electric field across interfaces; however, they also serve as catalysts, for example, the membrane enzymes [22, 23]. The role of surfaces should therefore also be investigated to support molecules that will serve as catalysts and to determine their effect on chemistry that occurs within a confined environment. In the process of developing "chemistry with no separation," surfaces and specifically membranes should be explored for their ability to selectively transfer reagents in and out of a compartment where the reaction occurs. Surfaces can also serve as the medium for aligning particles with magnetic properties; therefore, they can be used for spin-specific catalysis or for interacting with paramagnetic reagents.
4. *The role of the electric field.*
   An appreciation of the role of an electric field in protein function and structure has increased substantially in recent decades [24–28].Typically, one considers the electrostatic field resulting from the structure of the protein. However, every time two objects interact, an electric field is induced between them, as a result of their different electrochemical potentials. Even for two identical objects, a field is created due to the induced dipole interaction. Since all molecules and even proteins are polarizable objects, the field applied on them results in charge rearrangement and sometimes also in structural changes. Hence, besides the effect of a permanent electric field, there is a transient electric field that can also modify the outcome of reactions. Moreover, one object that interacts with a reactant can induce a field that will affect the reactivity of this reactant with another reagent. In biology, this is referred to as allosteric effect. However, in the past, allosteric effects were thought to be only related to structural changes; however, recently it was demonstrated that allosteric effects can result from charge reorganization [11, 29]. Besides inducing an electric field by interaction between molecules, a topic almost unexplored, it may be possible to induce specific reactions by an external magnetic field. For this to take place, the molecule and the electric field must be coupled to the same frame. Namely, the molecule can be attached to a surface, either directly or indirectly, or be within a membrane that is fixed in the molecular frame. However, electric field-induced reactivity has not been sufficiently explored; it can be important both for better

understanding biological systems and for fabricating a specific synthetic chemical.

In this short article, the aim was to point out the possibility of expanding the research topics in chemistry by aiming to transform molecules to systems, which, in principle, can be the building blocks for "artificial" life. The concept "from material to life" does not necessarily mean that evolution must be imitated. A "life-like material" may be achieved by a very different approach than that taken by evolution. Hence, the aim is not to produce "synthetic biology," but rather, to produce systems that can have biology-like functions even if they use different materials and structures. Even without reaching the ultimate goal above, this type of research may enrich chemistry in general and provide new methods and processes that will improve our well-being and our environment.

References

1. Whitesides GM. Reinventing chemistry. Angew Chem Int Ed. 2015;54:3196–209.
2. Zlenko DV, Zanin AM, Stovbun SV. Molecular self-assembly as a trigger of life origin and development. Orig Life Evol Biosp. 2022;52:105–11.
3. Hudder A, Nathanson L, Deutscher MP. Organization of mammalian cytoplasm. Am Soc Micro Molec Cell Bio. 2003;23:9318–26.
4. Qian H, Autzen E. A little engine that could: ATP-powered electrical battery and heater inside cells. Biophys J. 2012;103:1409–10.
5. Wang X, Liu S, Zhang Y, Lv S, Ni H, Deng Y, Yuan Y. A review of the power battery thermal management system with different cooling, heating and coupling system. Energies. 1963;2022: 15.
6. Yates MD, Goldena JP, Roy J, Strycharz-Glaven SM, Tsoi S, Erickson JS, El-Naggar MY, Barton SC, Tender LM. Thermally activated long range electron transport in living biofilms. Phys Chem Chem Phys. 2015;17:32564–70.
7. Bjerg JT, et al. Long-distance electron transport in individual, living cable bacteria. PNAS. 2018;115:5786–91.
8. Espinosa-Hoyos D, Jagielska A, Homan KA, Du H, Busbee T, Anderson DG, Fang NX, Lewis JA, Van Vliet KJ. Engineered 3D-printed artificial axons. Sci Rep. 2018;8:478.
9. Ariyaratne A, Zocchi G. Toward a minimal artificial axon. J Phys Chem B. 2016;120:6255–63.
10. Rubinson KA. Why proteins are big: Length scale effects on equilibria and kinetics. Protein J. 2019;38:95–119.
11. Banerjee-Ghosh K, Ghosh S, Mazal H, Riven I, Haran G, Naaman R. Long-range charge reorganization as an allosteric control signal in proteins. JACS. 2020;142:20456–62.
12. Minton AP. Excluded volume as a determinant of macromolecular structure and reactivity. Biopolymers. 1981;20:2035–285.
13. Zhou H-X, Rivas G, Minton AP. Macromolecular crowding and confinement: biochemical, biophysical, and potential physiological consequences. Annu Rev Biophys. 2008;37:375–97.
14. Prieto MJ, et al. Insights into reaction kinetics in confined space: real time observation of water formation under a silica cover. J Am Chem Soc. 2021;143:8780–90.
15. Weckhuysen BM, Kitagawa S, Tsapatsis M. Reactions in confined spaces. Chem Phys Chem. 2018;19:339–40.
16. Küchler A, Yoshimoto M, Luginbühl S, Mavelli F, Walde P. Enzymatic reactions in confined environments. Nat Nanotech. 2016;11:409–20.
17. Vantomme G, Crassous J. Pasteur and chirality: a story of how serendipity favors the prepared minds. Chirality. 2021;33:597–601.

18. Sang Y, Tassinaria F, Santra K, Zhang W, Fontanesi C, Bloom BP, Waldeck DH, Fransson J, Naaman R. Chirality enhances oxygen reduction. PNAS. 2022;119:e2202650119.
19. Naaman R, Paltiel Y, Waldeck DH. Chiral induced spin selectivity and its implications for biological functions. Annu Rev Biophys. 2022;51:99–114.
20. Bloom BP, Lu Y, Waldeck DH, Metzger T, Yochelis S, Paltiel Y, Fontanesi C, Mishra S, Tassinari F, Naaman R. Asymmetric reactions induced by electron spin polarization. PCCP. 2020;22:21570–82.
21. Ozturk SF, Sasselov DD. On the origins of life's homochirality: Inducing enantiomeric excess with spin-polarized electrons. Proc Natl Acad Sci U S A. 2022;119:e2204765119.
22. Pochapsky TC. Examining how enzymes self-organize in a membrane. PNAS. 2014;111:3659–60.
23. Sanders CR, Hutchison JM. Membrane Properties that Shape the Evolution of Membrane Enzymes. Curr Opin Struct Biol. 2018;51:80–91.
24. M.F. Perutz Electrostatic effects in proteins. Science. 1978;201:1187–91.
25. Warshel A. Electrostatic basis of structure–function correlation in proteins. Acc Chem Res. 1981;14:284–90.
26. Sharp KA, Honig B. Electrostatic interactions in macromolecules: theory and applications. Annu Rev Biophys Biophys Chem. 1990;19:301–32.
27. Warshel A, Sharma PK, Kato M, Parson WW. Modeling electrostatic effects in proteins. Biochim Biophys Acta. 2006;1764:1647–76.
28. Fried SD, Boxer SG. Electric fields and enzyme catalysis. Ann Rev Biochem. 2017;86:387–415.
29. Ghosh S, Banerjee-Ghosh K, Levy D, Riven I, Naaman R, Haran G. Substrates modulate charge-reorganization allosteric effects in protein-protein association. J Phys Chem Lett. 2021;12:2805–8.

Transforming the Future: A Look at the Latest Breakthrough Technologies

Thomas Herget

1 Introduction

Each year, a handful of innovations, breakthroughs, and discoveries push the boundaries of what is possible using technology. The most important are those that help to solve problems and overcome challenges that impact our planet or our everyday lives. The examples described here show significant steps forward and will have an immense effect on our lives in the future. In this chapter, we will highlight what they are and why they have the potential to make the world a better place. You'll recognize some; others might surprise you.

- **Disruptiveness of papers has dropped.**

Everybody will agree on innovations like the internet, genome editing, or quantum computing as being breakthrough technologies. Breakthrough or disruptive technologies are indeed not subjective but can be measured by the way of citation. Researchers at the University of Minnesota used information about citations from 45 million research articles and 3.9 million patents to develop an algorithm to measure how disruptive a study is, which they call the CD index. The index assigns values ranging from −1 (least disruptive) to +1 (most disruptive) to help measure the level of disruption caused by a particular study. When a study is very disruptive, other studies are less likely to refer to the sources that the disruptive study used, and are more likely to refer directly to the disruptive study itself. The number of technology- and science-related research papers published has climbed rapidly over the last few decades—but the "disruptiveness" of those papers has fallen over the last 60 years, according to their analysis published in NATURE [1].

T. Herget (✉)
Darmstadt, Germany
e-mail: Thomas.herget@merckgroup.com

U. A. Betz (ed.), *Curious Future Insight*,
https://doi.org/10.1007/978-3-031-41781-8_8

How does the concept of "disruptiveness" in research relate to scientific progress? Being disruptive—i.e., introducing new and groundbreaking ideas—is not always better than making incremental improvements to existing knowledge. It can be important to validate and replicate findings through additional research, which may not be particularly disruptive, but is still important for solid scientific progress. The authors note that it is difficult to determine why there has been a decline in the proportion of disruptive research over time. Being forced to publish in order to obtain ongoing funding may be one reason. On the other hand, the absolute number of highly disruptive studies has remained constant. This suggests that while relatively spoken there may be fewer groundbreaking discoveries being made, those that are being made are still having a significant impact on scientific knowledge.

This study, together with previously published other studies [2], offers a data-driven way to investigate how science changes and allows the establishment of new strategies for establishing the productivity of research in the future.

The following technologies are a selection envisaged as highly innovative and disruptive for environment and human health.

- **Batteries play a key role in the transition to renewable energy.**

The use of batteries is increasing globally, especially with the growth of electric vehicles (EVs) markets. With over 10% of global vehicle sales now being EVs (targeting 30% by the end of the decade), there is a growing recognition of the need to reduce carbon emissions from transportation and the potential of batteries to help achieve this goal. Governments around the world are introducing policies and incentives to accelerate the adoption of EVs and promote battery manufacturing. For example, the US government has introduced climate legislation that aims to invest billions of dollars in battery manufacturing and provide motivations for EVs purchases. Similarly, the European Union and several US states have announced plans to ban gas-powered vehicles starting in 2035, which will likely further drive demand for electric vehicles and battery storage systems.

Not surprisingly, battery technology is advancing rapidly, with improvements in energy density, durability, and cost. These improvements are making batteries more competitive with traditional fossil fuel technologies, and this trend is expected to continue as the cost of batteries continues to decline. Overall, the increasing use of batteries is a promising trend for achieving a more sustainable and resilient energy future, and the continued growth of electric vehicles markets and battery storage systems will play a critical role in achieving this goal.

However, currently most EVs use lithium-ion batteries, a technology that has been around for decades and is also used in laptops and cell phones. The years of development have helped to lower costs and improve performance, making EVs more affordable and able to travel longer distances between charges. While the focus for electric vehicle batteries is on making them smaller, lighter, and faster, the primary goal for stationary storage is to reduce costs. Lithium-ion batteries are also being utilized in electricity storage on the grid to help balance out intermittent renewable energy sources like wind and solar. Over 90% of global installations of electrochemical energy storage are using this technology nowadays.

Despite being commonly used for both, EVs and stationary storage, lithium-ion batteries are not the most ideal option for this purpose. The search for alternative battery chemistries is being driven by concerns about the availability of key battery materials like cobalt and lithium. This has resulted in a push to find new solutions that can replace the standard lithium-ion chemistry. While the focus for electric vehicle batteries is on making them smaller, lighter, and faster, the primary goal for stationary storage is to reduce costs. Therefore, different battery chemistries that are not as concerned with size and weight may be more suitable for grid storage.

As a result, there are now several promising alternatives to lithium-ion batteries emerging in the field of stationary storage. One of these promising alternatives is iron, and two companies in particular are making progress in this area.

Form Energy is a company that is developing a novel type of battery that uses a water-based electrolyte to store energy through reversible "rusting." This technology has the potential to provide low-cost, long-duration energy storage, which is particularly important for integrating renewable energy sources into the grid. Recently, Form Energy declared to build a USD 760 million manufacturing facility in West Virginia, building started in 2023. This facility will produce the company's iron-air battery technology, which has a capacity of up to 150 hours of continuous power [3].

ESS, Inc. is another company that is developing a different type of iron battery that employs similar chemistry. This battery has already entered manufacturing and is being used for long-duration energy storage. ESS, Inc.'s iron flow battery technology uses an iron electrolyte to store energy, which can be discharged over a long period of time [4].

Both Form Energy and ESS, Inc. are focused on developing low-cost, long-duration energy storage solutions, which are becoming increasingly important as the world shifts toward renewable energy sources. These companies' technologies have the potential to play a critical role in enabling the widespread adoption of renewable energy and achieving a more sustainable energy future.

Aqueous proton batteries are another promising technology for energy storage due to their high safety standards and potential for high-rate capability and long cyclability. Protons, which are simply hydrogen ions (H+), are the smallest and lightest charged particles, making them an ideal charge carrier for battery systems. Compared to traditional lithium-ion batteries, which use lithium ions as charge carriers as described above, aqueous proton batteries have several advantages. First, they use water-based electrolytes, which are abundant, inexpensive, and non-flammable, thus offering a higher level of safety. Second, because protons are smaller and more mobile than lithium ions, aqueous proton batteries can potentially deliver higher power density and faster charging rates. Finally, aqueous proton batteries have the potential to be more environmentally friendly than lithium-ion batteries, which require the mining and processing of lithium, a finite and non-renewable resource. Overall, the development of aqueous proton batteries could have a significant impact on the energy storage landscape, particularly in the areas of electric vehicles and grid storage in future [5].

In summary, battery energy storage systems (BESS) are one of the most popular and widely used energy storage technologies today. However, to expedite the deployment of BESS globally, there is a need for clear policy frameworks and energy storage targets. Governments can play a critical role in promoting the adoption of energy storage systems by implementing policies that encourage investment and incentivize the development of energy storage projects. Setting clear energy storage targets and including them in national energy plans can also help drive the deployment of BESS. In addition, the development of standardized regulations and technical standards can help streamline the process of integrating BESS into EVs and the grid and reduce regulatory barriers. All of these measures will help to accelerate the deployment of the energy storage industry, which is critical for achieving a more sustainable and resilient energy future.

- **Carbon removal factory in Iceland.**

Carbon dioxide is a greenhouse gas and responsible for contributing significantly to the climate change. The world's largest plant designed to suck carbon dioxide out of the air and turn it into rock has recently started running in Iceland. The companies behind the project are Switzerland's Climeworks [6] and Iceland's Carbfix [7]. The facility, outside Reykjavik, can capture 4000 metric tons of carbon dioxide out of the air every year. This Orca plant is named after the Icelandic word for "energy" and consists of four units, each made up of two metal boxes that resemble shipping containers. According to the US Environmental Protection Agency, the emissions from the Orca plant are equivalent to those produced by about 900 cars. The cost of building the plant is estimated to be around USD 15 million. Additionally, it has been reported that another larger plant is already planned.

To collect the carbon dioxide, the plant uses fans to draw air into a collector, which has a filter material inside. Once the filter material is filled with CO2, the collector is closed, and the temperature is raised to release the CO2 from the material, after which the highly concentrated gas can be collected. Then, Carbfix's process mixes the CO2 with water and injects it at a depth of 1000 meters into the nearby basalt rock where it is mineralized [8]. Carbfix says the CO2-water mixture turns to stone in about 2 years, and hydride of sulfur (H_2S), within 4 months. Both technologies are powered by renewable energy sourced from a nearby geothermal power plant.

Proponents of so-called carbon capture and storage believe these technologies can become a major tool in the fight against climate change. Critics however argue that the technology is still prohibitively expensive and might take decades to operate at scale. Of course, Carbfix will not solve the world's problem of greenhouse gas emissions. One severe limitation of the method described above is the need for substantial quantities of water and the presence of porous basaltic rock. Both are widely available on the continental boundaries, such as in Iceland and the Pacific Northwest of the USA, but are rare or absent in other parts of the world.

A growing number of individuals and companies, including Microsoft, Stripe, and Square, are already paying today's high costs to suck carbon out of the air as they strive to cancel out their emissions. That's providing crucial early revenue.

- **Heart transplant from pig to human.**

The shortage of organ donors is a major challenge in the field of transplantation, and many people are unable to receive the life-saving transplants they need due to the scarcity of organs. This has led to a growing interest in developing technologies that could enable an unlimited supply of transplantable organs. One potential solution that has been explored is xenotransplantation, meaning the transfer of organs between species. However, this approach has significant challenges, including the risk of hyper-acute rejection, where the human immune system attacks and destroys the transplanted organ within minutes or hours.

One animal that has been studied extensively as a potential source of organs for xenotransplantation is the pig. But tests have shown that pig organs are highly prone to hyper-acute rejection in humans, and significant advances in genetic engineering and immunosuppressive therapies will be necessary to make pig-to-human transplantation a viable option. In addition to the challenge of hyper-acute rejection, there is also a risk that pig viruses could be transmitted to humans during transplantation, potentially causing a pandemic. This risk highlights the importance of careful testing and monitoring of xenotransplantation procedures, as well as ongoing research into alternative strategies for addressing the shortage of organ donors.

Gene editing is one approach that is being explored to overcome the challenges of xenotransplantation and create pigs that can provide tolerable organs for human transplant. One company [9] that has been working on this approach is United Therapeutics, which has developed a list of ten gene modifications that it believes could make pig organs compatible with the human immune system. Three of these modifications involve "knocking out" pig genes that produce molecules that alarm the human immune system, while six involve adding human genes that would help to cover over differences between the pig and human immune systems. In addition, United Therapeutics has disabled a receptor that senses growth hormone in pigs to prevent the transplanted organ from growing too large in humans, who are smaller than pigs.

Many requests by the FDA (US Food and Drug Administration) had to be answered. Finally, David Bennett Sr. was near death when he received the genetically edited heart. It's important to note that while David Bennett Sr.'s transplant was initially successful and his new pig heart was functioning well, he unfortunately took a turn for the worse and passed away about 40 days after the transplant [10].

It's not entirely clear what caused his death, and an initial statement released by the University of Maryland School of Medicine did not provide any obvious cause. Later, it was disclosed, that Bennett's heart was affected by porcine cytomegalovirus, a preventable infection that is linked to devastating effects on transplants. The presence of the pig virus prompted a desperate effort to defeat it. The issue is now a subject of wide discussion among specialists, who think the infection was a potential contributor to Bennett's death and a possible reason why the heart did not last longer.

The findings of German researchers are consistent with the risk associated with the presence of the porcine cytomegalovirus in pig-to-human transplants. The virus, which is harmless to pigs, can cause serious health issues in humans. This highlights the importance of ensuring that the organs used in xenotransplantation have to be free from the virus to avoid potential complications and improve the success rate of the procedure [11].

The heart transplant was a major test of xenotransplantation, the process of moving tissues between species. But because the special pigs raised to provide organs are supposed to be virus-free, it now appears that the experiment was compromised by an unforced error. The biotechnology company that raised and engineered the pigs, Revivicor, a subsidiary of United Therapeutics, declined to comment and has made no public statement about the virus [12]. However, it's worth noting that xenotransplantation is a complex and still relatively untested field, and there are many potential risks and complications associated with transplanting organs between different species. While the use of genetically edited pigs for xenotransplantation holds promise for providing a nearly unlimited supply of transplantable organs, there is still much research and development that needs to be done to ensure the safety and efficacy of this approach.

While Bennett's death was a setback, the initial success of the pig-to-human heart transplant is seen as a major step forward in the field of xenotransplantation, and has given hope to millions of patients waiting for organs. United, eGenesis, Makana Therapeutics, and other companies working on xenotransplantation will need to address the issue of porcine viruses and other potential risks associated with the process, and work closely with regulatory agencies to ensure that future trials are conducted safely and effectively.

- **Anti-Alzheimer plaque drugs make their ways to approvals.**

Alzheimer's disease—also known as Alzheimer's dementia—is the most common form of dementia and an incurable disorder of the brain. Due to the death of nerve cells in the brain, people with Alzheimer's become increasingly forgetful, confused, and disoriented. According to Alzheimer's Disease International, there are around 50 million people, nearly 10 million just in the USA, with dementia worldwide. This number could rise to 152 million by 2050. Alzheimer's dementia is not yet curable. Finally, a few medications that can delay the loss of mental abilities and independence in mild and moderate dementia arrived. A new promising class of drugs was recently approved by the FDA to treat early-stage Alzheimer's patients. These experimental Alzheimer's drugs are therapeutic antibodies designed to bind to a protein known as amyloid beta. Research indicates these membrane-embedded proteins normally provide essential functions in the brain. But sometimes they misfold, which, over time, causes them to clump together and form a kind of sticky, toxic plaque that has long been thought to play a central role in the development of Alzheimer's.

The drugs Lecanemab from Biogen and Eisai, and Aduhelm from Biogen, are humanized monoclonal antibodies that appear to have a specific affinity for one form, called soluble protofibrils, that amyloid beta takes. This form is thought to be more toxic to neurons than others. Aduhelm (also known as aducanumab) and Lecanemab (also known as BAN2401 and Leqembi) represent a new category of drugs called amyloid beta-targeting agents, which work by recognizing protofibrils and prevent amyloid beta deposition in Alzheimer's brains.

Still, these drugs are not without suspicion. In the USA, Aduhelm was approved by the US Food and Drug Administration, subject to conditions (2021). The decision was highly debated due to the lack of clear proof of effectiveness. The FDA had linked the approval to the condition that Biogen prove the efficacy of the antibody with a further study. The European Medicines Agency (EMA) has not accepted Biogen's Alzheimer's drug Aduhelm in Europe. The EMA justifies the rejection of approval with the unproven effectiveness and the sometimes considerable side effects of the drug, such as brain swelling or even cerebral hemorrhages. In studies, Biogen was able to prove that Aduhelm effectively removes Alzheimer's-specific protein deposits from beta-amyloid in the brain. However, it could not be proven whether this improves the cognitive abilities of the patients [13].

The picture for the drug Lecanemab from Biogen and Eisai is much more convincing than for Aduhelm.

Eisai and Biogen's press release described what are known as topline results, indicating that the trial met its main and secondary goals and summarizing reported side effects [14]. That information was enough to frame Lecanemab as the most promising Alzheimer's drug to emerge from clinical testing in decades, and added billions of dollars to Eisai's and Biogen's respective market values. A NEJM paper showed that study participants given Lecanemab experienced significantly slower decline on four rating scales investigators used to gauge patients' cognitive and functional ability. In the trial of 856 patients with mild cognitive impairment or mild dementia, Lecanemab reduced clinical decline by 27% compared to placebo after 18 months [15]. As Lecanemab is meant to be taken chronically, there may be a potential benefit for more substantive changes in the disease's course beyond the trial's 18 months.

The FDA recently approved Lecanemab via the Accelerated Approval pathway for the treatment of Alzheimer's disease [16].

In summary, Aduhelm and Lecanemab represent a new category of medications permitted for treating Alzheimer's disease. The approval of these medications has been met with both excitement and controversy in the medical community, as the clinical trial data for Aduhelm has been somewhat mixed, and the drug's high cost has raised concerns about its accessibility for patients. Nonetheless, the development of amyloid beta-targeting agents represents an important step forward in the fight against Alzheimer's disease and provides hope for the millions of people affected by this devastating condition. These medications illustrate an important milestone in the ongoing fight to treat Alzheimer's disease by effectively removing plaques.

- **Human brain–computer interaction.**

Communication is an essential part of human interaction and is crucial for social and political inclusion, transmitting thoughts and ideas, and contributing to human cultures. For people with severe motor impairments, communication can be a major challenge, and technologies like brain–computer interfaces (BCIs) offer a promising solution. BCIs are devices that allow people to communicate with computers or other devices using their brain signals, rather than traditional methods like typing or speaking. These devices work by detecting and decoding neural signals from the brain, which can then be used to control computer interfaces, robotic devices, or other types of technology. Recent advances in BCI technology have led to the development of systems that can decode neural signals associated with speech, allowing people with severe motor impairments to communicate more effectively. An experimental device developed by researchers at Stanford University is one example of this technology [17].

BrainGate is an experimental system that uses implanted electrodes to allow people who are paralyzed to control computers and other devices using their thoughts. This technology relies on the detection and decoding of neural signals from the part of the brain that controls movement, which can be used to control computer interfaces, robotic devices, or other types of technology. In previous studies using BrainGate, participants had learned to control a computer cursor or robotic arm by imagining they were moving their hands. However, the recent study took things a step further, by having the participant imagine he was writing individual letters by hand while a computer monitored the electrical activity in his brain. The computer was then able to learn to decode the distinct pattern of activity associated with every letter of the alphabet, as well as several symbols, allowing the participant to "type" at a rate of about 90 characters per minute with 95% accuracy just by imagining he was handwriting letters on a sheet of paper. Because the new system relies on familiar thoughts, the participant was able to use it almost immediately. This incredible progress shows the potential of BCI technology to revolutionize communication for people with severe motor impairments. Since AI and machine-learning methods are rapidly improving, we can expect a promising path for future improvements. Also, the Stanford team is making its data set publicly available, which will accelerate advances further. While this technology is still experimental and has a long way to go before it can be widely used, the progress made so far is truly remarkable and offers hope for a better future for people with disabilities [18].

- **AI development: language model ChatGPT responds like a human being.**

Artificial intelligence (AI) refers to the ability of a computer or a robot controlled by a computer to perform tasks that typically require human intelligence and judgment. AI has become increasingly prevalent in our daily lives, with examples like Siri, Google Now, Amazon's Alexa, and Microsoft's Cortana being just a few of the most well-known examples. These digital assistants are designed to perform various tasks, such as checking schedules, searching the web, sending commands to other apps, and even controlling smart home devices. They use a combination of natural language processing, machine learning, and other AI technologies to understand and respond to user commands and queries. AI has the potential to transform many areas of our lives, from healthcare (see previous example: Human Brain—Computer interaction) and education to transportation and manufacturing.

OpenAI [19] recently introduced ChatGPT, a platform based on the GPT-3.5 model, which is one of the largest and most advanced language models currently available [20]. ChatGPT (Generative Pretrained Transformer) is designed for natural language processing tasks such as text generation and language translation, and it has the ability to generate human-like text responses to prompts. It became one of the most used apps within 3 months.

One of the key features of ChatGPT is its ability to understand and respond to natural language queries and prompts, making it useful for a wide range of applications. For example, it can be used to create chatbots for customer service, where it can answer frequently asked questions and provide helpful information to customers. It can also be used to generate responses to questions in online forums or to create personalized content for social media posts. Further potential applications of ChatGPT include text generation, language translation, text summarization, and sentiment analysis. To improve the text generation and the algorithms behind it, ChatGPT's AI is continuously trained by human feedback.

In order to use ChatGPT, you will need to have access to the OpenAI API (application programming interfaces), which requires an API key. Once you have the API key, you can use ChatGPT by submitting it a prompt in the form of a text block. The model will then generate a response based on the information it has been trained on.

The GPT-3 architecture, which serves as the basis for ChatGPT, is known for its ability to learn from large amounts of data. The model has been trained on a massive corpus of text data, which includes a diverse range of topics and styles. This has enabled the model to develop a deep understanding of language and a high degree of contextual awareness. As a result, ChatGPT is capable of generating responses that are highly relevant to the prompt and exhibit a level of knowledge and understanding that is similar to that of a human. The model is able to generate coherent and natural-sounding text that is appropriate for a wide range of natural language processing tasks.

Nevertheless, it's important to note that ChatGPT is a machine learning model and may not always produce accurate or appropriate responses. Like any other machine learning model, ChatGPT has certain limitations users should be aware of. It does not have a deep understanding of the world or the ability to reason like a human. As a result, the model may not be able to generate responses to complex or abstract questions, or to understand the context. Furthermore, users of ChatGPT should be aware of the potential for bias in the model's responses.

The ability of ChatGPT to generate natural-sounding text can have both positive and negative implications, e.g., in the field of education. On the one hand, the technology can be used to support students in their learning and help them to generate high-quality written work. For example, the model can be used to provide students with feedback on their writing, suggest revisions to their work, or even generate sample essays or reports to provide students with examples of good writing. However, on the other hand, the technology can also be misused for academic dishonesty. For example, students could use the technology to generate essays or other written work that appear to be their own but were actually generated by the model. This raises concerns about plagiarism and the authenticity of student work. Solutions need to be found to address these challenges and ensure that the use of ChatGPT and similar technologies in education is appropriate and ethical. This could include developing policies and guidelines for the use of the technology, as well as developing tools and techniques for detecting and preventing academic dishonesty like adding an electronic watermark.

Overall, ChatGPT is a powerful tool because of its large size, human-like responses, adaptability, and versatility. These features make it a valuable resource for anyone who needs to perform natural language processing tasks. OpenAI just released a new, more powerful version, called GPT-4, which can identify false news and additionally works with images. It can be foreseen that ChatGPT will change many aspects in our lives, like the way we will interpret data, do market research and analysis, train our students, or assess information. Due to its great success US technology giant firms like Google, Meta, Amazon, and Microsoft as well as Chinese tech giant Baidu are pursuing similar projects.

2 Outlook

Whether or not a new technology is a breakthrough with positive impact on mankind is often not immediately obviously but will show the future.

Such one example is fentanyl which started as a great medicine and then became a killer: Back in 1953, the Belgian doctor and chemist Paul Janssen set about creating the strongest painkiller he could. He believed he could improve on morphine, designing a molecule that was 100 times more potent but with a short duration. His discovery, the synthetic opioid fentanyl, would become the painkiller most widely used during surgery. However, today, fentanyl is setting grim records—it's involved in the accidental death of around 70,000 people a year in the USA, nearly 200 deaths a day. It's the leading cause of death in American adults under 50, killing more than car accidents, guns, and COVID together. Fentanyl kills by stopping your breathing. Its potency is what makes it deadly. Two milligrams can be a fatal dose. Janssen Pharmaceuticals, a division of Johnson & Johnson, played a role by making false claims about how addictive prescription opioid drugs were; making money while people got hooked on pills and patches. Janssen agreed now to pay a USD 5 billion settlement without admitting wrongdoing. Can things get worse? They can. US states are reporting a rapid increase in fentanyl deaths in young children who accidentally ingest pills.

The FTX cryptocurrency tokens [21], the "zero covid" policy in China, and Twitter terminating the site's governing policies are showing their negative effects rapidly. For others, like ChatGTP, we will learn both the positive and undesired effects in the future—as we do with fentanyl.

References

1. https://www.nature.com/articles/s41586-022-05543-x
2. https://doi.org/10.2139/ssrn.3822691
3. Form Energy: We are Transforming the Grid
4. Long-duration Energy Storage | ESS, Inc. (essinc.com)
5. https://link.springer.com/article/10.1007/s40820-022-00864-y
6. https://climeworks.com/direct-air-capture
7. https://www.carbfix.com/
8. https://www.usgs.gov/news/featured-story/making-minerals-how-growing-rocks-can-help-reduce-carbon-emissions
9. https://www.unither.com/index
10. Patient Who Received Groundbreaking Pig Heart Transplant Dies—The New York Times (nytimes.com)
11. Impact of porcine cytomegalovirus on long-term orthotopic cardiac xenotransplant survival—PubMed (nih.gov)
12. https://www.revivicor.com/
13. FDA approves Alzheimer's drug from Eisai, Biogen in closely watched decision | BioPharma Dive
14. Lecanemab confirmatory phase 3 clarity ad study met primary endpoint, showing highly statistically significant reduction of clinical decline in large global clinical study of 1,795 participants with early Alzheimer's disease | News Release:2022 | Eisai Co., Ltd.
15. Lecanemab in Early Alzheimer's Disease | NEJM
16. https://www.fda.gov/news-events/press-announcements/fda-grants-accelerated-approval-alzheimers-disease-treatment. 2023
17. High-performance brain-to-text communication via handwriting | Nature
18. https://neuroscience.stanford.edu/news/man-who-paralyzed-communicates-imagining-handwriting
19. https://openai.com/
20. ChatGPT—Wikipedia
21. https://www.reuters.com/business/finance/us-treasury-financial-watchdogs-companies-among-ftx-creditors-filing-shows-2023-01-26/

with management emphasizing the importance of concrete, measurable outcomes rather than overly controlling the process to achieve them. However, one critical element was in short supply: the availability of adequate resources. To qualify for this endeavor, it was mandatory to have previous hands-on experience in mission control. Becoming a member of the Cluster II flight control team provided me with firsthand exposure to the challenges, capabilities, and limitations of the existing mission control support tools and processes. This experience helped me identify an initial set of priorities for potential investigation and development.

To address the challenge of limited resources, we adopted a divided approach. A young team, freshly graduated from university, focused on exploring the potential of cutting-edge technologies like artificial neural networks, genetic algorithms, and fuzzy logic [1]. Their primary task was to create demonstration prototypes in our lab. As the team leader, my role centered on establishing a sustainable innovation strategy. I worked on building connections with stakeholders and identifying, adopting, and implementing efficient, state-of-the-art methodologies to maximize our team's limited resources and streamline our internal processes. Another significant challenge was motivating potential customers to come on board, considering the conservative nature of the environment we were working in.

We recognized the difficulties and obstacles but remained resolute in our belief that introducing an AI-based application to space operations, one that met the needs of flight control engineers and was productive with an acceptable level of risk, could open new opportunities in the European space landscape.

Our journey began with the introduction of a novel methodology called DSDM© (Dynamic System Development Methods) [2], which later played a foundational role in the development of modern Agile methodologies. From DSDM, we introduced two groundbreaking approaches: progressive delivery of prototypes in time boxes, featuring a snapshot of implemented features at each stage, and the flexibility to reorder or modify the primary functional and performance requirements of the application after each time box. This approach made it more manageable to navigate the uncharted territory we were exploring.

To strengthen the mission of the recently established AI & Operations Innovation Team (AI&OI) and to foster interest and dedication, an internal marketing campaign was initiated among the ESOC flight control teams. The campaign operated under these guidelines: It sought competitive proposals for defining use cases based on specific criteria such as maturity, data availability, impact, and urgency. The competition offered the winning teams the application free of charge, provided the problem they addressed was of a general nature. The investment made in internal R&D for this purpose was subsequently recovered by extending the use of the same AI-based application to other missions.

The workflow process was systematically developed and refined through iterations, emphasizing continuous improvement. However, the creative spark, the ideation process, often occurred in a more relaxed setting, such as a coffee shop. The team took pride in their ability to uncover novel needs from potential customers during informal coffee discussions. This laid-back environment, with fewer formal barriers compared to structured brainstorming meetings, proved highly effective in identifying and understanding what was of utmost importance to both flight and ground control engineers. Interestingly, some challenges were so complex that they were considered unsolvable using known technologies. In such cases, AI presented itself as a potential solution. It's also worth noting that brilliant and ingenious ideas for addressing these problems sometimes emerged at unexpected moments, like during a shower or in other unconventional settings.

One of the criteria for candidate use cases mentioned above was the availability of historical data. From this point of view, ESOC's mission control has always privileged the ability to archive historical telemetry data. This has proven to be a tremendous asset in accelerating the introduction and exploitation of AI in mission control. However, the process of getting that data out of the operational network firewall in a usable format was still complicated.

3 Threats and Opportunities

The catalyst for strengthening AI in space operations came from a recurring emergency on the Smart I [3] spacecraft. Smart I, launched in September 2003 with the ambitious goal of reaching the moon using electric propulsion, faced persistent anomalies. In response, the AI&OI team swiftly developed a novel platform called MUST [4] (Mission Utility and Support Tools). MUST was designed and implemented in record time to provide rapid retrieval, storage, and visualization of critical spacecraft data in an external and secure parameter archive, enabling remote access. This achievement had a profound impact as it allowed operators to remotely monitor the health of Smart I [5], alleviating the need for constant travel between home and the office to address new anomalies day and night. The MUST platform served a dual purpose and was soon adopted by other space missions, becoming the new standard for both remote monitoring and for data-intensive AI applications. It provided machine learning algorithms easy access to historical spacecraft data, empowering them to offer operational support in early anomaly detection, diagnostic assistance, and future predictions. With this robust platform in place, the AI&OI team could conduct various investigations and experiments, carefully test and validate innovative solutions, and collaborate with flight and ground operators to deploy effective tools for operational use.

To build trust and effectively manage risk during the adoption of AI applications, a concept called "operational prototypes" was introduced. This concept aimed to bridge the historical gap between early demonstrator prototypes and fully developed, often costly operational applications. The approach involved creating an intermediate stage where the operational prototype could be deployed alongside the existing standard application but in a non-invasive manner. This meant the prototype operated in the background without affecting critical processes. It underwent extensive validation, directly involving end users, before their ultimate acceptance. This staged approach helped ensure both trust and effective risk management in the AI adoption process.

4 Examples of AI Solutions for Space: Planning and Scheduling and Autonomy

Among the various applications implemented over the years, Mexar [6, 7] was a notable milestone. It made its debut in 2004 under the leadership of Simone Fratini and Nicola Policella. Mexar's purpose was to automate the daily data dumping sequences from various packet stores of the Mars Express, an ESA scientific probe orbiting Mars. This tool elevated the role of planners by handling mundane, repetitive tasks, enabling them to work at a higher conceptual level. Additionally, Mexar facilitated the creation of plans, exploration of alternative solutions, and the selection of the most robust plan for execution. The practical benefits included cost reduction in the planning process, minimized use of ground station time, and an overall increase in scientific productivity. This successful experience served as the catalyst for the development of a dedicated experimental platform for AI planning applications known as the Advanced Planning and Scheduling Initiative (APSI) [8].

As mission concepts became more complex, there was a growing demand for advanced capabilities in autonomous reasoning and resource management [9, 10]. These capabilities involve coordinating robotic tasks and synchronizing planning and scheduling both on the ground [11] and on board spacecraft [12]. This situation highlighted the necessity for more adaptable and sophisticated technologies that can facilitate the integration of action planning and resource allocation. To address this need, we initiated a collaboration with our colleagues at ESTEC, the European Space Agency's technology center [13].

In July 2013 Alphasat was launched into geostationary orbit as a public-private partnership, the biggest of its kind between ESA and UK operator Inmarsat. The AI&OI team created a system called TECO [14], which was built upon the APSI framework. This system has been in use since 2014 by the Technology Demonstration Payload ESA Coordination Office, and it has made a significant contribution by saving both time and human effort. TECO provides an automated service that optimizes the scheduling of payload activities. Importantly, this system remains in active operation today.

The APSI platform underwent another significant advancement in its modeling and problem-solving capabilities with the introduction of a new integrated planner and scheduler known as PLASMA [15], short for PLAn Space Multi-solver Application. PLASMA took the integration of symbolic planning and numerical resource management to the next level. It enabled the formulation of planning problems that directly specify resource requirements, enhancing the platform's overall functionality [16].

In addition to conventional planning tasks, there was a distinct focus on two key aspects: envisioning future scenarios and drawing inspiration from nature. A notable source of inspiration was the Disaster Monitoring Constellation [17], which marked the pioneering Earth observation constellation of budget-friendly small satellites. What set this project apart was its innovative approach to mission planning. The team designed a prototype planning system using a self-organizing multi-agent architecture, drawing inspiration from ant colony optimization algorithms [18, 19].

This approach resulted in a system that exhibits adaptability when faced with changes in the problem domain and the capability to harmonize satellite plans, thus preventing redundancy.

5 Examples of AI Solutions for Space: Spacecraft Health Caring and Its Spin-Offs

The initial precursor to the AI-driven approach aimed at improving monitoring and diagnostics took shape in 2001, following the development of demonstrator prototypes. This pioneering system relied on fuzzy logic and was specifically created to tackle the nutation issue [20] encountered during the Ulysses mission at NASA's Jet Propulsion Laboratory (JPL) in Pasadena, California. Ulysses, a collaborative mission between ESA and NASA, centered on solar exploration with a primary objective of studying the Sun across its entire range of latitudes, including the polar regions.

Another notable application of fuzzy logic occurred at ESOC in 2004. This time, it was employed to oversee one of the most crucial components on board a spacecraft, the 3-axis gyroscope. The gyroscope in question was a critical part of ESA's largest spacecraft, ENVISAT, an environmental satellite launched in 2002. ENVISAT operated in a sun-synchronous orbit at an altitude of 800 km and was equipped with a suite of 10 instruments, including optical and radar imaging payloads. The tool developed for this purpose was called the ENVISAT Gyroscope Monitoring Tool (EGM) [21], and it relied on fuzzy logic. EGM was specifically designed to conduct weekly health checks on the gyroscopes and to carry out health assessments during the annual maintenance of the backup gyroscopes. It had the capability to automatically detect faults such as performance degradation and provide diagnostic information.

Another significant milestone occurred in 2010 and 2011 when two patents were filed in the realm of enhancing onboard observability and analyzing spacecraft behavior. This accomplishment was the result of the vision and determination of pioneers in AI for space, led by Jose Martinez Heras. The first patent, referred to as "Novelty Detection," [22, 23] had the capability to identify the initial signs of a never-before-observed behavior before triggering an alarm. Notably, it achieved this with an exceptionally low rate of false alarms. The second patent, known as "Dr. MUST," [24] was focused on aiding in anomaly investigations and providing insights into elements that might be linked to the detected anomalous behavior. The adoption of the Novelty Detection and Dr. MUST tools by interested missions at ESOC marked a significant step forward [25] and paved the way for AI technology in the space domain, initially within the ground segment, with the potential for future integration onboard spacecraft. These achievements received recognition and appreciation not only within the community but also from NASA, as evidenced by their acknowledgment at the SpaceOps 2012 conference held in Stockholm, Sweden.

During this period, European national space agencies and the European space industry demonstrated a keen interest in understanding the potential of AI applications. Consequently, ESA initiated collaborative teams with researchers from CNES and DLR to facilitate knowledge transfer and further research efforts in AI for space.

The field of diagnostics is quite extensive, prompting our team to explore a new idea encapsulated in the motto: "from spacecraft health care to human health care." This led to the commencement of a new research initiative at ESA's European Astronaut Centre in Cologne, which opened up opportunities to apply AI research in support of astronaut health care and training [26]. To further promote these initiatives, various activities were organized, including open competitions tailored to address specific space challenges and hackathons like the Space2Health Hackathon, a collaborative effort between ESA and Merck in 2017. Building upon the success of an ESA academic open competition, the GalaxAI [27] machine learning toolbox was developed. GalaxAI is a versatile tool designed for efficient and interpretable end-to-end analysis of spacecraft telemetry data. It has been applied to two specific use cases related to orbiting spacecraft: analyzing and forecasting Mars Express thermal power consumption for planning purposes and predicting INTEGRAL's passages through the Van Allen belt.

6 Latest Developments, Elements Learnt and Conclusions

Over time, there have been significant advancements in computing technology and microprocessor capabilities. These advancements have paved the way for the emergence and deployment of new AI technologies, including Deep Learning (DL) [28] and Natural Language Processing (NLP). These developments led to the creation of four specific applications [29]: 1. Answering questions related to space mission and spacecraft design concepts. 2. Generating quizzes to aid in training for quality management and assurance in space science and engineering. 3. Extracting information for the long-term preservation of data in space. 4. Assisting in the evaluation of the innovation potential of ideas within the Open Space Innovation Platform of ESA (OSIP). These applications were the result of dedicated studies and research efforts, harnessing the power of AI technologies to address various needs in the field of space science and engineering.

ESA has now put in place a formal AI and automation roadmap for space operations. This roadmap serves as the foundational framework for advancing the development of applications that enhance and revolutionize space operations. Additionally, there are ongoing technical meetings involving AI technology specialists from various space domains within ESA. These meetings aim to foster exchange, collaboration, and the creation of synergies among experts in the field. Furthermore, a new standardization process has been initiated through the European Cooperation for Space Standardization (ECSS). The objective of this process is to produce a Machine Learning Qualification Handbook, which will provide guidance and support for the development of AI-based applications for space.

In summary, over the past two decades, the AI&OI team at ESOC, in collaboration with academic and industrial partners, has played a crucial role in supporting flight control and ground control teams. They have achieved this by developing specialized tools and promoting the integration of emerging technologies and products. These efforts have led to a transformation in how mission operations are conducted and supported in critical areas. The team's ability to listen, explore, and innovate has not only been instrumental in providing practical solutions but has also served as a source of inspiration and motivation. They have encouraged adaptability in the face of changing conditions and challenging situations. I am deeply grateful to my team and colleagues from ESA, industry, and academia for their unwavering curiosity and openness to new ideas, which have kept our journey exciting and enriching. One of the most remarkable moments in this journey was receiving a business phone call from astronaut Samantha Cristoforetti, seeking consulting support from our team for an upcoming AI-related project.

Throughout this journey, human factors and human potential have remained central. We have learned that we are capable of much more than we often realize, and circumstances do not always define our limits. It is evident that we possess untapped reservoirs of potential deep within us, and our mission is to unlock and utilize this potential to its fullest extent.

My passion now lies in leveraging the experiences gained from the challenges of space exploration to venture into new territory—facilitating the discovery of the untapped potential within each individual and making it easily accessible in the workplace. This journey is not just about personal growth but also about fostering deeper mutual trust and resonance within teams. An inspiring model for this transformation could be found in the harmony and energy of a choir. In this effort, freedom, trust, and unconditional love are three pivotal to cultivate that can help us navigate the next frontier—the space within ourselves.

Further Reading

1. Donati, A. Fuzzy logic applications for advanced space mission control functions. In Proceedings of the 10th Mediterranean Conference on Control and Automation-MED2002 Lisbon, Portugal. 2002.
2. Stapleton J. DSDM, dynamic systems development method: the method in practice. Cambridge: Cambridge University Press; 1997.
3. Camino O, Alonso M, Blake R, Milligan D, Bruin JD, Ricken S. SMART-1: Europe's lunar mission paving the way for new cost effective ground operations (RCSGSO). In Sixth international symposium reducing the costs of spacecraft ground systems and operations (RCSGSO), European Space Agency ESA SP-601. 2005, June.
4. Martinez-Heras JA, Baumgartner A, Donati A. MUST: mission utility & support tools. In DASIA 2005-Data Systems in Aerospace (Vol. 602). 2005 Aug.
5. Foing BH, Racca GD, Marini A, Evrard E, Stagnaro L, Almeida M, Koschny D, Frew D, Zender J, Heather J, Grande M. SMART-1 mission to the moon: status, first results and goals. Adv Space Res. 2006;37(1):6–13.
6. Rabenau E, Cesta A, Oddi A, Donati A, Léauté T. Using an artificial intelligence tool to perform science data downlink planning as part of the mission planning activities of mars express. In: Proceedings Fifth International Workshop plan. Scheduling Space; 2006.
7. Cesta SA et al. Mexar2: AI solves Mission planner problems. IEEE Intell Syst. 2007, July–Aug; 22(4): 12–19. https://doi.org/10.1109/MIS.2007.75.

8. Steel R, Niézette M, Cesta A, Verfaille G, Lavagna M, Donati A. AI techniques for space: the APSI approach. DASIA 2009-DAta Systems in Aerospace, 669. 2009. p. 29.
9. Donati A, Policella N. AI planning and scheduling infusion in space: ESA achievements and perspectives. In SpaceOps 2010 Conference delivering on the dream hosted by NASA marshall space flight center and organized by AIAA 2010.
10. Gao Y, Burroughes G, Ocón J, Fratini S, Policella N, Donati A. Mission operations and autonomy. Contemporary Planetary Robotics: An Approach Toward Autonomous Systems. 2016 Sep;6:321–401.
11. Fratini S, Policella N, Faerber N, Neves Madeira RN, Donati A, Teixeira De Sousa B. Ground station pass allocation in operations using AI. In 2018 SpaceOps Conference. 2018. p. 2726.
12. Fratini, S, Gorfer J, Policella N. On board autonomy operations for OPS-SAT experiment. In Advances and trends in artificial intelligence. From theory to practice: 32nd international conference on industrial, engineering and other applications of applied intelligent systems, IEA/AIE 2019, Graz, Austria, July 9–11, 2019, Proceedings 32. Berlin: Springer International Publishing, 2019. p. 182–195
13. Steel R, Hoffmann A, Niezette M, Cimatti A, Roveri M, Kapellos K, Donati A, Policella N. Innovative rover operations concepts-autonomous planner (IRONCAP)-concluding the adventure. In SpaceOps 2014 Conference. 2014. p. 1838.
14. Policella N, Benzi E. Alphasat TDP operations coordination via automated planning: an operational experience report. In 14th International Conference on Space Operations. 2016. p. 2514.
15. De Maio A, Fratini S, Policella N, Donati S. Resource driven planning with "PLASMA": the plan space multisolver application. ASTRA. 2015;
16. Fratini S, Nogueira T, Policella N. Integrating modeling and knowledge representation for combined task, resource and path planning in robotics. In KEPS 2017. 2017 Jun. p. 18.
17. Kameche M, Benzeniar H, Benbouzid AB, Amri R, Bouanani N. Disaster monitoring constellation using nanosatellites. J Aerosp Technol Manag. 2014;6:93–100.
18. Iacopino C, Palmer P, Brewer A, Policella N, Donati A. EO constellation MPS based on ant colony optimization algorithms. In: 2013 6th International Conference on Recent Advances in Space Technologies (RAST). Turkey: Istanbul; 2013. p. 159–64. https://doi.org/10.1109/RAST.2013.6581192.
19. Ntagiou E. Earth observation and data relay constellation missions' planning with ant colony optimization (Doctoral dissertation, University of Surrey).
20. Clivio R, McGarry A, Donati A. PHAEACIAN- Application of fuzzy logic to the Ulysses nutation problem. DASIA 2001-DAta Systems in Aerospace; Proceedings of the Conference. 2001.
21. Donati A, Bargellini P, Mantovani G, Ventimiglia L. ENVISAT gyro monitoring & early warning: operational assessment of an advanced prototype tool. InSpace OPS 2004 Conference. 2004. p. 118.
22. Martinez J. New telemetry monitoring paradigm with novelty detection. InSpaceOps 2012. 2012. p. 1275123.
23. Martínez-Heras JA, Donati A. Enhanced telemetry monitoring with novelty detection. AI Mag. 2014;35(4):37–46.
24. Martinez Heras JA, et al. DrMUST - a Data mining approach for anomaly investigation. Spaceops. 2012.
25. Donati A, Martinez J. Data mining for operations: expertise and services. In 14th international conference on space operations. 2016. p. 2302.
26. Martinez J, Donati A, Damann V, Biancat J, Brighenti C, Brighenti A, Ferrari F, Chientaroli R, Arcuri G, Silipo F, Trenti T. Data mining for astronauts medical autonomy. In 14th international conference on space operations. 2016. p. 2382.
27. Kostovska A, Petković M, Stepišnik T, Lucas L, Finn T, Martínez-Heras J, Panov P, Džeroski S, Donati A, Simidjievski N, Kocev D. GalaxAI: machine learning toolbox for interpretable analysis of spacecraft telemetry data. In 2021 IEEE 8th International Conference on Space

Mission Challenges for Information Technology (SMC-IT). New York: IEEE; 2021 Jul 26. p. 44–52.
28. Martinez J, Donati A. Novelty detection with deep learning. In 2018 SpaceOps Conference, 2018. p. 2560.
29. Gómez-Pérez JM, García-Silva A, Leone R, Albani M, Fontaine M, Poncet C, Summerer L, Donati A, Roma I, Scaglioni S. Artificial intelligence and natural language processing and understanding in space: four ESA case studies. arXiv preprint arXiv:2210.03640. 2022 Oct 7. https://ecss.nl/standards/

Part III

A Better World Driven by Curiosity and Action

Providing the Means of Action

Katja Becker

Ladies and Gentlemen,

Dear Colleagues,

Standing united by science for a better tomorrow, I wish to thank you very much for inviting me to shed some light on how science and research funding jointly provide the means for political action. Moreover, I am happy to address the role of science, technology and global research cooperation in solving today's greatest challenges in the age of multiple crises—pandemic, war, and climate change. Against this backdrop, I will focus on questions such as: What does it mean to be united by science? What is the role of research in this process? What is the role of research funding? And, most importantly, why do we need to stay curious to achieve this unity? Let me begin with the latter: after all, every research question, and therefore science as a whole, is initially ignited by curiosity.

1 Fostering Curiosity

Ever since Plato identified curious astonishment as key to human wisdom, curiosity has laid the foundation for all research activity, since it stimulates interest in generating knowledge and advancing innovation. Fostering scientific curiosity is therefore the ultimate goal of our endeavours in promoting free and knowledge-driven research as the backbone of the German science system.

Keynote Speech held at the Curious 2022—Future Insight Conference, 12 July 2022 (The text has been updated and revised in December 2022. The character of a speech was retained for stylistic reasons.).

K. Becker (✉)
Deutsche Forschungsgemeinschaft, Bonn, Germany
e-mail: katja.becker@dfg.de

U. A. Betz (ed.), *Curious Future Insight*,
https://doi.org/10.1007/978-3-031-41781-8_10

On the one hand, basic research contributes indispensably to overcoming the climate crisis, strengthening social resilience, and striving for technological sovereignty, among other things. On the other hand—as we continue to learn in the course of the COVID-19 pandemic and the war of aggression on Ukraine—basic research across its entire breadth helps us prepare for unforeseen challenges. In this sense, such research also lays the foundation for future innovations. This is why continuous investment in free and curiosity-driven research is the core of our research and innovation funding strategy.

The German Research Foundation (Deutsche Forschungsgemeinschaft, DFG) funds more than 30,000 basic research projects every year across the full range of academic disciplines—from archaeology to artificial intelligence, from law to atmospheric research, and many others. Since every funded project reflects an attempt to think in a new way, aiming to achieve innovation and establish unexpected creative interrelationships, the DFG regards itself as a catalyst for scientific ideas. This goal is fostered by our knowledge-driven funding procedures in which decisions are largely based on the originality of the research proposals, and hence derives indirectly from the researchers' own intrinsic curiosity. Each of these projects adds to the potential to change society overnight—frequently within complex societal, biological, or technological contexts and often on a global scale. This is also why proposals to the DFG can be submitted at any time, in any field, on any subject, and on any scale, because we firmly believe that scientific curiosity is the key to keeping up with the ever-changing world in which we live.

2 Knowledge Repositories

Facing up to the demands of the twenty-first century can make us all feel somewhat breathless as we adapt to the incessant stream of events that so rarely gives us a break. After all, the challenges our societies have to face are enormous. Among them, climate change and the consequences of the man-made changes affecting our living environment are issues we are all confronted with. Moreover, we might encounter the emergence, spread, and increasing resistance of pathogens. Finally, our societies often face challenges relating to migration and the scarcity of resources, including their uneven allocation, religious and cultural tensions, political upheaval, and armed conflict—right through to the whole kaleidoscope of political and societal challenges resulting from digitalisation.

These challenges continue to be highly urgent—and we do not yet know which of them will keep us particularly occupied in the near future. Nor do we know which elements of our scientific, technological, and social knowledge repertoire we might need to draw on to provide sustainable responses. But we do know that research can and should empower us to cope not only with today's pressing challenges but also with possible future crises that we are not yet able to anticipate. Only science fuelled by free and knowledge-driven basic research can provide us with knowledge repositories capable of meeting the unknown challenges of the future.

3 Corona Pandemic and mRNA Method

Allow me to highlight one of those repositories as evidence of why investing in basic research—the search for pure knowledge—is so important: the messengerRNA vaccination method, which has its origins in cancer research. This example demonstrates that the later application context of research results is not necessarily foreseeable at the time of funding. The DFG funded fundamental research into mRNA vaccination carried out by BioNTech co-founder Uğur Şahin more than 10 years ago in connection with individual projects and Collaborative Research Centres at the University of Mainz. This shows that excellent research can provide answers to questions that often do not arise until much later.

The messengerRNA vaccination method has significantly advanced our fight against the pandemic. Thanks to the progress made in basic biomedical research, clinical research and transfer to application, and thanks to the ongoing support for these accomplishments among policymakers and society at large, it has now been possible to offer an effective response to the coronavirus pandemic. Unfortunately, we must also accept that the cycles in which epidemics and pandemics can occur are becoming increasingly shorter. Firstly, this is due to the fact that we now perceive outbreaks more quickly and with greater precision than before. And secondly, our globally interconnected ways of life foster both the emergence of pandemics and our susceptibility to them.

It is our innate responsibility both as researchers and as research funders to consistently stay one step ahead of the next pandemic event. It is therefore crucial to reassess the risk factors relating to the emergence of pandemics and to counteract the enormous burden they place on our ecological and cultural systems. This can be called pandemic preparedness, or it can be linked to the notion of resilience, but it can also be defined somewhat more broadly. From my point of view, it is essential to maintain a reservoir of knowledge in the life sciences and natural sciences, as well as in the humanities, social sciences, and engineering sciences, so as to be able to tackle the challenges of the future.

4 Global Challenges and Global Science

Such future—and as yet unknown—challenges will very likely go beyond the perspective and scope of national research and national research funding. Hence, we should not only be thinking in ever new ways, but also engaging in global thinking and collaboration.

Once again, the pandemic has made it painfully clear what needs to be done. Confronted with the virulence of Sars-CoV-2, we have seen how important it is to think globally and take resolute action in order to lay the foundations for effective pandemic prevention and management. We should not remain attached to old boundaries but cast our eyes further afield to new interrelationships—based on an understanding of the real-life conditions of our planet Earth and geared towards its preservation for humans, animals, and plants.

In fact, the pandemic serves as a blueprint for what is yet to come and what remains to be done in both science and research funding. Such exceptional situations require scientific perception, political courage, and orientation towards global cooperation. Global challenges—be they societal, environmental, biomedical, or otherwise—need solutions provided by global science.

In view of this, the DFG has intensified its involvement in the Global Research Council (GRC) and has become even more active in shaping global research cooperation. As a global association of research funding organisations, the GRC takes on the role of a catalyst to enable funding organisations worldwide to share their experience and foster multilateral research and collaboration across continents.

As a shining example of such international scientific cooperation, I would like to draw your attention to our promising collaboration with scientists from Sub-Sahara Africa—namely, with the African Council for Higher Education (CAMES) and the UNESCO unit The World Academy of Sciences for the Advancement of Science in the Developing World (TWAS). For the past 12 years, TWAS and the DFG have jointly supported over 350 early-career investigators in pursuing research in Germany. Over a third of them are actually attending this conference and will be meeting tomorrow to start a TWAS/DFG alumni network. I am confident that this new network will lead to more joint research by Sub-Saharan and German scientists, more joint publications, and more German-African cooperation in Africa.

Also present at this conference is the "Research in Germany" initiative, launched by the German Federal Ministry of Education and Research (Bundesministerium für Bildung und Forschung, BMBF) in 2006—another example of successful collaboration dedicated to meeting global challenges. Under this initiative, several German research and funding organisations such as the DFG, the German Academic Exchange Service (Deutscher Akademischer Austauschdienst, DAAD), the Fraunhofer Society (Fraunhofer-Gesellschaft), and the International Office of the BMBF have joined forces to represent Germany abroad as a country of research and innovation. With this aim, the initiative has helped German researchers initiate and establish contacts worldwide.

5 The War in Ukraine

Sadly, this kind of international cooperation cannot be taken for granted, as we have come to learn in recent times. Science cannot stand idly by while a terrible war is raging in our European neighbourhood. The humanitarian catastrophe resulting from the Russian attack in breach of international law not only poses a threat to the values of a free and democratic Europe. It has also called into question core values of international research cooperation, attacking academic freedom, the free and open exchange of ideas, and even trust between scholars.

The DFG and the Alliance of Science Organisations in Germany therefore have condemned the invasion of Ukraine in the strongest terms possible and immediately suspended cooperation with Russian institutions. This far-reaching measure is a direct expression of our deep sympathy and solidarity with Ukraine. In addition, we must ensure that Ukrainian scientists and scholars are offered lasting perspectives. That is why we have opened several DFG funding lines and established tailored new lines of individual support, while also being strongly committed to preventing brain drain and rebuilding research structures within Ukraine, hand in hand with the National Research Foundation of Ukraine (NRFU).

With similar intentions, it is important to strengthen those forces within Russian academia that are so courageously calling for an end to the war. Although institutional dialogue may be temporarily interrupted for the time being, we strive to keep open individual channels of communication between German and Russian scholars in order to foster a swift re-opening of the previously stable bridges of science and scholarship to Russian society when peace is eventually restored.

Scholarly dialogue has always been about sharing knowledge and overcoming borders—not just intellectual borders but also those imposed by nations and political systems. As such, the return of nationalistic imperialism to Europe is actually prompting us to intensify our collaborative networks, both inside and outside Europe.

6 United by Responsibility

From the DFG's perspective, all these measures form part of our responsibility as research funders. The notion of responsibility also gives us an indication of what "united by science for a better tomorrow" could mean in this situation. Above all, it means that today—perhaps more than ever—we need to ask ourselves how we can organise science so that its findings contribute to increasing peace and prosperity.

First, this responsibility involves fact-oriented yet clear science communication. All the forces within society—science, civil society, policymakers, and businesses—need to communicate, explain, persuade, and engage in dialogue with each other more openly, more effectively, and more efficiently. It is through such dialogue that we will be able to leverage existing resources and create synergy effects.

However, to actually listen and not just talk to each other requires trust. This is why the trust scientists and researchers have gained in society at large over the past pandemic years presents us with a great opportunity: it is a gift for an enlightened, knowledge-based society. And this increase in trust has come about particularly due to the outstanding commitment demonstrated by individual scientists.

This trust is not something that can be taken for granted: it has to be earned anew every day. Living up to it is—again—a question of communication, which must be transparent, prudent, consistently objective and on an equal footing with its addressees. That said, it is also a matter of allowing multiple voices to be heard, and of weighing up findings and arguments within the research fields themselves. This diversity must be preserved—even when faced with the expectation of unambiguousness. This leads us to a second facet of scientific responsibility.

Apart from trusting communication, scientific responsibility also rests upon an understanding of politics and science as being different forms of rationality: in the political sphere, swift action is required, and democratic legitimacy has to be established based on majority support. By contrast, science seeks insights and pursues truth. In this way, science provides an indispensable basis for political deliberations and democratically legitimised actions. But by no means does it justify or authorise such actions. Here, responsibility must not be mistaken for accountability—for good reason, since the poles between science and politics cannot be broken down, nor do they have to be. Both science and politics are an integral part of our society.

Accomplishments such as the development of the coronavirus vaccine depend on how politics and society, science, and research funding work together and to what extent all the available dimensions of diversity and combinations of ideas, idea providers, and procedures can be activated to create something that is genuinely new.

The DFG takes account of the concerns of cutting-edge research and its global competitiveness. In doing so the DFG enables high-risk research in a positive sense while being mindful of the timeframes that such activity requires. By rigorously observing the criteria of scientific quality in its review and evaluation processes, the DFG helps ensure that scholars, policymakers, and society as a whole have a wide range of advanced scientific options at their disposal in case of acute need—an invaluable store of knowledge from which evidence-based solutions to concrete problems can swiftly emerge when needed.

In general, funding organisations like the DFG take on the role of an intermediary in this dynamic ensemble. They shelter and foster both scholarly curiosity and integrity, because scientific autonomy provides the best foundation for a responsible and successful interplay between science and politics, both nationally and internationally. And the DFG will continue to do so in the future.

Thank you very much!

Adding Years to Life and Life to Years

At Least Six Years of Higher-Quality Life for Everyone Is within Reach

Erica Coe, Martin Dewhurst, Lars Hartenstein, Anna Hextall, and Tom Latkovic

This is a decisive moment in the history of human health.

In many respects, health is a remarkable success story. Over the past century, life expectancy has dramatically increased in most parts of the world. But the portion of life we human beings spend in moderate and poor health hasn't changed, meaning we spend more years in poor health than at any point in history. Moreover, significant inequity continues to exist across and within countries.

We can do better—quickly.

Humanity mobilized against COVID-19 at a speed and scale previously unseen. While far from perfect, our success should inspire us to challenge what we think is possible. At its best, our response to COVID-19 demonstrates that when resources and motivation coalesce, scientific breakthroughs and large-scale behavior change are possible in very short periods of time.

It's time to set a new, more ambitious, more relevant goal for human health—a goal that galvanizes across continents, sectors, and communities to support everyone on the planet in adding years to their lives and life to their years. Humanity needs a goal that yields more time with loved ones, more accomplishments, and more time free from cognitive or physical impairment.

As a starting point for discussion, the McKinsey Health Institute (MHI) believes that over the next decade humanity could add as many as *45 billion extra years of higher-quality life—roughly six years per person on average, and substantially more in some countries and populations* [1].

Achieving this objective requires us, as a society, to challenge our beliefs about health and reorient material portions of public policy and the economy. It requires embracing a modernized understanding of health, including physical, mental, social, and spiritual health [2], and the full richness of factors that influence those elements of holistic health. It requires viewing health as an investment, not an expense. It

E. Coe · M. Dewhurst · L. Hartenstein (✉) · A. Hextall · T. Latkovic
McKinsey Health Institute, New York, USA
e-mail: lars_hartenstein@mckinsey.com

U. A. Betz (ed.), *Curious Future Insight*,
https://doi.org/10.1007/978-3-031-41781-8_11

requires scaling solutions that work, which could address 40% of the disease burden [3]. It requires dramatically more innovation and leadership from institutions outside of the traditional healthcare industry. It requires fully empowering individuals to steward their own health.

Every institution, every leader, and every person has an important role to play.

1 The State of Human Health: Great Achievement with Much More to Do

Good health underpins our ability to lead productive and enjoyable lives. Health enables social development and spurs economic growth. Year after year, experts cite social and physical components of health as two of the top three drivers of life satisfaction for individuals [4]. Global healthcare spending is often over (or about) $8 trillion each year and increases faster than GDP [5].

Over the past centuries, scientific progress, innovation, greater investment, and trade and exchange across the public, private, and social sectors have led to great advancements in prolonging and improving life. Between 1800 and 2017, average global life expectancy more than doubled, from 30 years to 73 years [6]. In some of the least advantaged global regions, life expectancy has increased by 10 years in just the past two decades [7]. Since 1900 in the United States, infant mortality has fallen by 90%, and maternal mortality has decreased by 99% [8]. Major breakthroughs in vaccination have enabled humanity to eradicate or suppress deadly infectious diseases such as smallpox and polio [9, 10]. Since 1990, we have seen breakthroughs in human genomics, a substantial reduction in cancer mortality, and improvements in smoking cessation [11].

While it's important to recognize and learn from our progress, we must also acknowledge how much more we have to accomplish collectively.

The share of our lives we spend in poor health has not diminished over time. On average, people spend about 50% of their lives in less than good health, including 12% in poor health [12–14]. The best available data suggest that this ratio has not changed much in the past 50 years. The upshot is that we spend more time in absolute terms in moderate and poor health than we have at any other point in history. The situation may be gradually worsening, particularly in high-income countries, where chronic conditions now afflict growing numbers of people for a significant portion of their lives (Fig. 1). Literature on life satisfaction shows that having a substantial health problem—defined as declining from "good health" to "poor health"—reduces life satisfaction twice as much as losing a job or becoming widowed, divorced, or separated and five times as much as losing half of one's income [15, 16].

Numerous known threats to human health remain insufficiently addressed. Infectious diseases still account for eight million deaths per year, and there are substantial unmet patient needs within oncology, diabetes, cardiovascular conditions, and brain disorders [17–21]. The prevalence of mental-health conditions has risen by 55% since 1990, and researchers anticipated 17% growth between 2020 and 2040 before

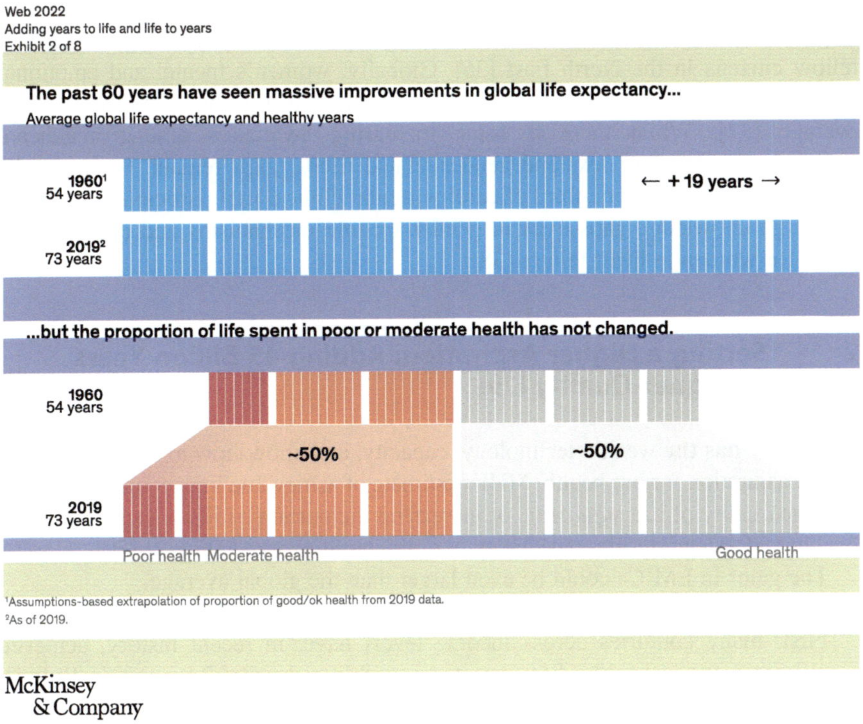

Fig. 1 Globally, lives have gotten longer but not healthier. Source: WHO and World Bank health and life expectancy data; country-level health system and survey data; McKinsey Health Institue analysis; and Remes, Jaana, et al. "Prioritizing Health: A Prescription for Prosperity," McKinsey Global Institute, 2020

COVID-19 [22–24]. The suicide rate in the United States has risen over the past 20 years to become the second-leading cause of death for people between the ages of 10 and 34 [25]. Global cases of dementia are expected to triple by 2050, afflicting more than 150 million people [26]. The rise in chronic, impairing lower-back pain is likely linked to obesity, which has tripled globally since 1975, from 4% to 13% of the world's population [27, 28]. There are also concerns about the potential negative effects of climate change on health, including challenges related to food security and infectious diseases. The impacts of climate change would disproportionately affect the health of vulnerable populations and people in low- and middle-income countries (LMICs), which is likely to exacerbate existing inequalities [29, 30].

Health inequity remains a major problem, with disparities in access and outcomes persisting between and within countries and regions and across gender, wealth, and other demographic identifiers. There is an 18-year gap in average life expectancy between low- and high-income countries, and a 30-year gap between the lowest and highest life expectancy [31, 32]. Maternal mortality rates are 50 or even 100 times higher in some low-income countries than in high-income countries [33]. Childhood cancer survival rates are above 80% in high-income countries but as low as 30% in LMICs [34]. In New York City, Black non-Hispanic women are eight times more likely to die of pregnancy-related complications than White women [35]. In England, the residents of London have a life expectancy 3 years longer than their fellow citizens in the North East [36]. Globally, women's mental and emotional health is at its lowest in 15 years and significantly below the overall population average [37]. While there is some increasing awareness among healthcare stakeholders regarding sex- and gender-related needs, gaps remain. For example, women make up only one-third of cardiovascular clinical trial participants globally, and large knowledge gaps exist in gender-specific mechanisms and optimal drug doses for women in heart failure [38].

2 Setting a Higher Aspiration: Adding 45 Billion Years of Higher-Quality Life

Humanity has the wealth, technology, capacity, and know-how to set and pursue a bolder aspiration for our health. MHI's estimate of achievable impact includes *lifting* average quality of life; *squaring*, or increasing the portion of life we spend in good health; and *extending* life expectancy over the baseline trend (Fig. 2) [39].

The gains in LMICs could be even larger than the global average.

What leads us to conclude such a feat is possible?

First, many countries across income levels have, in recent history, achieved significant gains in healthy life expectancy within a decade: 3.8 years in Bolivia,

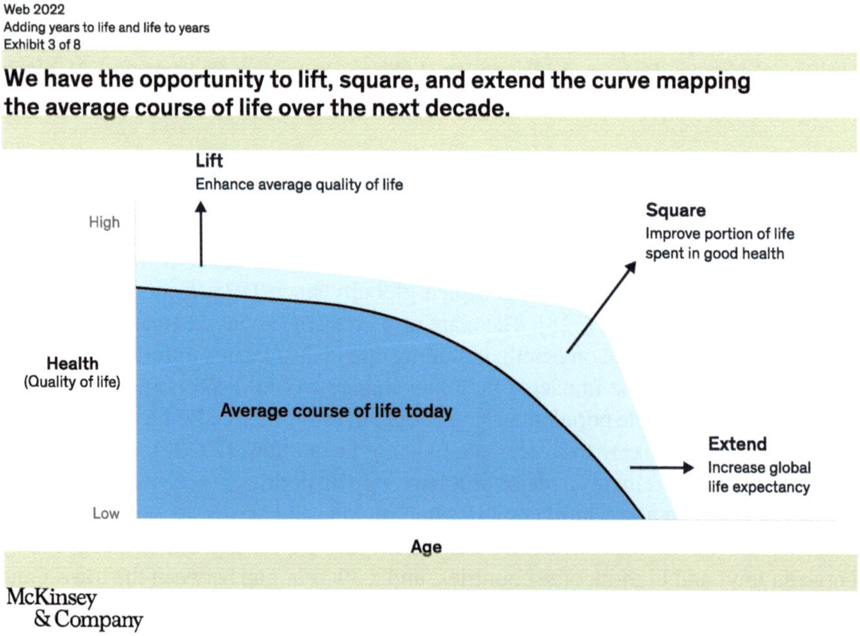

Fig. 2 We have the opportunity to lift, square, and extend the curve mapping the average course of life over the next decade

3.0 years in Ireland, 2.6 years in Oman, 6.5 years in South Africa, and 4.5 years in Thailand [40]. There are also countries and regions— the so-called blue zones—that appear to have achieved particularly high levels of longevity and stronger health in older age [41].

Second, tremendous untapped potential exists in the systematic, equitable, and extensive application of *existing knowledge*. Previous research led by the McKinsey Global Institute in 2020 concluded that applying existing and close-to-market interventions could eliminate about 40% of the current global disease burden by 2040 [24]. Health innovation in the visible pipeline could cut the disease burden by a further 6 to 10% [24, 42].

Third, the world's response to COVID-19 demonstrates that remarkable innovations and behavior changes can happen quickly when intense focus is combined with resources and collaboration across governments, for-profits, nonprofits, and communities. The world created not only multiple safe and effective vaccines but also numerous treatments and treatment protocols used by millions of clinicians across the world [43, 44].The pandemic also proved that it is possible for billions of people to rapidly modify their behavior in favorable ways (e.g., safeguarding protocols, handwashing, self-administered diagnostic testing) when they are convinced it's in their interest and are properly equipped. Even outside of the COVID-19 response, other industries (e.g., technology, mobile, and automotive) have also demonstrated significant progress in improving product output, safety, and quality at the same or lower cost over a single decade.

The remainder of this article describes the shifts MHI believes society will need to make to realize the possible gains in both life expectancy and quality of life.

> The world's response to COVID-19 demonstrates that remarkable innovations and behavior changes can happen quickly when intense focus is combined with resources and collaboration.

3 The Foundation: Embracing a Modernized Understanding of Health

Historically, society has defined health in terms of the presence or absence of disease. Someone is deemed to be "in good health" if disease has no impact on their life expectancy or physical function. To add up to 45 billion years of higher-quality life, MHI proposes that we embrace a broader definition of health that better aligns with individual aspirations and the latest scientific research. The World Health Organization (WHO) proposed just such a broad definition of health, with a greater emphasis on well-being, back in 1948: health is a "state of complete physical, mental, and social well-being and not merely the absence of disease or infirmity." The subcomponents of spiritual health [2] have also been recognized as relevant to health for decades [45–49]. MHI uses the term "spiritual" because it is the most common way for other healthcare institutions and leaders to refer to these concepts. Strong spiritual health does not necessarily imply the adoption of religious beliefs.

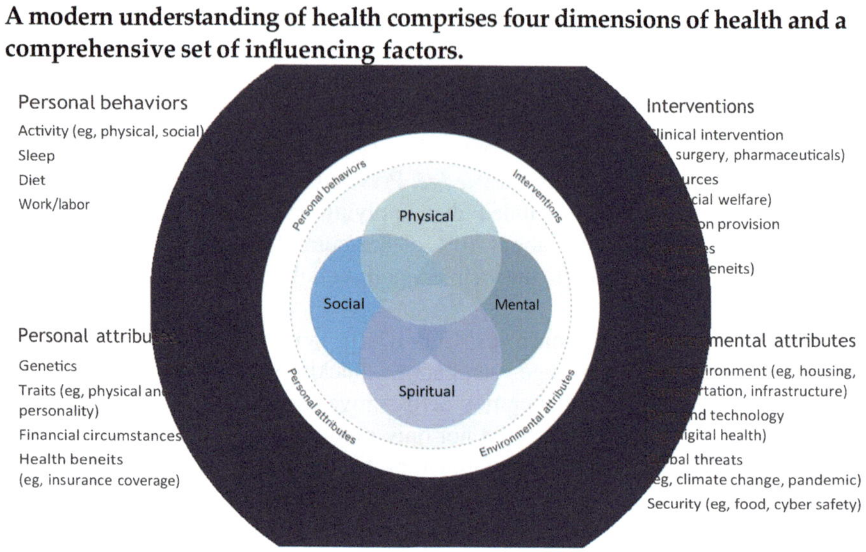

Fig. 3 A modern understanding of health comprises four dimensions of health and a comprehensive set of influencing factors

Unfortunately, a broad vision of health has not taken hold. MHI proposes that society fully embrace the proposed foundation and act on it.

MHI proposes an understanding of health with the following characteristics:

- ***Holistic***. It recognizes the relevance and interdependencies of physical, mental, social, and spiritual dimensions.
- ***Uses a positive frame***. The objective on each dimension is optimal health given an individual's physiological capacity, not simply the absence of disease.
- ***Anchored in function***. Health is relevant only to the extent it enables people to live fully—to build relationships, work or volunteer, and contribute to society while also enjoying pursuits.
- ***Affected by a multitude of influences***. This understanding of health recognizes the vast set of factors that affect health, including personal attributes, personal behaviors, interventions, and environmental attributes.
- ***Objective***. It is measurable across time, geographies, health systems, and cultures.

MHI proposes four interconnected dimensions of health in greater detail (Fig. 3):

Physical health is the extent to which an individual can competently perform physical tasks and activities without substantial discomfort. It includes the capacity to move through the environment in which one lives with confidence and independence and to control one's interactions with the physical world via fine motor control. People with good physical health have sharp sensory capacities with keen senses of touch, vision, hearing, taste, and smell. Physically healthy individuals are full of energy and vitality, free from the twin scourges of debilitating pain and fatigue [50].

Mental health is an individual's cognitive, behavioral, and emotional state of being. Mental health is needed for an individual to understand and interact with the world through memory and language. Mental health allows us to experience joy, direct anger, limit harmful impulsive behavior, and avoid serious depressive episodes. Mentally healthy individuals have the resilience to cope with normal stresses and adverse events while maintaining a positive and realistic sense of self [51, 52].

Social health represents an individual's ability to build healthy, nurturing, genuine, and supportive relationships. People in good social health have the capacity to form meaningful connections with others, to both receive and provide social support. Social health gives people a strong sense of belonging to a community [53, 54].

Spiritual health enables people to integrate meaning in their lives. Spiritually healthy people have a strong sense of purpose, belonging, or identity. They feel a broad sense of connection to something larger than themselves, whether to a community, a calling, or a form of divinity. Spiritual health helps people feel rooted and mindful in the present moment [55–58].

Strong anecdotal and empirical evidence suggests that these four health dimensions collectively contribute to both longevity and quality of life.

Individuals often suffer harm when their health fails along even one of these dimensions. For example, global data indicate that severe mental health disorders can reduce life expectancy anywhere from 10 to 25 years [59]. On the social health dimension, loneliness and social isolation are associated with higher risks of heart attack and strokes [60]. In fact, research shows that loneliness and social isolation can be as damaging to an individual's health as smoking 15 cigarettes per day [61], which is especially concerning when we note that up to 29% of elderly people report feeling lonely [62]. A lack of social connections has been associated with an increase in inflammation at the same magnitude as physical inactivity in adolescence, and in old age, the effect of social isolation on hypertension exceeded that of clinical risk factors such as diabetes [63]. On a more hopeful note, for older American adults, greater purpose in life has been linked with a lower risk of stroke [64].

A more complete understanding of human health also includes acknowledging the extensive set of factors that affect it. These ***influencing factors*** fall into four groups: personal attributes, personal behaviors, environmental attributes, and interventions. Personal behaviors refer to individual actions such as sleep, diet, exercise, and adherence to treatment regimens. Personal attributes are individual characteristics such as genetics, education, and relationships that typically cannot be modified, at least in the short term. Environmental attributes are factors that shape the health of all individuals within a given context and include the context's political and economic system as well as global threats such as climate change. Interventions refer to deliberate actions intended to bring about change, such as clinical interventions, financial support, or incentives.

4 From Possibility to Reality: Six Shifts

Our research suggests that adding up to 45 billion years of higher-quality life would require at least six material shifts in societal mindsets and actions (Fig. 4). These shifts are highly interdependent and mutually reinforcing, and adopting them would represent a material reorientation of public policy and the economy.

4.1 Invest More, Disproportionately on Prevention and Promoting Optimal Health

Unlocking up to 45 billion years of higher-quality life will require individuals, governments, and private institutions to invest more financial and human capital as a percentage of GDP to improve health. Greater investment in health-related interventions is justified in areas traditionally considered part of the healthcare system and in other parts of the economy—such as education, nutrition, and agriculture, consumer products, financial services, and technology—that have the potential to improve health.

The case for greater investment in health is economically sound and responsive to citizen and consumer preferences. McKinsey Global Institute's *Prioritizing health* report estimates that improving the health of the global population by effectively scaling known interventions could generate an ROI of two to four times, even when only considering economic benefits. Individuals consistently place a high priority on improving health [24]. Year after year, experts cite social and physical elements of health as two of the top three drivers of life satisfaction for individuals [65]. A significant positive correlation exists between greater investment in health and improvements in living standards and wealth across countries and time. MHI believes a causal link exists. High-income countries do not appear to have reached a point of diminishing returns [66–68].

> Improving the health of the global population by effectively scaling known interventions could generate an ROI of two to four times, even when only considering economic benefits.

MHI further proposes disproportionately investing in disease prevention and health promotion in addition to treatment to help people truly thrive (Fig. 5). Promotion is defined here as actions that help individuals achieve and sustain the best possible physical, mental, social, and spiritual health given their intrinsic biological capacity.

Currently, health spending is heavily tilted toward curative care. OECD countries spend just 2.8% of their health budgets on organized prevention programs such as vaccinations, disease screenings, and health education [69, 70]. Low-income economies spend ten times as much (as a proportion of budget) as OECD countries do on preventive measures, which amounts to 20 to 35% of these low-income economies' healthcare budgets [71]. At the moment, most investment and innovation in achieving optimal health come from the private sector. The global

Fig. 4 Six shifts are needed to reach the full potential for human health

wellness market, for example, is valued at $1.5 trillion, roughly four times the size of what governments and NGOs spend globally on preventive care and health promotion [72].

Fig. 5 Society should empower people to thrive, going beyond treatment to promoting a healthy way of life for all

Finally, MHI proposes disproportionately investing in underresourced populations as both a moral imperative and a good financial decision. Improving the health of populations with historically reduced access to care or worse-than-average health outcomes can give societies a potent economic boost. Better health is associated with improved labor productivity and higher income. For example, 11% of economic growth in LMICs from 1970 to 2000 resulted from reduced rates of adult mortality [73, 74].

As a first step to realizing this shift, governments might consider conducting a "health opportunity assessment" across all government agencies, including those outside of healthcare (e.g., environment, education, treasury, trade, agriculture, housing, and infrastructure) to understand the most critical links to health and to unearth potential high-ROI investments for consideration, including the net effect of laws, regulations, and government activity on private-sector activity [75].

4.2 Improve Measurement of a Modernized Understanding of Health with Better Data

Attaining the aspiration requires improving global standards and systems to measure a modernized understanding of health, collecting significantly more comparative data on each element and dramatically increasing transparency.

The rationale is straightforward: measurement is foundational to improvement. Measurement helps us understand what works and what doesn't so we can allocate resources effectively. Conversely, weak (or nonexistent) measurement leads to waste and prevents us from investing more in what might be promising innovation. Overly generalized measurements are also ineffective; for example, the absence of a diagnostic framework for autism specifically for women leads to underdiagnosis [76, 77]. It's clear that researchers need subgroup-specific data collection and measurement initiatives.

MHI estimates that fewer than 5% of factors that influence and measure a modern understanding of health are defined consistently (or at all), captured systematically, and made broadly available as data. Most elements of a modern definition of health have very modest, if any, reliable measurement standards or data, including social and spiritual health and large categories of influencing factors such as the role of nutrition, employment, housing, and sleep. An estimated 75% of studies related to health can be classified as having primary endpoints that measure physical health, while just 12% address mental health, 6% address social health, and 1% examine spiritual health [78]. Most countries have very limited or no data regarding mental health, and data that are comparable across regions are lacking. For example, more than 100 countries have no data about mental-health conditions among adolescents [79]. And of the 23 mental-health indicators measured by OECD, only two—life satisfaction and death by suicide—were available in more than 90% of OECD countries [80]. Finally, there is only a modest attempt to define or consistently measure the extent to which individuals have achieved optimal possible health.

Even physical-health measurements, the area with the best standards and data, offer considerable room for improvement. High-income countries capture primarily disease prevalence and select health interventions (e.g., prescriptions, office visits, and procedures). Few data are systematically captured or available with respect to function itself or health status such as mobility, toxin levels, pain, and sexual function. Challenges in lower-income countries are more fundamental and include undigitized patient records, siloed reporting due to unintegrated care, and lack of access to diagnostic technologies [81, 82]. Positive examples, such as India, have shown how LMIC economies can still build robust nationwide health registries over time [83].

An example of the tangible downside associated with weak measurement is the challenge high-income countries face in subjecting nonpharmaceutical interventions to rigorous empirical analysis. Since 2007, only 11% of food-related health claims—a large component of the nonpharmaceutical-intervention category— have met the EU criteria for robust scientific evidence [84–86]. Of the estimated 20,000 mental-health apps available for download on personal computers and smartphones, five have been formally vetted and approved by the US Food and Drug Administration [87].

4.3 Scale What Works

McKinsey Global Institute's *Prioritizing health* report estimates that applying existing and proven interventions could reduce the global disease burden by about 40% (Fig. 6) [24]. Scaling known interventions could reduce child mortality by 65% and help a typical 65-year-old to be as healthy as a typical 55-year-old today. These numbers only account for improvements in physical and, to some extent, mental

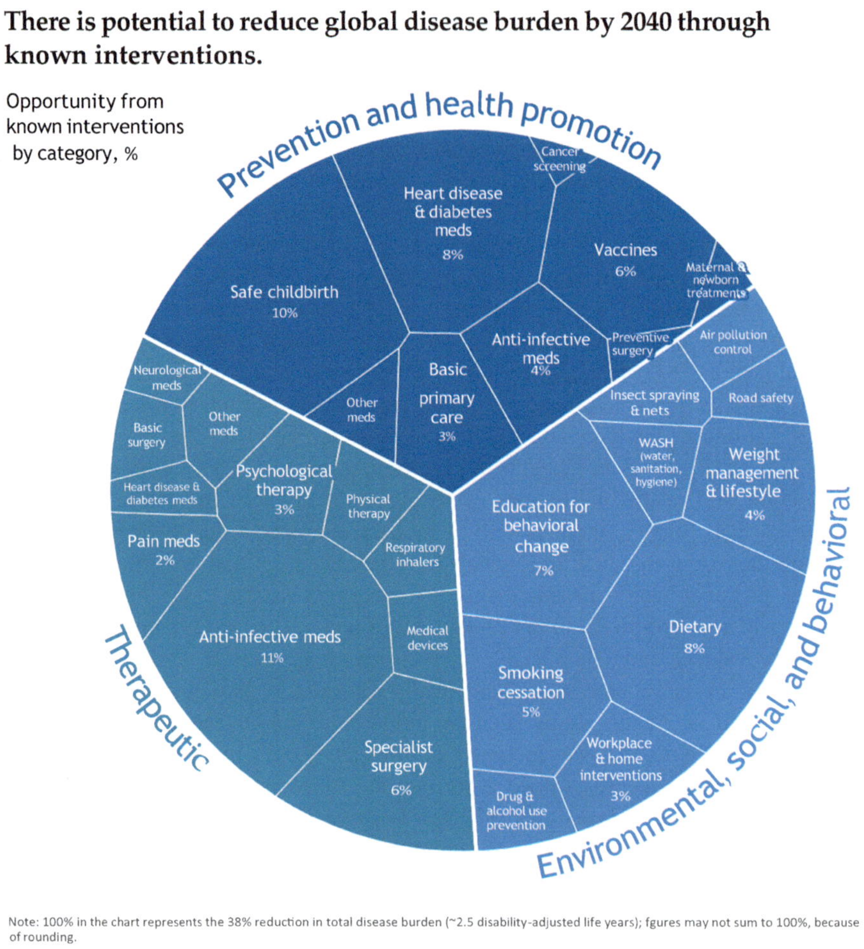

Fig. 6 There is potential to reduce global disease burden by 2040 through known interventions. Note: 100% in the chart represents the 38% reduction in total disease burden (~2.5 disability-adjusted life years); figures may not sum to 100%, because of rounding. Source: *Prioritizing Health: A prescription for prosperity*, McKinsey Global Institute, July 8, 2020; McKinsey Health Institute analysis

health. The potential for improved outcomes when health is considered more broadly is even greater. For high-income countries, most of the potential is in addressing diabetes, cancer, and cardiovascular disease. Low-income countries would benefit from more investment in known, basic health infrastructure [24]. Furthermore, scaling up largely preventive solutions and targeting infectious diseases and nutritional disorders as well as maternal, neonatal, and child health are high-potential improvements.

Past experience shows that scaling existing technologies can be very challenging—the global rollout of hepatitis B vaccines took more than 17 years, and survival after a diagnosis of type 1 diabetes in some LMICs is as low as 1 year, often due to the unaffordability of insulin [88–90].

To scale what works, societies need awareness and a commitment from people, governments, payers (including employers), and healthcare providers to adapt. Each group has an opportunity to overcome myopia and biases and establish strategies, policies, and incentives to increase the application of known solutions with high ROI. In many cases, such adaptation also requires greater net investment, as previously stated.

Another key to scaling what works is applying proven interventions across contexts, including countries and populations. Literature extensively describes differences in health outcomes for the same diseases by health provider, even within the same geography [91–95]. At their worst, healthcare systems and especially providers can be overly focused on their own systems. For example, the Aravind Eye Care System in southern India developed a successful and efficient system for delivering high-quality and low-cost cataract surgery [96]. Despite the potential for meaningful application in other parts of the world, including high-income countries, this system remains primarily anchored in the local context.

Increasing innovation will require more foundational research, better collaboration, and quicker and more effective nurturing and scaling of the most promising concepts.

4.4 Innovate More, and More Quickly

More innovation in all forms will be required, including in such areas as business models, government policies, incentive schemes, pharmaceuticals, medical devices, clinical standards, mobile apps, medical products, process improvements, and novel applications of existing technologies such as artificial intelligence. Increasing innovation will require more foundational research, better collaboration to break through silos, and quicker and more effective nurturing and scaling of the most promising concepts.

Innovation, along with greater investment in health, has been the driving force behind most advances in health. Once scientists identified the link between diabetes and insulin production in the pancreas a century ago, a continuous chain of discovery and commercialization has transformed the lives of diabetes patients [97]. Cancer mortality in the United States fell 32% from 1991 to 2019 [98], in part due to new surgical techniques, detection tools, and targeted therapies [99]. The exceptional pace of development and deployment of vaccines for COVID-19 is illustrative of the opportunity to drastically accelerate progress in public health when societal will, incentives, and resources are aligned. COVID-19 vaccines progressed from discovery to global distribution of the finished product in less than 12 months. Vaccine development is a high-risk process that traditionally takes more than a decade to complete. Prior to COVID-19, the fastest vaccine ever developed—for mumps in 1967—took 4 years from discovery to market launch.

Yet there are many areas where innovation has not occurred as hoped, often in the context of market failures. Despite the urgent threat of antimicrobial resistance (AMR), innovation in new antibiotics and alternatives has been weak, due in part to lack of incentives to invest in research and development [100, 101]. The antibiotic pipeline is substantially weaker than those for other therapeutic areas, such as treatment for cancer. One comparison: between 2017 and 2020, 1751 immuno-oncology drugs were in preclinical and research phase, and 45 received approval. In the same period, just 292 antibiotics were in preclinical and research phase, and 11 received approval [102, 103].

In addition to even more innovation from the life sciences industry, humanity would benefit from substantially more innovation from other industries and domains, especially via the application of emerging technologies [104] such as applied artificial intelligence, next-generation computing, and distributed computing. In addition, society has yet to realize the full potential of digital health innovation to improve health outcomes. Good momentum

for catalyzing new or greater innovation exists, especially among venture- and private-equity-backed companies. Venture capital investments in digital health reached an all-time high of $29 billion in 2021, nearly doubling 2020's investments [105]. Promising concepts include wearables (e.g., activity tracking, cardiac anomaly detection, patches for blood sugar monitoring, physiotherapy for pain relief with body-posture detection), telesurgery, and AI-enabled diagnostics.

Telemedicine exploded worldwide in 2020 due to COVID-19; in the United States, it grew by 3800% in 2020, and some consumers continue to indicate a preference for it [106, 107].

Finally, capturing the full potential impact from innovation will require a broader rethinking of innovation. This should involve creating an ecosystem of innovation by looking at all processes and influencing factors and engaging all stakeholders who affect a desired health outcome. This kind of innovation allows for better implementation mechanisms, end-to-end coordination, and collaboration to effectively and efficiently support sustainable behavior change and improve patient pathways. One such example is the Stop TB Partnership, which aims to end tuberculosis worldwide by 2030 [108]. The organization defined the optimal end-to-end care flow for tuberculosis, identified key interventions, and mapped stakeholders against each. It then aligned all contributors on the same data standards and coordinated their efforts, which resulted in an ecosystem that is both effective and scalable. Another example of an effective ecosystem can be found in Amsterdam, where multiple city departments (beyond the health department) work together in a coalition to achieve highly ambitious health and well-being targets [109–111].

Impactful innovation requires integrated, purposeful action across governments and the private sector. Governments might consider how to adapt investments, budgets, regulations and regular processes, intellectual property schemes, reimbursement principles, and laws to better catalyze innovation in areas with the greatest societal needs. Government action may be particularly relevant where the risk of market failure is high. Individuals and private payers, including insurers and employers, might consider how to adapt their consumption patterns and reimbursement policies to seek and reward promising innovations. Finally, greater innovation requires businesses from outside the healthcare sector to view health as an emerging or core market to enter, disrupt, and pursue.

4.5 Unleash the Full Potential of all Industries

Achieving the set aspiration requires institutions outside of the traditional healthcare industry to pursue health-related business opportunities much more aggressively, better enable and empower their employees, and better define and honor health-related environmental, social, and governance (ESG) commitments.

Health is deeply relevant to every business in the world, at minimum because employers affect the health of their employees, and the health of employees affects their performance. Sixty-eight percent of companies say employee well-being and mental health is a top strategic priority [112]. Poor employee health costs about $3.5 trillion annually [113]. In the United Kingdom, 17.9 million working days were lost to mental-health illnesses from 2019 to 2020, which was more than half of the total working days lost to poor health [114]. For many businesses, health is also part of their ESG commitments.

Yet, while 66% of Fortune 500 companies publish sustainability reports, only about 4% published health impact reports prior to COVID-19 [115]. Fewer have developed or implemented systemic policies to improve employee health.

Even more critically, MHI estimates that 40 to 45% of companies in the S&P 500 that are not considered part of the healthcare industry deliver products or services that directly affect the health of individuals (Fig. 7), sometimes favorably and sometimes unfavorably. Industries with highly health-relevant products or services include food and nutrition, consumer products, social media, transportation, gaming, travel, consumer financial services, non-health-related insurance, housing, and heating and cooling systems. Companies in these industries have an opportunity to leverage their relevance much more aggressively to explore even more significant, more disruptive entry into healthcare in both traditional and emerging domains. In some cases, companies face an imperative to understand and mitigate the potentially unfavorable effect of their products or services.

Individual health behaviors are the single biggest driver of an individual's health.

Moreover, companies that don't have a direct or strong link to health still have the opportunity to empower employees to improve their health and honor health-related environmental, social, and government commitments.

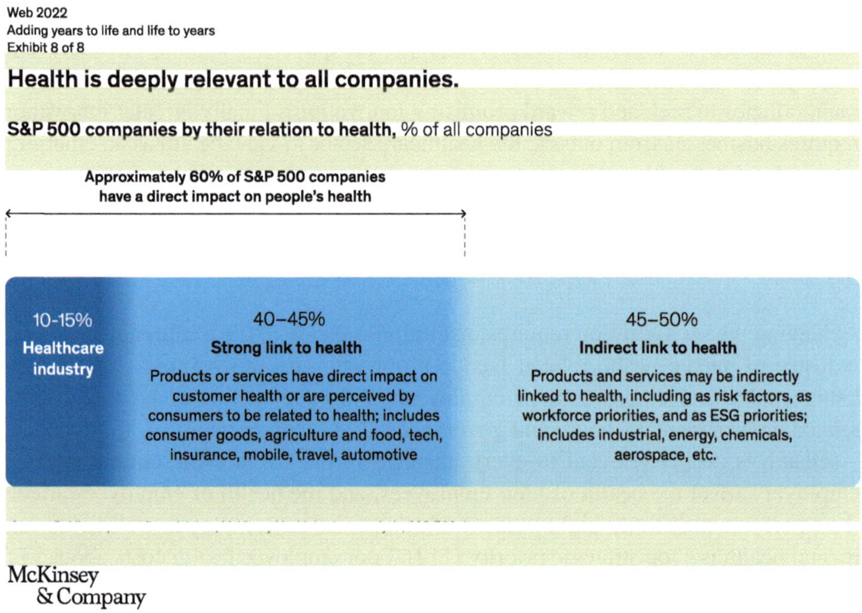

Fig. 7 Health is deeply relevant to all companies. Source: Public corporate finance data, McKinsey Health Institute analysis, S&P 500 data

4.6 Empower Individuals to Steward Their Own Health

To unlock up to 45 billion years of higher-quality life, individuals need to be empowered as the primary stewards of their own health and their loved ones' health. Individual behaviors are the single biggest driver of an individual's health. Multiple studies have shown that modifiable behaviors—including unhealthful diet, activity levels, sleep practices, medication use, and tobacco use—contribute to as much as 60% of deaths worldwide [116]. Individual mindsets and behaviors are likely even more relevant when considered through the lens of a modernized, holistic understanding of health that also includes the role of mental, social, and spiritual functioning. Almost every aspect of our lives and most of our choices affect our health.

Consumers across cultures and countries are increasingly demanding more empowerment. Their expectations are shaped by experiences of speed, convenience, personalization, and access to information from other industries, and they expect the same with respect to their health. Trust in institutions is declining in many regions, increasing the felt need of many individuals to be more self- directed [117, 118]. Finally, in many parts of the world, healthcare remains largely self-funded, which puts an even greater onus on the individual.

Recent experience shows that massive changes of behavior are possible. Billions of people significantly changed their daily actions, such as wearing face masks and social distancing, to protect themselves and their families during the COVID-19 pandemic. Insights from behavioral economics also show how to motivate and empower individuals so they are most likely to make decisions that are in their own best interest. Nudging—which behavioral science defines as a means of altering people's behavior without restricting their choices or changing their economic incentives—can influence the behavior and decision-making of individuals and groups. For example, nudging has been shown to improve hand hygiene among healthcare workers to decrease the number of healthcare-associated infections [119]. In addition to purposeful nudging, social networks and their associated culture, mindsets, and behaviors are highly correlated with individuals' health-relevant decisions and behaviors.

Empowering individuals will require action from the government, more innovation, and adaptation from traditional healthcare stakeholders. Governments can consider how they might adapt laws, regulations, incentives, and government-operated healthcare services to better empower people. Governments might also consider how to adapt (or create) environments, structures, or systems to make it easier for consumers to make healthy choices.

Empowering individuals will require substantially more technology-based innovation, including mobile or remote access that optimizes exercise, sleep, caloric intake, posture, and the like.

Innovation that empowers individuals is exploding across low-, middle-, and high-income countries. In rural India, for example, smartphones enable individuals to have greater access to healthcare providers [120]. Approximately 200 million Chinese consumers have used Ping An's Good Doctor mobile platform to receive consultation, referrals, and appointments [121]. In 2019, 21% of Americans reported using smart watches or fitness trackers [122].

Finally, healthcare stakeholders, especially provider institutions and clinicians, might consider how to embrace many patients' desire to be self-directed. Providers have an opportunity to operate more as coaches who help patients define and execute a coauthored care plan. Providers also have an opportunity to embrace many of the previously described technology-based solutions emerging outside the traditional healthcare industry.

Addressing Health Inequity

As highlighted in this report, addressing health inequities across countries and populations is both the right thing to do and a requirement to reach the aspiration. MHI acknowledges that this is an arena where more research and work could be critical, especially to achieve impact at scale. Moreover, equity priorities will vary considerably across communities based on contexts and cultures. That said, research suggests a few foundational actions are needed. The first is to acknowledge and measure inequity to encourage accountability, motivate action, and better understand root causes of poor health. The second is for governments, businesses, and social institutions to be even more

(continued)

intentional in creating focused strategies to address inequities. Finally, MHI believes a focus on equity can and should be considered and integrated into all other actions to improve health, both to uncover potential opportunities to reduce inequity and to mitigate the risk of unintentionally exacerbating it.

5 Entering the Arena

Dramatically improving our health requires an ecosystem approach—exchanging ideas, aligning around standards, working across multiple stakeholder silos. It will require unprecedented collaboration to shift society's mindsets and actions enough to realize possible gains in life expectancy and quality of life.

McKinsey Health Institute is an enduring, non-profit-generating global entity within McKinsey that strives to catalyze actions across continents, sectors, and communities to extend and improve lives. MHI is committed to contributing to this collective ecosystem. MHI is fostering a strong network of organizations committed to this aspiration through a range of collaboration types—convening and enabling leaders, advancing research, creating and promoting open-access data assets, and stimulating innovation. MHI is sharing resources, innovations, data, and findings in the public domain so others can replicate what proves effective and looks to its ecosystem partners to commit to the same.

MHI welcomes connection with committed organizations interested in building out this ecosystem together. MHI is actively seeking opportunities to collaborate across its identified six shifts as well as seven initial key focus areas: brain health, healthy living, infectious diseases, equity and health, sustainability and health, aging, and healthcare-worker capacity.

Dramatically improving our health requires an ecosystem approach— exchanging ideas, aligning around standards, working across multiple stakeholder silos.

6 About the Authors

Erica Coe is a coleader at the McKinsey Health Institute (MHI) and a partner in McKinsey's Atlanta office; Martin Dewhurst is an alumnus and former coleader of MHI; Lars Hartenstein is a coleader at MHI in the Paris office; Anna Hextall is a consultant at MHI and is based in the London office; and Tom Latkovic is an alumnus and former MHI coleader of MHI.

The authors wish to thank Jesse Bradford, Jacqueline Brassey, Kana Enomoto, Soheil Eshghi, Anne-Marie Jennen, Jack Manley, Daphne Maurel, Lorenzo Pautasso, and Eike Pfefferkorn for their contributions to this article.

References

1. Calculated in a scenario-based simulation sizing three effects: increase of life expectancy ("extending"), decrease of lifetime spent in poor health, and decrease of lifetime in moderate rather than good health ("lifting" and "squaring the curve"). Data sources comprise WHO HALE data, health system data, life-expectancy data for more than 20,000 administrative regions globally, and the McKinsey Global Institute's report *Prioritizing Health: A prescription for prosperity*, July 8, 2020. The McKinsey Health Institute's aspiration to add six years of higher-quality life on average is going significantly beyond delaying the average onset of disease from, for example, 55 to 65 years mentioned in the latter report.
2. Spiritual health includes meaning, belonging, purpose, and identity. It does not necessarily require or include religious belief.
3. Prioritizing health: A prescription for prosperity. McKinsey Global Institute. 2020, July 8.
4. *World happiness report 2021*, Sustainable Development Solutions Network, 2021; social health is the ability of individuals to form healthy and rewarding interpersonal relationships with others.
5. Global spending on health: Weathering the storm. World Health Organization. 2020, December.
6. Gapminder.org: Life Expectancy; McKinsey Global Institute analysis.
7. Ortiz-Ospina E, Ritchie H, Roser M. Life expectancy. Our World in Data. 2019, October. World Bank country and lending groups, World Bank, 2022.
8. Healthier mothers and babies. Morbidity and Mortality Weekly Report, Centers for Disease Control and Prevention. 1999, October 1.
9. Sharrow D, et al. Levels and trends in child mortality. United Nations Inter-agency Group for Child Mortality Estimation. 2020.
10. Ten great public health achievements–United States, 1900–1999. Morbidity and Mortality Weekly Report, Centers for Disease Control and Prevention. 1999, April 2.
11. "1990: Launch of the Human Genome Project," National Human Genome Research Institute, May 6, 2013; Global Data Health Exchange, Institute for Health Metrics and Evaluation, 14 February 2022; deaths from diarrhea, lower respiratory infection, and other common infectious diseases declined by 60 percent between 1990 and 2019. Globally, the annual number of new HIV cases decreased by 39 percent since its peak in 2000. Polio cases decreased 99.99 percent since 1980. Deaths attributed to smoking per 100,000 decreased by 40 percent, from 146 in 1990 to 90 in 2017. Globally, suicide rates per 100,000 decreased 35 percent from 1990 to 2017. The global age-standardized cancer mortality rate decreased 15 percent between 1990 and 2017.
12. An individual in "good health" is defined as a person without a relevant chronic medical condition impacting quality or duration of life, but who may have very well managed chronic conditions or acute injury with high potential for swift and complete recovery. "In moderate health" defines a person suffering from one or a combination of several acute or chronic conditions that are impacting quality and/or duration of life, but without relevant impairment of activities of daily living. An individual in "poor health" is defined as suffering from one or more acute or chronic conditions requiring constant or frequent attention by carers and/or impacting activities of daily living significantly and/or reducing quality of life or shortening life expectancy significantly.
13. McKinsey analysis; Life Expectancy at Birth, Total (Years) Database, World Bank Data. Accessed 13 December 2021.
14. Global Burden of Disease Study 2019, Demographics 1950–2019, Institute for Health Metrics and Evaluation. Accessed 13 December 2021.
15. Kapteyn A, et al. Life satisfaction within and across countries; The role of societal capital and relative income. University of Southern California. 2016, August.

16. Frijters P, et al. A happy choice: Wellbeing as the goal of government. Centre for Economic Performance discussion paper number 1658, October 2019, cep.lse.ac.uk; World happiness report 2021.
17. Global Burden of Disease Collaborative Network Databases, Institute for Health Metrics and Evaluation. Accessed 16 February 2022.
18. Global Burden of Disease Study 2017, Institute for Health Metrics and Evaluation. 2018.
19. World Bank; Global Burden of Disease 2017.
20. 10 global health issues to track in 2021, World Health Organization. 2020, December.
21. The top 10 causes of death, World Health Organization. 2020, December.
22. Global, regional, and national burden of 12 mental disorders in 204 countries and territories, 1990–2019: A systematic analysis for the Global Burden of Disease Study 2019. Lancet. 2022, February; 9(2).
23. Mental health fact sheet. WHO. Accessed 20 January 2022.
24. Prioritizing health. 2020, July.
25. Centers for Disease Control and Prevention (CDC) WISQARS Leading Causes of Death Reports, 2019.
26. Estimation of the global prevalence of dementia in 2019 and forecasted prevalence in 2050: An analysis for the Global Burden of Disease Study 2019. Lancet. 2022, January; 7(2).
27. Freburger JK, et al. The rising prevalence of chronic low back pain. Arch Int Med. 2009, February; 169(3).
28. "Obesity and overweight," (fact sheet). WHO. Accessed January 20, 2022.
29. The state of food security and nutrition in the world 2021: Transforming food systems for food security, improved nutrition and affordable healthy diets for all. Food and Agriculture Organization of the United Nations. 2021, October.
30. Watts N. et al. The *Lancet* Countdown on health and climate change: from 25 years of inaction to a global transformation for public health. Lancet. 2018, February; 391(10): 210.
31. Uneven access to health services drives life expectancy gaps. World Health Organization. 2019, April.
32. Life Expectancy. 2019, October.
33. Goldenberg RL, McClure EM, Saleem S. Improving pregnancy outcomes in low- and middle-income countries. Reproductive Health. 2018, June; 15:88.
34. Childhood cancer factsheet. World Health Organization. 2021, December.
35. De Blasio administration launches comprehensive plan to reduce maternal deaths and life-threatening complications from childbirth among men of color. City of New York. 2018, July 20.
36. Life expectancy for local areas of the UK: between 2001 and 2003 and 2017 to 2019. Office for National Statistics. 2020, September 24.
37. Women's emotional health is at its worst in 15 years. Global Women's Health Index, Hologic. 2021, September.
38. Lam CSP, et al. Sex differences in heart failure. European Heart J. 2019, December; 40(47), 3859-36858.
39. On current trajectory, life expectancy is expected to increase 2.6 years on average over the next decade disproportionately in LMICs.
40. Healthy Life Expectancy at Birth Years Database, World Health Organization, 20 January 2022; Population (Total) Database, World Bank, 20 January 2022; HALE at birth over the periods 2000–2010 (for Ireland and Thailand) and 2010–2019 (for Oman, South Africa, and Bolivia): WHO income data.
41. Herm A, Pes G, Poulain M. The Blue Zones: Areas of exceptional longevity around the world. Vienna Yearb Popul Res. 2013; 11: 87-108.
42. Westervelt A. The medical research gender gap: How excluding women from clinical trials is hurting our health. Guardian. 2015, April 30.
43. COVID-19 Treatments Database. European Medical Agency. 2022, February 2.
44. Know your options for COVID-19. US Food & Drug Administration. 2022, January 27.

45. Chaturvedi SK, Dhar N, Nandan D. Spiritual health, the fourth dimension: A public health perspective. WHO South-East Asia Journal of Public Health. 2013, January; 2(1): 3-5
46. Chirico F. Spiritual well-being in the twenty-first century: It is time to review the current WHO's health definition. J Health Social Sci. 2016, March, 1(1): 11
47. Puchalski CM. Integrating spirituality into patient care: An essential element of person-centered care. Polish Archives Internal Medicine. September 2013; 123(9): 491-7
48. Lucchetti G, et al. Spirituality and health in the medical schools in Brazil. BMC Medical Education. 2012, August; 12:78
49. Spiritual aspects of health: Global strategy for health for all by the year 2000. South-East Asia Advisory Committee on Medical Research, World Health Organization. 1984, March.
50. The World Health Organization quality of life (WHOQOL) - BREF, 2012 revision. World Health Organization. 2012.
51. Galderisi S, et al. Toward a new definition of mental health. World Psychiatry. 2015, June; 14(2): 231–233.
52. Patel V. The Lancet commission on global mental health and sustainable development. Lancet. 2018, October;392(10):157.
53. Waite LJ. Social well-being and health in the older population: Moving beyond social relationships. Future directions for the demography of aging: proceedings of a workshop. 2018, June.
54. Keyes CLM. Social well-being. Soc Psychol Quarterly. 1998, June; 61(2).
55. Personal well-being in the UK quarterly, April 2011 to September 2020. Office for National Statistics. 2021, February.
56. Kim ES, et al. Purpose in life and reduced incidence of stroke in older adults: 'The health and retirement study'. J Psychosomatic Res. 2013, May; 74(5), 427-32.
57. Edmonds GW, Hampson SE, Hill PL. A purposeful lifestyle is a healthful lifestyle: Linking sense of purpose to self-rated health through multiple health behaviors. J Health Psychol. 2017, May; 24(10): 1392-1400.
58. Levine GN, et al. Meditation and cardiovascular risk reduction: A scientific statement from the American Heart Association. J Am Heart Assoc. 2017, September; 6(10): e002218.
59. WHO, Information sheet: Premature death among people with severe mental disorders.
60. The Loneliness Epidemic. US Health Resources and Services Administration. 2019, January.
61. Chawla K, et al. Prevalence of loneliness amongst older people in high-income countries: A systematic review and meta-analysis. PLOS One. 2021, July.
62. Kim ES, et al. Purpose in life and reduced incidence of stroke in older adults: The Health and Retirement Study. J Psychosomatic Res. 2013, May.
63. Yang C, et al. Social relationships and physiological determinants of longevity across the human life span. PNAS. 2016, January; 113(3): 578-583.
64. Purpose in life. 2013, May.
65. *World happiness report 2021*; social health is the ability of individuals to form healthy and rewarding interpersonal relationships with others.
66. In 2019, EU countries devoted on average 8.3 percent of their GDP to healthcare; this figure has remained largely unchanged since 2014 as growth in health spending remained broadly in line with overall economic growth. In the US, the share of national health expenditure has increased six percentage points over the last two decades, in line with the pace of economic growth. Health Expenditure in Relation to GDP. Health at a Glance: Europe 2020: State of Health in the EU Cycle, 2020
67. Current Health Expenditure (% of GDP). World Health Organization Global Health Expenditure Database, World Bank. 2021, January 19.
68. Historical: National Health Expenditure Data. Centers for Medicare & Medicaid Services. 2021, January 19.
69. Countries need to up spending commitments on health promotion, urge think tanks. International Longevity Centre. 2021, November.

70. Health spending in most OECD countries rises, with the U.S. far outstripping all others. OECD. 2004, March.
71. Global spending on health: A world in transition. World Health Organization. 2019, December.
72. McKinsey analysis; global wellness-market figure pulled from McKinsey's *Future of wellness* report and encapsulates products ranging from personal health trackers to nonsurgical aesthetic procedures.
73. Jamison DT, Lau LJ, Wang J. Health's contribution to economic growth in an environment of partially endogenous technical progress. In: López-Casasnovas G, editor. Health and economic growth: findings and policy implications. Cambridge, MA: MIT Press; 2005. p. 67–91.
74. Blake C. Alkire, et al. The economic consequences of mortality amenable to high-quality health care in low- and middle-income countries. Health Affairs. 2018, June; 37(6).
75. A similar perspective with a UK lens is explored in *Whose Health Is It, Anyway?*, by Sally C. Davies and Jonathan Pearson-Stuttard, published by Oxford University Press in 2021.
76. Golding J, Russell G, Steer C. Social and demographic factors that influence the diagnosis of autistic spectrum disorders. Social Psychiatry Psychiatric Epidemiol. 2010, October; 46.
77. Dworzynski K, et al. How different are girls and boys above and below the diagnostic threshold for autism spectrum disorders? J Am Acad Child Adolesc Psychiatry. 2012, June; 51(8), 788-97.
78. Sample analysis based on PubMed.
79. Erskine HE. et al. The global coverage of prevalence data for mental disorders in children and adolescents. Epidemiol Psychiatric Sci. 2017, August; 4.
80. Key findings and recommendations. A new benchmark for mental health systems: Tackling the social and economic costs of mental ill- health. OECD. 2021, June.
81. Ngugi P, et al. Development of standard indicators to assess use of electronic health record systems implemented in low-and- medium-income countries," PLOS One. 2021, January.
82. Mayhew SH, Mays N, Mounier-Jack S. Integrated care: Learning between high-income, and low- and middle- income country health systems. Health Policy Planning. 2017, November; 32: iv6-iv12.
83. Hay K, Mohanan M, Mor N. Quality of health care in India: Challenges, priorities, and the road ahead. Health Affairs. 2016, October; 35(10), 1753-1758.
84. EU Register of Nutrition and Health Claims Made on Foods Database (v.3.6). European Commission. 2022January 7.
85. Mobile health approaches to weight management: Food for thought, Organisation for the Review of Care and Health Applications (ORCHA). 2021, January.
86. Mernin A. App warning in addressing cancer care backlog. Health Tech World. 2021, April 14.
87. Gold J. In a murky sea of mental health apps, consumers left adrift. California Healthline. 2021, June 21.
88. Chen D-S. Hepatitis B vaccination: The key towards elimination and eradication of hepatitis B. J Hepatol. 2009, April; 5(4): 805-16.
89. Chen D-S, Kao J-H. Global control of hepatitis B virus infection. Lancet. 2002, July; 2(7): 395-403.
90. Yudkin JS. Insulin for the world's poorest countries. Lancet. 2001, March; 355(9).
91. Daw P, et al. A systematic review of provider-and system-level factors influencing the delivery of cardiac rehabilitation for heart failure. BMC Health Serv Res. 2021, November, 21: 1267.
92. Saposnik G, et al. Hospital volume and stroke outcome: Does it matter? Neurology. 2007, September, 69(11), 1142-51.
93. Armstrong BK, Jong KE, Vale PJ. Rural inequalities in cancer care and outcome. Med J Aust. 2005, January; 182(1), 13-4.
94. Laura K. Stein et al., Correlations between physician and hospital stroke thrombectomy volumes and outcomes: A nationwide analysis. Stroke. June 2021; 52(1)
95. Vijay G. Profitable audacity: One company's success story. Harv Bus Rev. 2012, January 25.

96. Our story. Aravind Eye Care System. 2022, February 26.
97. Kjeldsen T, Kurtzhals P. A hundred years of insulin innovation: When science meets technology. Diabetes. 2021, September;70(9):e1–2.
98. Seigel RL, et al. Cancer statistics, 2022. CA Cancer J Clin. 2022, January;72(1):7–33.
99. Advancing cancer therapy. Nature Cancer. 2021, March, 2, 245-246.
100. Antimicrobial resistance: Invest in innovation and research, and boost R&D and access. IACG. 2018, June.
101. Global shortage of innovative antibiotics fuel emergence and spread of drug-resistance. World Health Organization. April 2021.
102. FDA, New Drug Therapy Approvals (2021, 2020, 2019, 2018, 2017) and Biologic Approvals by year; approvals of anti-infectives and antibacterials between July 1, 2017, and September 30, 2020. Only FDA approvals counted. Count of drugs, not indications. Approvals given for total/new molecule.
103. 2020 antibacterial agents in clinical and preclinical development: An overview and analysis. World Health Organization. 2021, April.
104. Corbo J, Henke N, Ostojic I. The top trends in tech. McKinsey. 2021, June.
105. Evans B, Krasniansky A, Zweig M. 2021 year-end digital health funding: Seismic shifts beneath the surface. Rock Health. 2021, January 10.
106. Bestsennyy O, Gilbert G, Harris A, Rost J. Telehealth: A quarter-trillion-dollar post-COVID-19 reality? McKinsey. 2021, July 9.
107. Cordina J, Levin E, Stein G. COVID-19 consumer healthcare insights: What 2021 may hold. McKinsey. 2021, November 19.
108. Re-imagining TB Care. Stop TB Partnership. Accessed February 26, 2022.
109. Dutch health targets for future. Open Access Government. 2017, December 11.
110. The Amsterdam healthy weight approach: Investing in healthy urban childhoods: A case study on health diets for children. UNICEF. 2020, November.
111. Aiming for zero new HIV infections in Amsterdam by 2026, Fast-Track Cities. 2020, May.
112. Meister J. Five strategic HR priorities for 2021. Forbes. 2021, May 28.
113. McKinsey analysis, WHO, WTO Integrated Benefits Institute.
114. Working days lost in Great Britain, 2019–20. WHO Health and Safety Executive. Accessed 14 December 2021.
115. Jha AK, Sands P. Most CEOs don't have a global health strategy. That needs to change. Fortune. 2019, March 25.
116. Kushner RF, Sorenson KW. Lifestyle medicine: The future of chronic disease management. Curr Opin Endocrinol Diabetes Obes. 2013, October;20(5):389–95.
117. Perry J. Trust in public institutions: Trends and implications for economic security. UN Department of Economic and Social Affairs. July 20, 2021
118. Esteban Ortiz-Ospina and Max Roser, "Trust," Our World in Data, 2016.
119. Caris MG, et al. Nudging to improve hand hygiene. J Hosp Infect. 2017, September;98(4): 352–8.
120. Smart doctor: How smartphones hold the key to improving rural health in India. Nanyang Technological University. July 16, 2020.
121. Technology empowers Ping An's health care amid coronavirus. Ping An. August 13, 2020.
122. Vogels EA. About one-in-five Americans use a smart watch or fitness tracker. Washington, DC: Pew Research Center; 2020, January 9.

From Talk to Action: How Biopharma and Medtech Can Drive Sustainability

Charline Wurzer, Elia Tziambazis, and Elisabeth Richenhagen

Abstract

"There is a rapidly closing window of opportunity to secure a livable and sustainable future for all," the latest IPCC report concludes. To keep global warming below an increase of 1.5 °C compared to pre-industrial levels, rapid reduction of almost half of current greenhouse gas emissions is needed by 2030. Currently, we are not on track to achieve this target, as projections indicate an 11% increase of emissions by 2030. At the same time biodiversity loss is accelerating, water scarcity is increasing, and plastic pollution is becoming ubiquitous. Undoubtedly, fighting human-induced climate change and the broader impact of human activities on the environment is one of the defining challenges of the twenty-first century, calling for bold action from the public and private sector globally. Healthcare has also a key role to play with the sector accounting for 5% of global emissions.

"There is a rapidly closing window of opportunity to secure a livable and sustainable future for all," the latest IPCC report concludes.[1] To keep global warming below an increase of 1.5 °C compared to pre-industrial levels, rapid reduction of almost half of

[1] Synthesis Report of the IPCC Sixth Assessment Report. Summary for Policymakers (2023), pg. 25.

C. Wurzer (✉) · E. Tziambazis · E. Richenhagen
The Boston Consulting Group, Inc., Boston, MA, USA
e-mail: Wurzer.Charline@bcg.com; Tziambazis.Elia@bcg.com; Richenhagen.Elisabeth@bcg.com

E. Tziambazis
The Boston Consulting Group, Inc., Boston, MA, USA

E. Richenhagen
The Boston Consulting Group, Inc., Boston, MA, USA

U. A. Betz (ed.), *Curious Future Insight*,
https://doi.org/10.1007/978-3-031-41781-8_12

current greenhouse gas emissions is needed by 2030. Currently, we are not on track to achieve this target, as projections indicate an 11% increase of emissions by 2030. At the same time biodiversity loss is accelerating, water scarcity is increasing, and plastic pollution is becoming ubiquitous. Undoubtedly, fighting human-induced climate change and the broader impact of human activities on the environment is one of the defining challenges of the twenty-first century, calling for bold action from the public and private sector globally.

For the past 200 years, biopharma and medical technology (medtech) companies have been a beehive of innovation, improving diagnostics, patient treatments, and clinical outcomes across a vast array of disease areas. These breakthroughs cover the widest potential range of healthcare possibilities: everything from anesthetics, antibiotics like penicillin, vaccines, cancer treatments, implants and protheses to MRI scans, transcatheter heart valves, genetic mapping, organ transplants, fertility treatments, and COVID tests. But despite this unmatched record of scientific advances, an additional task for both biopharma and medtech companies arises more and more clearly: reducing their impact on the environment.

Unfortunately, biopharma and medtech companies also impact the environment in several ways:

- First, healthcare accounts for 5% of the total global emissions; if healthcare were a country, it would be the fifth largest CO_2 producer. These emissions are driven by products and supply chains (50%), including drugs and medtech devices; patient care (35%) in hospitals and outpatient care settings as well as ambulance trips; direct patient emissions (10%); and R&D (5%).[2]
- Second, healthcare generates a large amount of waste, primarily through single-use products, packaging, and product expiries. To get a tangible sense of the sheer volume, look at Dutch artist Maria Koijck's eye-opening and moving 1 ½ minute YouTube video [1] in which she spreads out the vast troves of single-use materials generated during her breast replacement surgery—the plastic gowns and globes, the spent syringes, the clamps and tape, and everything else—in an ever-widening circle on the floor. She proceeds to lay in the center of this remarkable accumulation of waste. The World Health Organization estimates that high-income countries produce on average 3.5 kg waste per hospital bed per day, while low-income countries generate on average 1.5 kg. About 15% of this waste is considered hazardous, either infectious, chemically toxic, or radioactive [2]. Other sources say that the negative impact is even higher. Practice Greenhealth, a healthcare environmental solutions organization in the USA and Canada, contends that US hospitals produced about 13 kg of hospital waste per patient per day pre-COVID [3].

[2] Arup & Health Care Without Harm/Arup (2019); World Bank & Health Care Without Harm (2017); UK National Health Service & Lancet (2020); The Lancet Countdown on Health and Climate Change (2020, 2018); Environmental Research Letters: International Comparison of Health Care Carbon (2019); Health Affairs: Health Care Pollution And Public Health Damage In The United States: An Update (2020); BCG analysis based on literature review case experience.

- Third, healthcare also has an impact on biodiversity. Many drugs and diagnostic tests use animal derivatives and other natural ingredients. For instance, some diagnostic tests contain horseshoe crab blood, which is used to determine if certain drugs and vaccines contain endotoxins, bacterial substances that can cause fevers and even be fatal to humans. Other drugs contain palm oil, and most packaging contains cardboard. The production of both ingredients leads to deforestation and ultimately biodiversity loss.
- And finally, healthcare activities consume a lot of water. The biopharma sector alone uses about 530 billion liters of water annually, an amount equal to the annual water consumption of all Swedes.[3] This is primarily used during the pharma manufacturing process, which requires purified water to dissolve ingredients like sugar and to clean stainless steel tanks. Only about 27%[4] of this water is recycled. And downstream, several pharmaceutical products—IV fluids, disinfectants, and the like—contain water, and various medtech devices, including dialysis machines and sterilization equipment for surgical instruments, need water to operate.

Healthcare's negative environmental impacts are not the result of malicious intent. Rather, they have evolved from fundamental decisions made to provide the best and highest-quality products and services for patients and providers, in many cases during periods when environmental concerns were not considered an urgent priority. These well-intentioned decisions must be examined closely by healthcare companies now to design alternative approaches that meet the same performance goals but are less harmful to the environment and to efforts to reduce climate change.

The healthcare choices that bear new consideration now, in particular their effect on the environment, include:

- *Single-use products.* Over time the healthcare industry has grown to favor medical equipment and instruments that are used only once in a hospital, clinic, or patient's home and then discarded. Disposable items were initially developed to advance patient and provider safety. Single-use medical supplies, such as syringes, test kits, and scalpel blades, prevent germs and viruses from spreading from one patient to another and are essential in infection control and reducing hospital bacteria. Disposables—including masks, plastic gloves, and body protection—reduce incidents of infection among healthcare personnel while protecting patients from germs carried by providers on their skin or clothing.
  Single-use equipment also increases the efficiency of hospital and outpatient physician-patient interactions, which is increasingly important as shortages of healthcare personnel widen. By reducing the burden of sterilization and disinfection after each medical intervention, single-use items allow healthcare providers to see more patients and spend more time with them.

[3]Refinitiv, Water consumption from public companies within pharmaceutical industry (2021).

[4]Refinitiv, Water consumption from public companies within pharmaceutical industry (2021).

In addition, some single-use equipment, such as auto-injectors for drugs like insulin, enable patients to take care of their health at home. This releases some of the strain on the healthcare system, while improving the patient's quality of life since doctor visits are not required for every insulin shot.

- *Packaging*. Both biopharma and medtech products tend to come in a lot of packaging, ultimately waste, primarily to ensure that the items are sterile and intact for patient use. Often included in this packaging are lengthy paper leaflets with instructions and ingredients to ensure that the products' functions and capabilities are clearly understood and that they are used correctly by providers and patients.
- *Ingredients*. Researchers have made substantial progress over the decades in curing or treating simple and complex conditions and diseases using natural active ingredients and substances. Hence, many different medicines and vaccines—or specific formulations of a medicine, such as tablets, capsules, creams, or mixtures—contain animal derivatives and/or natural ingredients. As noted before in the references to horseshoe crab blood, palm oil, and cardboard, the use of these ingredients may negatively impact biodiversity.

Although these facets of healthcare products have a deleterious effect on the environment, sustainability initiatives need to be pursued by healthcare, biopharma, and medtech companies with a delicate balance in mind. Obviously, while enhancing their environmental record is important and becoming more dire, healthcare companies must still prioritize patient safety and improved outcomes. Ultimately, they must stack up sustainability requirements against patient benefits and impact on the healthcare system—and attempt to reduce environmental damage with new ideas and innovation so that both sides of the equation can be satisfied.

This tricky balance is expressed well by Johnson & Johnson in its mission statement: "We believe our first responsibility is to the patients, doctors, and nurses, to mothers and fathers and all others who use our products and services. In meeting their needs everything we do must be of high quality. We are (also) responsible to the communities in which we live and work and to the world community as well.. . .We must maintain in good order the property we are privileged to use, protecting the environment and natural resources." [4] And Belén Garijo, CEO of Merck KGaA, stresses in the company's 2022 annual report to shareholders that Merck "continued to care, pioneer, and outperform for. . .[their] patients and customers while creating high impact across the Group for society as a whole." [5].

1 The Business Case for Sustainability

The pressure on healthcare companies to adopt substantive sustainability disclosure, goals, and strategies is increasing. Regulators—most prominently, the EU—are introducing transparency and reporting requirements for environmental impact disclosures. The EU's Corporate Sustainability Reporting Directive provides a template for most businesses to report on carbon emissions and other relevant environmental data, as well on broader ESG matters (e.g., human rights). These rules will take effect between 2024 and 2026 depending on company type.[5]

The EU has also introduced specific sustainability requirements for products and operations. For instance, RoHS and REACH are focused on restricting the use of harmful substances, the Medical Device Regulation contains a clause on reprocessing of single-use devices, and the proposed EU Packaging and Packaging Waste Regulation covers packaging size and materials and recycling.[6]

In addition, customers and shareholders are focusing more and more on the way companies view their responsibilities to the environment and other ESG criteria—and they are rewarding businesses that demonstrate their commitment to reducing carbon emissions and limiting their impact on the environment more broadly. Further, there is growing evidence that minimizing environmental impact can be a catalyst for other significant advantages for healthcare companies; among them, cost reduction, product differentiation, cheaper capital access, higher valuation and shareholder returns, business resilience, employee attraction and retention, and improved reputation.

- *Gain preferential position in tenders*. Although prices continue to be a substantially more important factor in tenders and non-tender based procurement decisions than sustainability, especially in the pharmaceutical market, companies across the healthcare sector embracing positive sustainability principles are beginning to enjoy tangible benefits.
  For instance, the UK's Procurement Policy Note (PPN) 06/20 [6] established "social value" requirements when agencies assess purchasing bids, including environmental and climate change goals. In addition, starting in 2027, the NHS will only purchase from suppliers that are committed to net-zero by 2040 [7]. Similarly, the 2021 Nordic joint tender procurement initiative, involving Denmark, Iceland, and Norway, sets down qualitative environmental criteria that must be met for companies to win government contracts [8].
  Some private sector players are also starting to use sustainability as a determinant in procurement. For example, Kaiser Permanente, an American healthcare consortium and one of the biggest healthcare payer/providers to join in these efforts, mandates that for all major strategic and critical purchasing decisions, a variety of specific environmental criteria be considered along with product performance and cost effectiveness [9].
- *Drive product differentiation*. Reducing emissions and the environmental impact can also enable companies to tap into new revenue streams, such as climate-friendly (energy-efficient and lower-waste) product lines. These ecolines can be

[5]EU, Corporate Sustainability Reporting Directive: 1 January 2024 for companies already subject to the NFRD (reporting due from 2025); 1 January 2025 for large companies that are not presently subject to the NFRD (reporting due from2026); 1 January 2026 for listed SMEs, and small non-complex credit and captive insurance undertakings (reporting due from 2027).

[6]Contact sensitive plastic packaging of medical devices and contact sensitive plastic packaging of in vitro diagnostics are exempt.

developed by focusing on reused raw materials, using more sustainable materials, reducing overall raw material volume, and minimizing energy consumption during use. All these features can separate new products from traditional ones—and show that a company is serious about adopting a more positive environmental agenda.

Two illustrations: Swedish MedTech company Mölnlycke, which specializes in wound care and surgical supplies, has launched a set of surgical drapes with bio-based raw materials instead of plastic.[7] Philips has made sustainability an essential element for all new products with its EcoDesign line, which focuses on improvements in energy efficiency and recyclability. Philips hopes that 100% of its new products will meet EcoDesign standards by 2025.[8]

- *Realize cost reduction.* Many CO_2 and waste reduction programs have been shown to actually make operations more efficient and less costly. If planned right. Abating 20% to 30% of emissions could generate a net cost saving—and cutting avoidable emissions by 60–80% could be cost neutral.[9] The cost reductions would come from new logistics strategies (including shifting from air transport to ocean and road to rail), material efficiency, reduced business travel, and increased efficiency in production (especially for raw materials like plastics).
- *Improve access to capital.* There is also deepening investor interest in ESG. Public market investors are increasingly examining ESG scores when making investment decisions and leading private equity players are setting up dedicated green/impact funds to invest in sustainable companies. On the debt side, sustainability-linked and green bonds enjoy significant investor interest. For instance, Amgen's $750 million green bond issued in 2022, intended to advance the environmental goals in Amgen's ESG framework, was six times oversubscribed, according to the company [10].
- *Drive business resilience.* A by-product of sustainability efforts is often increased supply chain resilience, which companies are striving for in light of recent supply disruptions due to the pandemic and ongoing international conflicts. For instance, to reduce transport emissions and to start sourcing from firms with a greener energy mix, companies start looking at near sourcing, i.e., shifting to suppliers that are located close to their own manufacturing operations and end markets. Moreover, as was clearly seen during the COVID chip shortages, companies with more reprocessing/recycling capabilities were able to lessen the pain of shortfalls by using repurposed materials and parts.
- *Increase employee attraction and retention.* Having a strong sustainability ambition and narrative is also becoming increasingly important for employee attraction and retention, especially concerning young talent; 40% of millennials consider ESG elements when selecting a job or employer [11].

---

[7] Mölnlycke, Sustainability Report (2021), pg. 64.

[8] Philips, Integrated Annual Report (2022), pg. 51.

[9] BCG analysis and case experience.

- *Protect reputation.* Using sustainable packaging and sustainable product designs shields companies' reputations and provides a positive counterpoint to the notion that healthcare companies are not sufficiently sensitive to the needs of the environment.

2 Moving to Action

It is one thing to understand the contours of a problem—clearly, healthcare, like most industries, can do more to reduce its impact on the environment and pressure to act is increasing. Perhaps that is the easy part. The more difficult thing is to move into real action. Here we will break down each of the crucial categories of environmental impact areas for healthcare companies and offer practical advice on how to create impact. Across all categories, we have identified four essential steps: (1) measure impact and create a solid baseline, (2) set a target, (3) define the levers to reach this target (including near term goals), and (4) relentlessly track and report on progress.

3 Carbon

As mentioned earlier, healthcare companies contribute about 5% of the total global emissions [12], the equivalent of the fifth highest-emitting country in the world after China, the USA, India, and Russia.[10] To understand key emission drivers, healthcare companies must quantify where the lion's share of their emissions are coming from. That means healthcare companies first need a clean baseline, an accurate and carefully drawn overview of emission levels and sources.

Let's first look at the industry level: As illustrated in Fig. 1, about 50% of emissions come from products and supply chains, including ingredients for drugs (so-called APIs, or active pharmaceutical ingredients, and glucose); plastic, metals, and glass for medtech devices; and transport of raw materials and finished goods. Another 35% is the result of patient care, including electricity and heat for hospitals, ambulance rides, and anesthetic gases. And the rest is direct patient emissions (10%) from product use and patient travel and R&D (5%).[11]

This view includes emissions from the entire value chain including suppliers, biopharma and medtech companies, providers, and patients.

On the company level as well, full value chain emissions must be considered. They are usually clustered into three scopes (see Fig. 2), i.e., Scope 1 and 2;

[10] Arup & Health Care Without Harm, Health care climate footprint report (2019).

[11] Arup & Health Care Without Harm/Arup (2019); World Bank & Health Care Without Harm (2017); UK National Health Service & Lancet (2020); The Lancet Countdown on Health and Climate Change (2020, 2018); Environmental Research Letters: International Comparison of Health Care Carbon (2019); Health Affairs: Health Care Pollution And Public Health Damage In The United States: An Update (2020); BCG analysis based on literature review case experience.

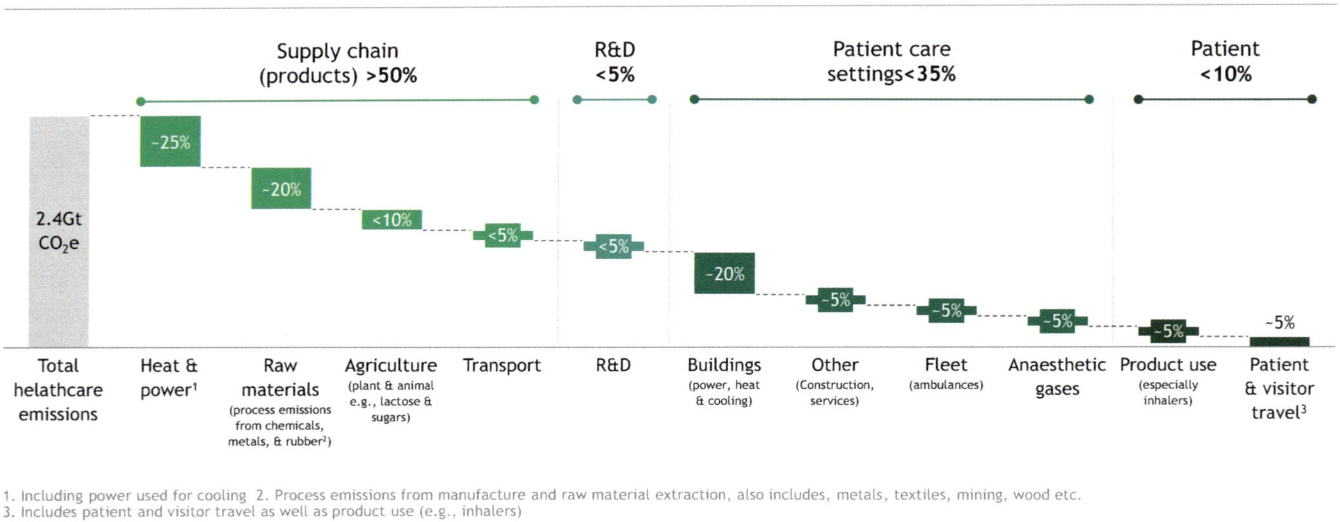

Fig. 1 50% + emissions generated from products and supply chains. Source: Health Care Without Harm/Arup (2019); Health Care Without Harm/World Bank (2017); UK National Health Service/Lancet (2020); The Lancet Countdown on Health and Climate Change (2020, 2018); Environmental Research Letters: International Comparison of Health Care Carbon (2019); Health Affairs: Health Care Pollution And Public Health Damage In The United States: An Update (2020); BCG analysis based on literature review

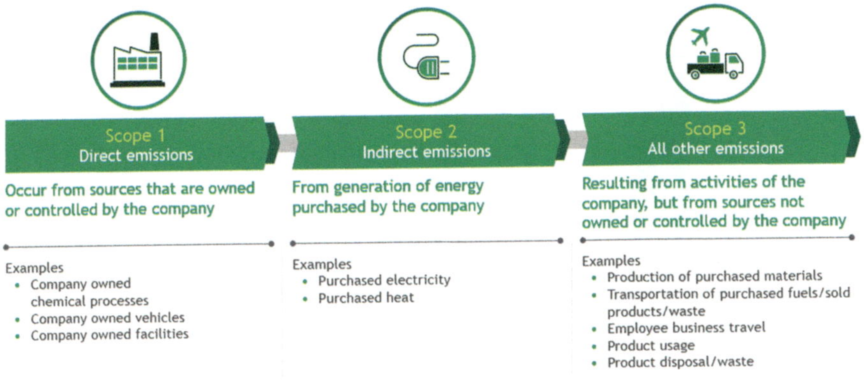

Fig. 2 Emissions are categorized into three scopes. Source: Health Care without Harm/Arup (2019), Health Care without Harm/World Bank (2017); UK National Health Service (2020); The lancet countdown (2020, 2018); Environmental research letters (2019); Health affairs (2020); BCG analysis

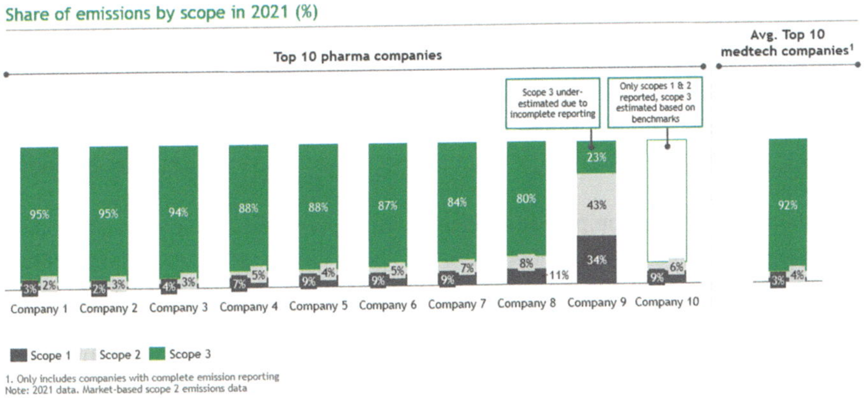

Fig. 3 80% + emissions in scope 3 across pharma and medtech companies. Source: Companies 2021 Sustainability Reports, Refinitiv ESG data, BCG analysis

emissions from the company's own activities and energy use; and Scope 3, emissions that are generated by suppliers (called "upstream") and by customers (called "downstream").

As illustrated in Fig. 3, across biopharma and medtech companies, Scope 3 emissions account for more than 80% of the total company CO_2 output. Consequently, to fully reduce their footprint, healthcare companies will need to work with both suppliers and customers.

The next step is to set a target. To limit global warming to a temperature rise below 1.5 °C, healthcare companies need to set net-zero targets for all three emission scopes. This is currently considered to be the gold standard for corporate climate action. But what is the best path to meet these targets? The Science Based Targets

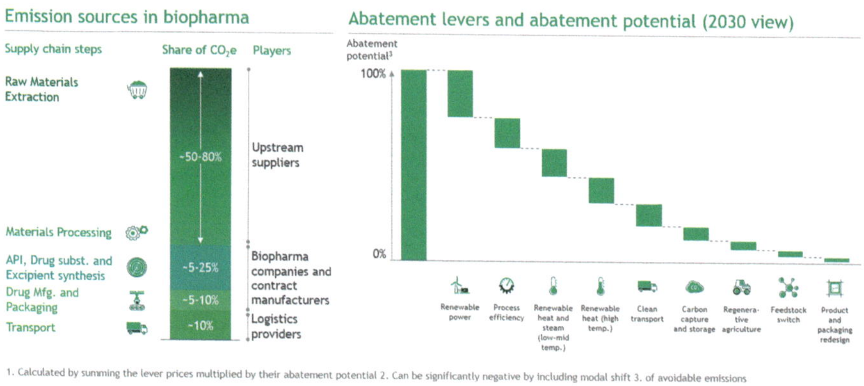

Fig. 4 Most biopharma emissions created by suppliers—renewable power and heat, process efficiency, and clean transport as key abatement levers. Note: Analysis on the left focused on product supply chain (small molecule and biologic drugs only, not medical devises and equipment). Source: BCG analysis based on literature review and case experience; CDP, company sustainability reports, CO2.AI, IEA, IPCC, WEF, ETC, MPP, Agora, academic papers

initiative (SBTi), a partnership between CDP, the United Nations Global Compact, World Resources Institute, and the World Wide Fund for Nature, offers this approach to reach net-zero: (1) prioritize rapid, deep emission reductions by setting near-term targets, halving emissions by 2030; (2) set long-term targets and cut all possible emission by 2050; and (3) neutralize residual emissions, via carbon removal and storage [13].

Among healthcare companies, biopharma firms have been the most ambitious in taking aim at their environmental impact. Nine of the top ten pharmaceutical companies have adopted emission reduction targets across all three scopes, and seven have established net-zero goals across all three scopes. Compared to biopharma companies, medtech is further behind; only 40% of the top players included Scope 3 emission targets and only three of the top 20 medtech companies have set net-zero targets.[12]

With a CO_2 baseline and target in hand, companies can consider which reduction levers are most relevant and may be the most effective for them (see Figs. 4 and 5 for an overview of biopharma and medtech emission sources and reduction levers). Indeed, across pharmaceutical and medtech companies, more than 60% of emission reductions can be achieved by improvements in four areas: sustainable power, sustainable heat, green logistics, and efficiency gains.

Let's start with companies' own emissions. Although these emissions may account for 20% or less of the total, companies can directly impact them and, hence, more quickly make a dent in CO_2 output than with any other aspects of the emissions equation. Levers to reduce company carbon emissions include sourcing of renewable energy and heat, increasing energy efficiency especially in manufacturing

[12] BCG analysis on top 10 biopharma and top 20 medtech latest ESG reports, as of March 2023.

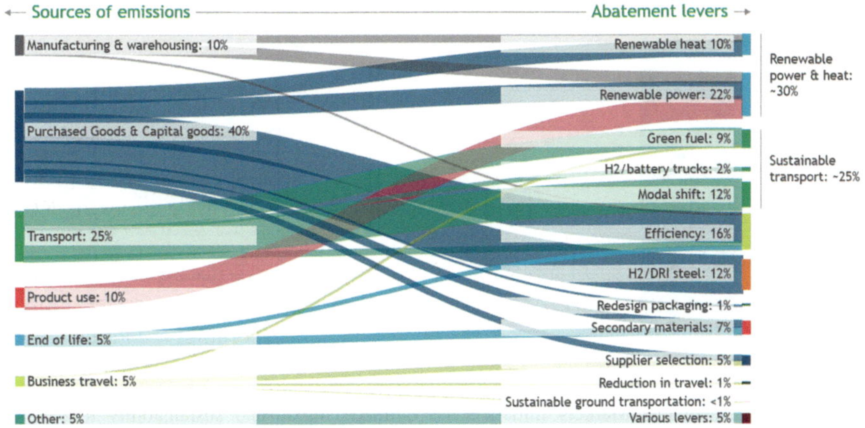

Fig. 5 Medtech with similar emission sources and abatement levers as biopharma. Source: BCG case experience and analysis. Note: Only shows avoidable emissions

processes, switching the vehicle fleet to electric vehicles or other more sustainable fuel sources, and aiming to reduce car usage. Some companies have already made headway on these improvements. For instance, Philips reached 100% renewable electricity in 2020 and 77% renewable energy in 2022.[13] Meanwhile, Astra Zeneca is one of the few companies to report progress on moving to a green fleet. Ten percent of the biopharma company's owned vehicles are EVs while 63% could be called "green" (EV, hybrid, or plug-in hybrid).[14]

But "upstream" supplier emissions from raw materials, APIs and components tend to account for the largest share of emissions. Therefore, it is crucial to work with existing suppliers to decrease the emission intensity of procured goods and services and build in sustainability criteria in supplier selection going forward. To streamline requirements for suppliers as well as to accelerate supply chain decarbonization, a group of biopharma players (Merck KGaA, Astra Zeneca, Roche, Samsung Biologics, GSK, Sanofi, and Novo Nordisk) along with other healthcare companies have joined the Sustainable Markets Initiative, which in part is formulating supplier ESG requirements. Among these supplier targets are rules for assessing and disclosing Scope 1, 2, and 3 emissions by 2025; commitments to set near-term, science-based goals aligned to limiting global warming to 1.5 °C; commitments to switch to at least 80% renewable power by 2030; a clear effort to explore options for shifting to green heat by 2030; and commitments to set carbon abatement goals targets for their own suppliers (see Fig. 6) [14].

On top of those carbon reduction approaches for supplier emissions, additional opportunities (especially for medtech companies) to reduce these include switching to more sustainable materials—for instance, from virgin plastics to biobased

[13] Philips, Integrated Annual Report (2022), pg. 57, 60.

[14] Astra Zeneca, Sustainability Report 2022, pg. 21.

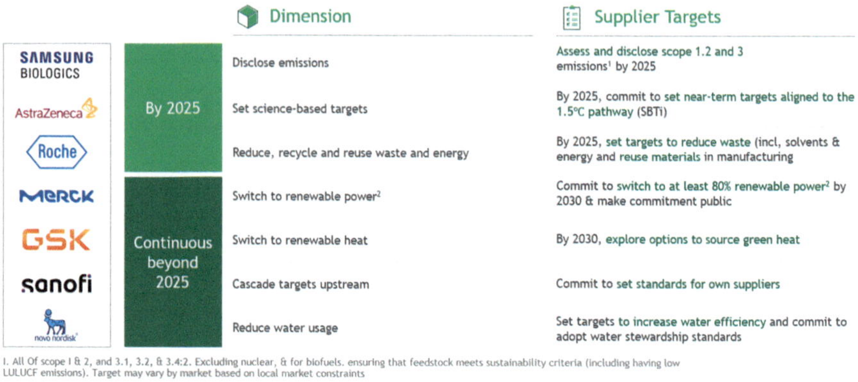

Fig. 6 The Sustainable Markets Initiative has defined climate and sustainability targets for suppliers. Source: Sustainable Markets Initiative. Note: These supplier targets apply only to suppliers of the healthcare business of each private sector member

materials and recycled materials)—product miniaturization and increasing material efficiency. One example is device manufacturer Dexcom's latest glucose monitoring product, the G7. It contains 20% less plastic, and its packaging uses 56% less plastic and paper than the previous version.[15]

Another area where companies can realize sizable emission reductions is logistics. As mentioned earlier, companies can consider switching to more sustainable transportation modes (moving from air to ocean or road to rail) and using green fuels. Or they can increase transport efficiency via improved routing or near sourcing and reductions in product packaging sizes, which may reduce the weight and number of required shipments. Merck KGaA is exploring several of these ideas in pilot programs.[16] And Novo Nordisk has partnered with transport company Kuehne +Nagel to explore the development of sustainable aviation fuels, which are not yet available [15].

4 Product-Related Waste

Healthcare generates a sizable amount of highly visible waste, primarily through single-use products, packaging, and product expiries. This section examines reducing product-related waste through circular product design, an increasingly popular concept in which products or its components are kept in circulation for as long as possible.

Besides minimizing product-related waste, circularity can help companies reduce emissions and save costs. In a teardown analysis of three products, we found that

[15] Dexcom, Sustainability Report (2022), pg.47.

[16] Merck KGaA, Sustainability Report (2022), pg. 159.

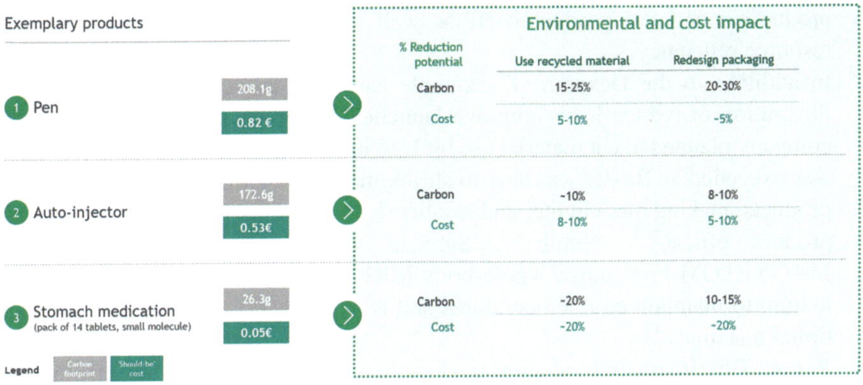

Fig. 7 Teardown analysis shows positive carbon and cost impact of circular design and packaging. Source: BCG Inverto analysis

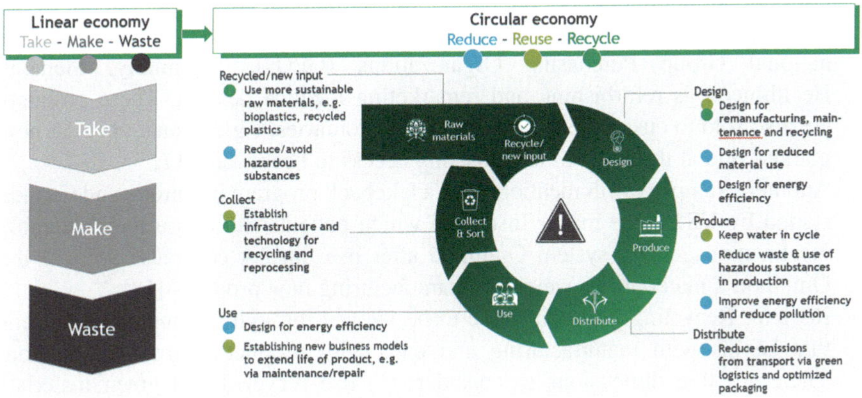

Fig. 8 Replacing the linear "take, make, waste" business model with a circular "reduce, reuse, recycle" approach. Source: BCG case experience and analysis

companies can achieve 25% in CO_2 reduction and 20% in cost savings from use of recycled materials alone (see Fig. 7).

A 2021 study [16] demonstrated that increased recycling and remanufacturing in making electrophysiology catheters can result in CO_2 savings of as much as 50%, largely driven by less new plastic production.

It may represent a dramatic shift for many companies but to curtail waste, while saving carbon and costs, replacing the linear "take, make, waste" business model with a circular "reduce, reuse, recycle" approach can be considered (see Fig. 8, including an overview of levers).

- *Reduce*. This can be achieved by making a product smaller and its walls or components thinner as well as eliminating waste in manufacturing as much as possible. In addition, this entails minimizing the use of hazardous substances in

production and the product itself as well as increasing product energy and resource efficiency.
In addition to the Dexcom G7 example earlier in this chapter, another good illustration of reduce is an initiative launched by Baxter in 2020, in which the company planned to cut material use by 15% in three essential products. That goal was exceeded as Baxter was able to shrink material use through redesigns in six products, making them lighter and smaller (by as much as 47%) without affecting product efficacy.[17] Similarly, Siemens Healthineers has launched the MAGNETOM Free.Star, a whole-body MRI scanner that uses significantly less helium to maintain cool temperatures and is more energy efficient than conventional machines.[18]

- *Reuse*. This involves producing devices that can be returned after their initial one-time use and refurbished by the manufacturer to be used again. US medtech firm Stryker created a new business unit, Stryker's Sustainability Solutions (SSS), to provide reprocessing and remanufacturing services for single-use medical devices. Based on the financial benefits of buying reprocessed devices (hospitals can save up to $one million per year), Stryker has inked contracts with several national Group Purchasing Organizations (GPOs).[19] Similarly, Siemens Healthineers is refurbishing and remarketing medical devices. These products are also sold to customers in less developed countries, e.g., Ghana, opening new markets and at the same time increasing access to healthcare [17].
  Another example worth mentioning is a takeback program in Europe and Canada started by US device maker Insulet in which patients can return their wearable insulin management system Omnipod after use. Insulet can reuse 95% of the Omnipod's materials by weight in manufacturing new products [18].[20]
- *Recycle*. Recycling programs have to be viewed through a comprehensive lens that begins with manufacturing and ends with product return or collection. There are five dimensions to consider: (1) use recycled and green materials from the outset; (2) consider modular design approaches to facilitate material decomposition and separation of hazardous and non-hazardous waste; (3) limit the material mix—it tends to be easier to recycle products made of fewer material types; (4) draw up the most efficient way to take back or collect products; and (5) recycle material waste from production.
  Several recycling pilots are already underway in the healthcare orbit [19].
  For instance, Johnson & Johnson has setup a pilot program with the Asklepios clinic chain in Germany to recycle surgical instruments, aiming to recycle 80% of materials after sterilization [20]. Other interesting recycling projects in healthcare include Baxter's recycling program for anesthetic gases; and GE Healthcare's recycling program for iodine-based contrast media used in computer tomography

[17] Baxter, Sustainability Report (2020), pg. 17.
[18] Siemens Healthineers, Sustainability Report (2022), pg. 36.
[19] Stryker, Sustainability Report (2022), pg. 51.
[20] Insulet, Sustainability_Report (2021), pg. 46/47.

(CT scans), available in eleven European countries as well as Canada and the USA [21]. And Sartorius, a German bioprocessing player, has setup a partnership with its supplier Südpark to recycling manufacturing waste of its ultrapure barrier films.[21]

Reducing materials volume and product weight and size can be explored for most of healthcare devices and equipment. But decisions about *reuse* and *recycling* depend on a product's economic value and whether the product's level of contact with patients is so high that protecting against infecting the next patients using the equipment is an essential consideration.

More specifically, reuse (refurbishing and remanufacturing as well as repair and maintenance) tends to be more viable for high value equipment with low patient contact levels, such as imaging devices and patient monitors.

For products that are high value but have significant patient contact levels—such as endoscopes—sterilization and reprocessing is preferred. And for lower value products—gloves, oxygen masks, and surgical instruments—recycling is the best option.

5 Packaging

Another substantial driver of waste in healthcare is packaging of biopharma and medtech products. However, companies can relatively easily reduce the amount and size of packaging to limit waste and waste-related emissions and realize cost savings as illustrated in Fig. 7—that is, when local regulations, usually involving patient safety, do not hinder these moves.

Medtronic's packaging reduction initiative is an apt example of how much of a difference this strategy can make. By simply redesigning the packaging layout for catheter trays and lids, the device maker was able to cut back materials use and improve use of space for more efficient transportation and storage. The reduction of unit sizes slashed annual truck shipment volumes in half and eliminated 155,000 pounds of waste [23]. Boston Scientific has also been active in this regard, eliminating paper instructions for products sold in countries where downloadable directions are allowed by regulators.[22] Zoetis, an animal health company, switched to temperature-controlled trucks, which allows more than 90% of shipments to be made without styrofoam coolers.[23]

Merck KGaA realized a reduction in packaging and transport emissions for its ZooMAb® antibodies through product innovation. Most traditional antibodies need to be shipped at temperatures between 2 °C and 8 °C, using specific insulated shipping containers with wet ice bricks. This results in lots of packaging waste

[21] Sartorius, Sustainability White Paper (2023), pg. 17.

[22] Boston Scientific, Sustainability Report (2022), pg. 53.

[23] Zoetis, Sustainability Report (2021), pg. 45.

and increased transport emissions. To overcome this, Merck redeveloped its ZooMAb® antibodies as a freeze-dried product, allowing them to be shipped at ambient temperatures and eliminating the need for expanded polystyrene (EPS) coolers and ice bricks.[24]

But perhaps the most common tactic explored by healthcare companies is shifting to recycled or more sustainable materials. In Astra Zeneca's case, 97.5% of its paper-based product packaging is slated to be comprised of sustainable materials.[25] Novartis aims to eliminate polyvinyl chloride (PVC) in its packaging by 2025 and to be plastic neutral by 2030, which means that the weight of Novartis plastic packaging entering the environment for disposal will be approximately the same as the weight of plastic being recovered for recycling. Novartis says that in 2022, nine out of ten sites in this program's scope have already disposed of PVC in packaging.[26]

Some healthcare players like Zoetis, CVS, and Walgreens have joined the sustainable packaging coalition, a group of companies from multiple industries teaming up to design more environmentally friendly packaging ideas. Zoetis backed up its participation in this group by adopting new policies to ensure the packaging of every new and existing product is evaluated for its environmental impact.

6 Biodiversity and Bioderivatives

Although it often goes unrecognized, healthcare's impact on the environment extends to the potential to adversely affect biodiversity. This can occur through the use of natural and animal-based raw materials for everything from diagnostic tests to drugs, and by the release of harmful substances, mainly via patient use and from plant wastewater, that can disrupt animal and plant health.

To mitigate the potentially negative impact on ecosystems, healthcare companies can consider sourcing natural ingredients and bioderivatives sustainably, find safe chemical alternatives, and adhere to practical higher standards for improving ecopharmacovigilance (reduced adverse effects of pharmaceutical products on the environment). At the same time, implementation of effective water stewardship practices (for their own operations and compel suppliers to do the same) are key to limit the potentially dangerous ingredients that company operations are discharging into water supplies.

In 2022, the UN Convention on Biological Diversity set a goal for large businesses and financial institutions to disclose their impact and dependency on biodiversity and ecosystems by 2030. Some biopharma companies have joined the pledge for mandatory disclosures [24].

[24] Merck KGaA, Sustainability Report (2022), pg. 96.

[25] Astra Zeneca, Sustainability Report (2022), pg. 7.

[26] *https://www.gsk.com/en-gb/responsibility/environment/water/* accessed April 30, 2023.

GSK is one of the first companies to create a pilot program using the Science Based Targets Network for Nature (SBTN) methodology to better understand its impact on nature. SBTN is a consortium of global experts, business associations, and consultants formed to design blueprints that companies can follow to assess their effect on biodiversity and limit their harm to the planet. GSK has used SBTN's methodology to create a baseline of biodiversity calculations for 80% of its sites and its wider value chain. Using this baseline, GSK has also already promulgated two main biodiversity targets: (1) achieve a positive impact on biodiversity at all sites by 2030, and (2) ensure that 100% of agricultural, forestry, and marine-derived materials are sustainably sourced and deforestation free by 2030.[27]

To meet these goals, GSK is investing in site action plans for improving habitats, protecting species, and enhancing soil and water quality. The company also created a new Sustainable Sourcing Standard encompassing environmental, social, and ethical requirements which must be met by 2030 in each supply chain. As a first stage, GSK is directing this initiative toward twelve high-risk materials and has developed a roadmap for reducing their use (e.g., by packaging redesign, portfolio consolidation, and other types of efficiencies) or avoiding them altogether (e.g., by transitioning to synthetic alternatives where appropriate). However, the company concedes that especially moving to synthetic alternatives may take some time since safety and regulatory hurdles could stand in the way.

Other healthcare businesses are also making strides in this direction. For instance, Astra Zeneca is working on identifying tools to address the environmental risks of pharmaceuticals and is collaborating with experts to adopt leading forest restoration techniques in communities where the company is affecting tree growth and survival. Astra Zeneca has promised that agricultural, forestry, and marine-derived materials used in its products and research activities will be sustainably sourced by 2028.[28]

Focusing on the issue of wastewater, Merck KGaA plans to reduce potentially harmful residues to below the no-effect threshold by 2030. To this end, Merck has completed the first step—identifying the relevant wastewater sites. Next, Merck will conduct a risk assessment to determine toxic substances in the water and the degree of deviation from the no-effect threshold. Following that, the company will implement improvement actions.[29] Astra Zeneca has adopted a goal to improve water quality by reducing pollution, eliminating dumping, and minimizing release of hazardous chemicals and materials by 2030. In 2022, 100% of Active Pharmaceutical Ingredients (API) discharges from Astra Zeneca sites and 92% of discharges from direct suppliers followed this target.[30]

[27] GSK, Sustainability Report (2022), pg. 18/19.

[28] Astra Zeneca, Biodiversity Position Statement (2023), pg. 4.

[29] Merck KGaA, Sustainability Report (2022), pg. 163.

[30] Astra Zeneca Sustainability Report (2022), pg. 7.

7 Water Consumption

As a large consumer of water resources, the biopharma industry, in particular among healthcare companies, must take water shortage issues seriously. The first step, as is also true with any environmental impact, is to measure the effect on the environment and then design strategies that directly diminish its extent. Indeed, reducing water usage is just one interim answer; moving to water usage neutrality—no drawdown of resources through recycling, especially in water stressed regions—is essential.

Among the most promising water saving efforts underway at biopharma companies, Novartis is monitoring water usage at all its facilities with the aim of reducing water consumption in operations by 50% in 2025 compared to 2016 and becoming water neutral by 2030. GSK is committed to a 20% drop in overall water use by 2030 and to become water neutral in both its own operations and at key suppliers in water stressed areas.[31] Merck KGaA is assessing water-related risks across product lifecycles, in part by reviewing site-specific water management practices and creating a water intensity score at each of its operating locations.[32]

8 Conclusion

There is no doubt that healthcare companies, like companies across all other industries, have their work cut out for them in fully understanding their impact on the environment, defining improvement targets, and implementing initiatives to meet these targets. But the benefits of taking these difficult steps far outweigh the difficulties: a more positive customer perception, new revenue streams, lower operating costs, higher employee retention, improved supply chain resilience, staying ahead of regulations—and, of course, saving the planet from climate change.

Given the stakes, healthcare companies really have no choice. Nor do we as individuals in our private and professional lives. Each of us can make a difference. Students may focus on sustainability in a research project; scientists may design more sustainable, efficient, and clean chemical processes or more sustainable materials and products; a business leader can drive sustainability by establishing a green agenda for sourcing, manufacturing processes, logistics, product circularity, and by integrating the topic in day-to-day decision-making; and investors can fund green technologies and sustainable companies, products, and solutions.

Ultimately, healthcare companies are not alone in the need to focus on protecting the environment. We all have our job to do. As Ernest Hemingway once said, "The Earth is a fine place and worth fighting for."

[31] *https://www.gsk.com/en-gb/responsibility/environment/water/* accessed April 30, 2023.

[32] Merck KGaA Sustainability Report (2022), pg. 163.

References

1. Koijck M. This is the waste of one operation, my operation ...... https://www.youtube.com/watch?v=w28M7wjIr6E. Accessed 30 Apr 2023.
2. World Health Organization, Health-care waste. 2018. https://www.who.int/news-room/fact-sheets/detail/health-care-waste. Accessed 30 Apr 2023.
3. Practice Greenhealth, Waste. https://practicegreenhealth.org/topics/waste/waste-0. Accessed 30 Apr 2023.
4. Johnson & Johnsohn. Our Credo. https://www.jnj.com/credo/#:~:text=We%20believe%20our%20first%20responsibility,costs%20and%20maintain%20reasonable%20prices. Accessed 30 Apr 2023.
5. Merck KGaA. Annual report. 2022. https://www.merckgroup.com/en/annualreport/2022/to-our-shareholders/letter-from-belen-garijo.html. Accessed 30 Apr 2023.
6. NHS Procurement Policy Note. 2020. https://assets.publishing.service.gov.uk/government/uploads/system/uploads/attachment_data/file/921437/PPN-06_20-Taking-Account-of-Social-Value-in-the-Award-of-Central-Government-Contracts.pdf. Accessed 30 Apr 2023.
7. NHS. Suppliers. https://www.england.nhs.uk/greenernhs/get-involved/suppliers/. Accessed 30Apr 2023.
8. Sykehusinnkjop. New joint Nordic tendering procedures. 2021. https://www.sykehusinnkjop.no/en/nyheter/nyheter-2021/new-joint-nordic-tendering-procedures-published/. Accessed 30 Apr 2023.
9. Kaiser Permanente, Sustainable Procurement. https://supplier.kp.org/impactspending/what-we-do/sustainable-procurement/. Accessed 30 Apr 2023.
10. Bloomberg Law, Amgen Mulls more green bonds after 'robust interest' in debut. 2022. https://news.bloomberglaw.com/antitrust/amgen-mulls-more-green-bonds-after-robust-interest-in-debut. Accessed 30 Apr 2023.
11. Fast Company. Most millennials would take a pay cut to work at an environmentally responsible company. 2019. https://www.fastcompany.com/90306556/most-millennials-would-take-a-pay-cut-to-work-at-a-sustainable-company. Accessed 30 Apr 2023.
12. N. Watts, M. Amann, P. N. Arnell et al., The 2020 report of the Lancet Countdown on health and climate change: responding to converging crises (2020), Lancet, vol. 397, pg. 129–170.
13. Science based targets, The Corporare Net-Zero Standard. https://sciencebasedtargets.org/net-zero. Accessed 30 Apr 2023.
14. https://a.storyblok.com/f/109506/x/1f15c6a6ef/hstf-supplier-targets.pdf. Accessed 30 Apr 2023.
15. https://newsroom.kuehne-nagel.com/kuehnenagel-and-novo-nordisk-expand-global-partnership-into-low-carbon-fuels/#:~:text=Novo%20Nordisk%20chose%20the%20Kuehne,around%2030%2C000%20tonnes%20of%20CO2. Accessed 30 Apr 2023.
16. https://www.mdpi.com/2071-1050/13/2/898. Accessed 30 Apr 2023.
17. https://www.siemens-healthineers.com/perspectives/refurbished-imaging-systems.
18. Insulet. Sustainability report. 2021. https://www.insulet.com/sites/insulet/files/2022-12/Insulet_2021_Sustainability_Report.pdf.
19. https://www.novonordisk.com/sustainable-business/zero-environmental-impact/can-you-recycle-an-insulin-pen.html. Accessed 30 Apr 2023.
20. https://www.jnjmedtech.com/de-DE/news-events/innovation-made-hamburg-ethicon-startet-recycling-projekt-fur-medizinprodukte-am. Accessed 30 Apr 2023.
21. https://www.gehealthcare.com/insights/article/recycling-iodine-how-hospitals-are-keeping-contrast-media-in-the-circular-economy. Accessed 30 Apr 2023; *Bringing recycled anaesthetic gas to Europan hospitals* (*medicaldevice-network.com*) accessed April 30, 2023.

22. G.M. Kane, C.A. Bakker, A.R. Balkenende Towards design strategies for circular medical products, Resour Conserv Recycl, 135, (2018), pg. 38–47.
23. https://www.medtronic.com/ph-en/our-impact/product-stewardship.html. Accessed 30 Apr 2023.
24. Convention on Biological Diversity Press Release. 2022. https://www.cbd.int/article/cop15-cbd-press-release-final-19dec2022. Accessed 30 Apr 2023.

Contribution from Digital Minister Professor Dr. Kristina Sinemus for Curious 2022

Kristina Sinemus

It has certainly been a long time since a new technology has changed our lives, our society, as dramatically as digitalisation has. It has brought not just profound change to all areas of life, but also extremely fast change. Keeping pace with this and extensively harnessing the many opportunities digitalisation harbours poses a challenge in all kinds of situations—because alongside making things far easier in many cases, the digital transformation also entails risks and fears. All of this needs to be addressed. This is why, since 2019, there has been a dedicated portfolio for Digital Strategy and Development in the Hessian Government pooling and coordinating all its digital activities and managing the digital budget totalling more than EUR 1.3 billion. In 2021, as a kind of roadmap for the vision of Hesse in 2030, the Hessian State Government published the strategy "Digital Hesse—Where the future begins"—because we need to tap into the potential harboured by digitalisation for social progress, greater value creation and sustainable development, and ensure that digitalisation serves people and not the other way around. The aim is for citizens to use digital technologies naturally and confidently, for companies to develop and produce these innovations, and for science to harness digital progress for its research.

The basic prerequisite for and foundation of any and all digitalisation is a high-performing and secure digital infrastructure. Hesse is doing well here nationwide thanks to its proactive market-driven network buildout—in fixed line and mobile communications alike. The goal is to provide Fibre-to-the-Building connections throughout Hesse by 2030. As we progress towards this, we will forge ahead with the gigabit connection of homes, schools, hospitals and businesses to make sure now that the necessary bandwidths are in place. Hesse is supporting the fibre buildout in all of these areas with huge subsidies.

Hesse is also making rapid progress in evolving the mobile communications infrastructure, especially when it comes to densifying the LTE and 5G networks. As

K. Sinemus (✉)
Hessische Ministerin für Digitale Strategie und Entwicklung, Wiesbaden, Germany
e-mail: kristina.sinemus@digitales.hessen.de

U. A. Betz (ed.), *Curious Future Insight*,
https://doi.org/10.1007/978-3-031-41781-8_13

part of our digital strategy, we will continue to vigorously support the expansion of 5G, especially in rural areas, in order to make our vision of full 5G coverage in Hesse a reality and to be ready for the 6G network from 2030 onwards. Here, too, stimulating the market also has a special role to play. The "Pact on the Future of Mobile Communications for Hesse" in cooperation with the mobile network operators carries on seamlessly from the successful Mobile Communications Pact of 2018 with a view to further accelerating the expansion. Of course, the State Government is not just promoting the commercial buildout, though, is also deploying targeted funding from its own mobile communications funding programme for mobile masts in areas where the market is failing. Furthermore, the "Digitale Dorflinde" (Digital Village Green) funding programme allows WLAN hotspots to be put in place at central sites in local communities in Hesse, so that many people can access the Internet locally.

Digitalisation is not an end in itself and is more than technology alone. Digitalisation is also a social process that we need to actively design and shape if we want to promote the sustainable and successful digitalisation of our society. That is why strengthening digital literacy and ensuring participation in the opportunities digitalisation offers are also very much at the fore. But what is most important is making sure the benefits for people and society always come first. With the Council for Digital Ethics, we have established a body where experts from science, business and society tackle issues arising from the digital transformation. Examples include the challenges the digital transformation poses for the stability of democracy and the issue of trust in artificial intelligence (AI). The Centre Responsible Digitality (ZEVEDI) addresses ethical and legal issues of digitalisation, adopting an interdisciplinary, dovetailed and dialogue-oriented approach—because the digital transformation can only succeed if there is sufficient trust in individuals and institutions. The use of AI in particular needs to be transparent and trackable in order to build trust in AI in the long term. The vision of the Hessian State Government is therefore to design responsible AI which benefits people. In our digital strategy, we have stated our ambition of making "AI made in Hessen" our trademark. At the beginning of the next decade, the aim is for Hesse to be connected worldwide with innovative and trustworthy AI. The strategic framework for this is the Hessian AI Agenda for the Future, which is a sub-strategy of the Digital Strategy. This inter-ministerial agenda defines five fields of action and three fields of innovation, pooling existing AI measures and presenting new projects that will further bolster the vibrant AI ecosystem in Hesse. The spectrum of actors in the AI ecosystem ranges from universities to companies, start-ups and social initiatives to the public administration.

The Hessian Centre for Artificial Intelligence hessian.AI is becoming an important crystallisation point for "AI made in Hessen" whose start-up phase from 2020 to 2024 is being supported with EUR 38 million in state funding. Thirteen universities of different types are behind hessian.AI, allowing them to combine their strengths. The centre offers cutting-edge research, applied research, transfer to business and society and opportunities for young researchers. To provide scientific institutions and companies—especially start-ups and SMEs—with low-threshold access to AI technology, an AI innovation lab is being established in the scope of hessian.AI which is being funded by the state with a further EUR 10 million and which offers access to high-end AI computing infrastructure. To promote the quality of AI systems and make it verifiable, the Hessian Digital Minister has initiated the AI Quality & Testing Hub together with the VDE (German Association for Electrical Electronic & Information Technologies) which brings relevant research and development, standardisation, testing methods and infrastructures as well as experimental spaces together under one roof. These are just a few examples of how "AI made in Hessen" can be used innovatively and responsibly—and become a globally recognised trademark.

Innovative is also the watchword for the programme "Distr@l Digitalisierung stärken—Transfer leben" (Strengthening digitalisation and transfer in practice). Since the end of 2019, the Hessian State Government has been providing targeted support for digital research and development projects that demonstrate a high degree of innovation. The programme funds feasibility studies, digital product and process innovations, knowledge and technology transfer all the way to providing spin-off funding at universities and growth funding at start-ups. The focus is on application-related projects. Examples include a life assistance system based on artificial intelligence that analyses household consumption of electricity and water, recognises household activities and dangerous situations and automatically calls for help to allow people to continue living as long as possible in their own homes worry-free. Or a communication and tracing app for HIV-infected people that is one of a kind in Germany and is designed to significantly improve the medical care of HIV patients and, in the future, other chronically ill patients too.

These examples showcase how digital innovations and developments create added value for our state. That is why we want to continue to make every effort to achieve the goals we have set ourselves so that everyone can participate in the opportunities digitalisation harbours.

For further information on the "Digital Hesse—Where the future begins" strategy, visit https://digitales.hessen.de/Digitalstrategie.

Still Curious: When Science Meets the Public

Shirley M. Malcom

Abstract

Humans are born curious, and that curiosity fuels our growth and development. For scientists and engineers that curiosity remains a way of life. Thus, curiosity is the bridge that can support interactions between scientists and the many non-scientist publics. But the relationship between science and society requires understanding of and attention to the needs of many different public audiences who support the enterprise as well as commitment and skills in reinforcing those relationships. The American Association for the Advancement of Science (AAAS) has been committed to supporting these interactions since its founding. The challenges of delivering on that commitment change over time with changing circumstances, societal values, and norms; so too has the organization evolved to meet these challenges. The lessons learned are instructive for the larger science community.

1 Science-Society Interactions

What happens "When Science Meets the Public?" I have reflected on this theme in writings and presentations multiple times over the past 30 years. This reflection has been stimulated in part by the organizational arrangement of my position at the American Association for the Advancement of Science (AAAS) whose education, diversity, and public engagement programs I headed for decades. The advisory committee that focused on science and society became a component aspect of the unit for which I had responsibility. In part, the reflection was also related to concerns about the many "marginalized" communities of color such as the one in which I was

S. M. Malcom (✉)
AAAS, Washington, DC, USA
e-mail: smalcom@aaas.org

U. A. Betz (ed.), *Curious Future Insight*,
https://doi.org/10.1007/978-3-031-41781-8_14

born and raised—a segregated community in Alabama that had not been a target for science engagement.

I find some irony in the fact that world events in science finally touched children even in the segregated schools of Birmingham. I came to science because of one of the most visible moments in the history of science and technology—the successful launch of Sputnik in 1957 by the then Soviet Union, and the United States' response to being "scooped" by a social, political, and ideological rival country. News headlines and television programs about the satellite and the "Space Race" and attention to science and mathematics education in schools over the following years led many of us to careers in science and engineering, though we were probably considered unlikely "recruits."

These experiences convinced me that there were opportunities for engagement with communities everywhere, beyond those whose members formed the mainstream of science representation and attention. But, for the most part, these opportunities had not been pursued, even when science reached into all of our lives and communities. There have always been (and still are) many science-interested audiences with whom we might engage. The story of science meeting the public is all too often a story of missed (or neglected) opportunities.

Since many people encounter science through technology or because of changing societal or environmental circumstances, my presentations over the decades have always been different as the times, the needs and the contexts have changed. The strategy for effective engagement with science has been (and continues to be) helping people see the connections to their personal, social, political, and economic lives, meeting different public audiences on their own terms, as recent events have clearly demonstrated.

2 Spotlight on a Pandemic

One of the more recent opportunities for engagement occurred in early 2020. As a novel coronavirus made its way across the globe, the word "pandemic" entered our lexicon. In 2020, both *Merriam-Webster* and *Dictionary.com* declared "pandemic" as the word of the year. *Merriam-Webster* noted that on March 11 when the World Health Organization (WHO) officially declared COVID-19 a "pandemic," searches for the word spiked 115,806% compared with the same day in the previous year. Clearly that term and associated terms (as well as the opportunity to explain the science) were top of mind for public audiences [1]. As the pandemic wore on the challenges of getting the science right and getting the public health messaging right were of paramount importance. The science community achieved the first goal and stumbled badly on the second. But the other issue that was being confronted was the pervasiveness of disinformation such as that conveyed on social media. In the USA, all of these communications challenges were happening within a backdrop of politics and political messaging. The greatest danger emerged as the politics threw the announcement of the development of a vaccine effective with SARS-CoV-2 into a chaotic mix. What and whom to believe?

The earlier concerns I had when first confronting the idea of public engagement with science were all played out: the need to understand that there are many publics who needed appropriate messages; seeing that those publics likely needed different messengers and media strategies as well; developing partnerships with and enlisting and amplifying the voices of trusted messengers as many leading scientists and political leaders didn't know how to talk about the pandemic or vaccines and just added to the confusion. The science and public health messages were also caught in the "swirl" of political and social media messaging. How does one disentangle these as the death toll from COVID-19 continued to rise? And in the USA, how does one deal with health disparities that were laid bare, showing greater impacts on persons from minoritized communities?

The pandemic and the world's response to it underscored how woefully unprepared the science community was when facing a public in need of straight talk from trusted voices. Would that the science community have started "trust-building" efforts sooner. Would that the science community have conveyed not only the facts as they emerged but also information about how science works.

3 What Scientists Get Wrong About Science-Public Interactions

My interest in scientist-public interactions as an object of research dates back to that AAAS committee and the proceedings of a 1991 workshop they had held that bore the title, *When Science Meets the Public* [2]. The volume was largely complete when I was asked to add some thoughts. I had previously shared some thoughts verbally with the committee based on more than a decade of experiences focused on sharing science with persons from marginalized communities, especially communities of color. As I read the manuscript, I saw the need to "unpack" the idea of who those public audiences might include and how they might be engaged.

Too often the challenge of engagement as seen from the lenses of those from the science community emerges from assumptions about different publics that are not borne out by experience or which do not hold up under scrutiny. Concerns about funding levels for science and support for research have often prompted many of these conversations. In a 1997 interview, physicist Neal Lane, then director of the US National Science Foundation (and later science advisor to President Clinton and head of the White House Office of Science and Technology Policy) spoke of the "reservoir of public goodwill" that science enjoyed across the public. Surveys supported by the foundation indicated that over 40 percent of the public indicated strong interest in science and technology, but he continued, ". . . .only one in 10 surveyed believes that he or she is well informed about science and technology, and only one in four has some knowledge of science. And the vast majority of people have no understanding of the scientific process--98 percent of them don't know what research means. To me this gap is very troubling: two thirds laud the value of science, but very few understand the enterprise."

Lane emphasized the importance of genuine dialogue with public audiences as an aspect of scientists' professional responsibility. He noted, "The climate for science has changed forever. While it is necessary to increase public understanding of science and technology, it is equally important for scientists to deepen their understanding of the public" (Interview with Neal F. Lane—Scientific American [3]).

Surveys of public and science audiences conducted by the Pew Research Center and published in 2015 showed incredible agreement on the part of both these groups as to the importance of science in the society and the economy and the need for broad support of science. But there were strong differences of opinion between these populations on a number of science-related issues (e.g., the relationship between climate change and human activity, requirements for vaccines), indicating that they are seeing these issues through very different lenses [4].

Results from Pew Research Center studies published in 2020 continued to show strong support for science, but with some science-related issues, differences based on education, family income and party affiliation emerged. While most people see positive net benefits of science, the degree of agreement differed by race as well as by science knowledge, with Blacks and Hispanics showing lower levels of enthusiasm, along with persons with less science knowledge [5].

Scientists are, for the most part, viewed favorably by different publics, especially when compared with other sectors of society (e.g., political leaders, journalists). But too often these publics are not viewed positively by scientists who approach engagement with many non-scientist audiences from deficit models.

Outside of the school or college classroom or lab, when most people encounter the ideas of science, they are often bound around or embedded in technology. While those inside the science community make a big deal of saying this is not the same as science, most "regular" people don't distinguish between these two. Occasionally people will be put into a position of having to deal with science directly, such as if placed in the circumstances that arise where they might be considering DNA evidence (not just in a trial, but, e.g., when they are looking at genealogy) or when exploring specific treatment options for a health condition. But much of the time science is just part of the background noise of our lives until something happens. There may be a "blip" in attention for stunning pictures of a black hole, return pictures from a space mission, a volcanic eruption, a severe weather event, or the outbreak of a pandemic. But otherwise, no.

A focus of the AAAS committee's discussions had been largely on the communication of science and the places where people encounter it—the media of engagement. Unfortunately, while noting audience differences and needs, we still see a focus on facts and media rather than experiences and dialogue. That focus too often disrespects the audiences and the curiosity and knowledge they can bring. We must reshape the who, how, when, where, and why science meets the public. And on whose terms do these encounters occur.

4 Exploring One's World

Every human's relationship with science begins with incredible promise—curiosity and exploration as a fundamental human trait. That idea is reflected in the title of one of my favorite books, *The Scientist in the Crib* [6]. The book explores cognitive development and focuses on early learning by pointing out the ways in which infants and children explore and learn about themselves and their environments as they engage with "everyday physics," "everyday biology," and "everyday psychology." And if we are really lucky to be born into the "right" family and in the "right" country and at the "right" time, that good early start can be maintained! A FaceTime call from one of my daughters was made on behalf of our three-year-old grandson who wanted to know "how do magnets work." Walking through the house he can explore where magnets "stick" and where they do not. He does not yet want or need a deep scientific explanation of dipoles. So, the content is only part of the story; the other part is the audience, the context, and the opportunity for and experience of exploring [7].

With such promising beginnings—curiosity as a foundation to science engagement—what can be done to encourage and support that? How do we enhance science's interactions and engagement with the larger society and society's interactions and engagement with science? Not only are humans curious; so too are many animals. My own research related to imprinting in birds and showed how, over time after hatching, those birds move beyond the mother to explore and learn the larger environment as they grow and develop.

Gopnik et al. [6] argue in their book that scientists provide an example of a group that maintains that curiosity that human babies display as they grow and develop. Science helps us make sense of the things around us, and scientists are not just seekers but also have opportunities to be guides.

5 AAAS and the Challenge of Public Science Literacy

There is a rich history to the AAAS role in public science literacy, one that threads through the stories of the Association and its journals, ever seeking new and better ways to support connections among science and people.

AAAS was established as a professional society in 1848 around some central ideas: to be a forum of science across disciplines and to support democratic principles by advancing knowledge and connecting science to people. At the time of its founding there was strong interest in science among the educated public. This was driven in part by the westward expansion of the United States. The history of the Association that was published for the organization's sesquicentennial in 1998, *The Establishment of Science in America: 150 Years of the American Association for the Advancement of Science,* [8] records some of this history, including public response to science during those early years. It should be noted that this response was largely that of the educated non-scientist public whose enthusiasm for science led them to attendance at lecture halls and purchase of periodicals, pamphlets, and texts. Science was an aspect of "culture," and it attracted that small but fervent audience. Small colleges that sprang up expanded their curricula to include science, including women's colleges such as Mt. Holyoke and Vassar. As the Annual Meeting of the organization moved each year from town to town, it attracted not only the scientists but also the community elite who enjoyed scientific talks as well as the interaction with scientist attendees [9].

Science and the association were affected by the larger socio-political forces in the communities and in the nation. Just as education in the 1850s and 1860s was not accessible to all, based on economic status, class, race, and gender, neither was access to the formal study of science available to all. AAAS, unlike some other societies, never restricted participation, for example, by women, but women's lack of greater access to science due to restricted educational and career opportunity meant that they were less prominent in the life of the organization for the first 120 years of its history. The significant population of Black Americans who were enslaved in the country did not have access to any education. But stories suggest a strong interest in and use of local science knowledge such as in navigation (e.g., the Underground Railroad used by enslaved people to escape to the North and into Canada) and invention. After emancipation many historically Black colleges were established for the education of Blacks. Though most provided education starting at a very basic level, the curricula eventually moved beyond applied areas such agriculture and mechanics to include study of the basic sciences. Howard University, for example, became the home of basic researchers such as developmental biologist, Ernest Everett Just. While they had received doctorates from white research universities, they were not hired by any of them. And while some free Blacks had access to formal education, even before emancipation, they were unable to pursue careers in science also because of lack of employment opportunities. Edward Bouchet was the first African American to earn a Ph.D. from any American university, completing his doctorate in physics from Yale University in 1876. But his employment prospects were quite different from the White graduates in his class.

So which scientists and which publics were able to engage with an organization like AAAS was based, not only on the willingness and openness of the association, but also on the socio-historical context of the times. As it weathered the period before, during, and after the Civil War, the organization tried to do so in a way that welcomed all, a real challenge of "science diplomacy!" Some socio-political events were just too monumental to be navigated, however. AAAS did not hold Annual Meetings during the Civil War or during WWII. Negative encounters at the 1955 Annual Meeting in Atlanta with the reality of "Jim Crow" laws that enforced racial segregation led to a resolution not to meet in the segregated South—this, at a time when other professional societies continued the practice. Differences in perspectives and values related to interactions with society are reflected in these different choices—whether science is open to all or if societal or regional "norms" are allowed to trump openness in science.

The challenges for engagement during that time were significant:

- How does an organization maintain an open yet focused forum for scientists to exchange information about their research?
- How does it foster communication among disparate scientific disciplines and promote interdisciplinarity?
- How does it embody principles for democracy and popular access while advancing scientific knowledge that seemed increasingly esoteric to many people and where others did not have access or the background to understand its connection to their lives?

6 *Science:* A New Tool to Support Engagement

The journal *Science* was founded in 1880 by John Michels with support from Thomas Edison and later from Alexander Graham Bell. The journal was not formally connected to AAAS until 1900, but clearly this connection was good for both: expanding the journal's reach and enhancing the organization's ability to attract members and fulfill its mission: to advance science and support the exchange of research information across all fields of science. *Science,* editorially independent though published by AAAS, could also serve to amplify the stories of science discovery. Over the years as the journal matured and grew in prestige and circulation, it could also become a source for news of scientific discoveries that could reach non-scientists through the popular press and other media. As it enjoyed the stewardship of capable editors and publishers, it too evolved to address the needs of the time, for example, improvement of peer review processes; developing a strong news component; becoming a digital as well as a print product; expanding its titles to become the *Science* family of journals; adding digital media and other enhancements to support expanded audiences and different preferences for engagement and more.

Science is a "big megaphone," and editors, publishers, and leaders have used this voice over the years to speak to concerns about and needs of the science enterprise, including the need for openness and inclusion in science.

7 Science Meets the Public: Imagining a Perfect World

While we can outline the deficiencies of today's current environment for science interactions with our many publics, it is important as well to articulate and work toward establishing the conditions in which these interactions can thrive.

What might be considered "heaven" in terms of promoting the interaction between science and its many publics?

Such a goal might include conditions where:

- *The curiosity that children are born with is encouraged and strengthened over time.* This means that within the home and larger community, there are opportunities to promote curiosity as a natural part of healthy growth, development, and learning.

- *Science is an everyday part of people's lives (like sports and music).* Science has become a part of the larger social and cultural landscape of people's lives, not a compartmentalized section reserved for experts and elites. In some ways, weather forecasting and reporting have already reached this level of acceptance and integration into people's lives.
- *Children have rich instruction in science throughout their formal schooling and often engage (along with their families) in science activities informally.* Science experiences are available from the earliest point of formal education, including pre-school, and quality education in science includes both school and out of school experiences in community places of science. This also assumes having schools available to all where science is valued and taught well by highly qualified teachers in ways that support learning by all and cultural connection for all.
- *There are many opportunities to engage with science at all stages of life.* Completion of formal education is not the end of opportunities for engagement. While there are books, online videos, articles, museums, science centers, and more, engagement is a two-way street. Science is accessible in ways that you can find it, and it can find you through the everyday activities in which you participate.
- *Scientists are active partners with members of the public and comfortable interacting with them.* There are no barriers between scientists and public audiences. There is comfort and pride to be found in these interactions. Scientists are appreciated and recognized for this work as an aspect of fulfilling their professional and civic responsibilities.
- *Scientists are visible members of communities as citizens and as scientists.* Scientists are seen in their roles as citizens such as when they use their knowledge in support of evidence-based decision-making, serve in advisory roles where their knowledge and perspectives can add value, or participate as elected or appointed officials, office seekers, and office holders.
- *Science is broadly representative—with talent that includes all demographics; not just the domain of the well-off (accessible and egalitarian) or of those who have historically participated.* The science and engineering communities look a lot like the communities around them and actively engage with and build trust with those communities, bringing scientific and technical knowledge to help address the challenges of those communities. Their lived experiences and personal identities inform their work, and they add value to the work and support excellence in the enterprise.
- *Science is well supported as a public investment, and people are fully engaged with and curious about the science, the enterprise, and its impact.* Research and development continue to be supported as a public investment because people see the contributions that science and technology make to them as individuals, as communities, to countries and to the health of the planet. They support the investment because they are strongly vested in the payoff.
- *Scientists are respected in the roles they play in the larger society—from working to understand the way the world works, to addressing problems that emerge.*

While there is currently strong support for scientists and the work they do, these views expand to communities that have been less supportive because of the increasing diversity of the science community, improved engagement with these publics, greater openness and transparency about the science, and willingness to discuss and resolve historical barriers across these communities.
- *People consume scientific knowledge and ideas and understand and value the role of science in their lives.* There is strong demand for science, and members of the public seek to determine the attitudes toward and positions related to support of science by those seeking public office.
- *People use scientific findings and scientific ways of thinking to help them make personal decisions.* There is demand for scientific information around issues such as vaccine uptake; and where questions remain, people actively seek science-based explanations rather than mindlessly embracing misinformation/disinformation found on social media.
- *People can contribute to science (citizen science).* People are actively invited to assist in the work that scientists do, such as in data collection. And because of increased trust and appropriate engagement, people are willing to participate in clinical trials and to become partners in the research enterprise.
- *People are attracted to and interested in studying science and engineering and following scientific and technical progress much as they do sports.* As with those who anxiously followed the pace of testing during clinical trials of COVID vaccines and those who follow release of stunning pictures from the recently launched Space Telescope, people are excited to see science advances and to learn about the processes and people behind them.
- *The public endorses and policymakers seek science informed-solutions to policy challenges.* While acutely aware that science cannot be the only consideration when developing policies, there is increasing expectation that it should inform those policy decisions.

While my list may not mirror that of the reader, I have tried to offer a vision of what I consider important goals to discuss. I urge anyone interested in the challenges of science-society interactions to devote some time thinking about what you might want those interactions to be and to accomplish.

8 Science Meets the Public: Facing the Realities

The biggest problem with my list is that it does *not* reflect real-world findings.

- Science is NOT universally encouraged for and taught to young children, in spite of the role of S&T in our world and the natural curiosity that children display.
- While some children have access to rich instruction in science in their schooling, most do not. Often disparities are related to budgets for education; other times these might be related to the educational and income levels of families. In other cases, they relate to cultural expectations of who can and cannot do science.

Historic barriers still exist for girls and women and for those from countries' minoritized populations.
- Scientists are not necessarily active partners with their communities nor comfortable interacting with them. And science is often poorly communicated, even when it is a matter of life and death. Communication and engagement strategies are not generally included in the education and training that scientists receive; and too often, those who undertake these efforts are marginalized among their peers.
- The demographics of the science communities are not especially reflective of the communities in which they live or the publics that support their work.
- And increasingly there is an erosion of trust in science and scientists overall, but more by marginalized communities than by others.
- Publics are increasingly distanced from science and from those who create it.
- Too often science does not factor into either personal or public policy decision-making.

We have all too many recent examples of these failures of the science-public connection.

These challenges have not just developed. They have been around for a very long time. Perhaps the gravity of the pandemic made them more visible and made it more urgent that we address them.

So given all of this, what do we DO?

9 Advancing Science, Serving Society

An organization such as AAAS can only remain vibrant for 175 years if it continues to consider its relevance and re-set its role within the scientific enterprise. AAAS has undergone such discussions and actions over the years, often in response to challenges emerging in the larger society. For example, the movement for civil rights, women's rights, and social justice led to a major re-set in the early 1970s as the Association established advisory structures and programs in support of diversity, equity, and inclusion which have been prominent in the organization for some 50 years. Also, the need for greater input for science within policymaking structures led to the establishment of the AAAS Science and Technology Policy Fellowship Program, again 50 years ago. *Science* expanded its engagement with media to amplify its reach to more public audiences.

But even with these key programs in place, circumstances within the larger society can create new or renewed urgency for action, such as the disruption caused by the pandemic, the murder of George Floyd, and the stark realization of societal inequities revealed in the population disparities of who contracted COVID-19, who died from COVID-19, and who did and did not have access to quality care and to vaccines. A hard look within the research enterprise revealed differential funding, where Black biomedical scientists were less likely to receive support for their research, especially where this research was focused on understanding disparities in their own communities, leaving dangerous knowledge lacuna. How could we claim research excellence in our work when diverse perspectives and diverse voices were not being included?

This was the backdrop for an AAAS governance modernization process begun over 2 years ago. It represented the first time in over 70 years that the organization had considered its structure and processes to align them to the needs and issues of today, such as the increased role of interdisciplinarity and need for responsiveness and nimbleness in addressing current and future science-related challenges. The mission of AAAS remained the same: advance science, engineering, and innovation throughout the world for the benefit of all. But, in re-visiting our place in the scientific enterprise AAAS had also to consider the strategic goals that would take the organization into the future. It emerged with these:

- *Advance scientific excellence and achievement*
  AAAS recognizes, inspires, and enables a robust research ecosystem that drives discovery and innovation and prepares future scientists and engineers.
- *Foster equity and inclusion for scientific excellence*
  AAAS fosters the diverse, equitable, open, and inclusive scientific enterprise that is essential for scientific excellence.
- *Build trust among scientists and communities*
  AAAS builds trust among scientists and engineers and broader communities and is a valued source of accurate scientific information that is foundational to countering misinformation.
- *Catalyze progress where science meets policy*
  AAAS provides actionable evidence for public policy that serves society and promotes policies that enable quality science.

The organization arrived at these goals through broad consultation and discussions that incorporated diverse member voices, guided by diversity in board and staff leadership. It is critical to note the alignment of these strategic goals with the concerns of and need to repair structural weaknesses in current science-society interactions [10].

10 Programming to Support Strategic Goals

While it is important to have goals, it is equally important to align goals and actions, to measure impact and effectiveness, and to hold ourselves accountable for meeting the goals. A few recent and historical examples are relevant here:

- In response to the pandemic, *Science* committed itself to publishing the best research and to making that research freely available to all as quickly as possible. Review and editorial processes were accelerated to achieve that.
- The news presented through *Science* was amplified to help counter misinformation.

- SciLine identified a diverse group of scientists who made themselves available to speak with reporters who were not specifically trained in or responsible for covering science.
- Established in 1975 the AAAS Mass Media Science and Engineering Fellows has offered 10-week summer media internships to some 800 advanced science, engineering, and medical students and postdoctoral scholars. While not designed to promote career transition to science journalism, the program has played that role for many and enriched the pool of skilled science and technology communicators among journalists and scientists.
- Awards have been effective tools to give more visibility to policymakers to the importance and impact of basic research (Golden Goose Award championed by Representative Jim Cooper); they also recognize the work and value to the science enterprise of active scientists and engineers who carry out excellent work in engaging with public audiences (AAAS Mani L. Bhaumik Award for Public Engagement with Science and AAAS Early Career Award for Public Engagement with Science).

Over the years, AAAS programs have reached out specifically and served to engage Members of Congress, agencies and departments of government (AAAS Science and Technology Policy Fellows); clergy (Dialogue on Science, Ethics and Religion); members of minoritized communities, especially through organizations based in communities (Black Churches, Historically Black Colleges and Universities, community-based organizations); youth-serving groups and much more.

11 "Advance Science, Engineering, and Innovation Throughout the World for the Benefit of All."

In the USA and indeed in many of the countries of the globe, the challenge is not just about doing excellent science; it is also about whether the science that gets done is accessible to all, so that they can both use science knowledge and contribute to science knowledge.

A final story highlights the need for diverse inputs into the science and the need also to engage with diverse communities in respectful ways. The AAAS Early Career Award for Public Engagement with Science was established in 2010, to recognize early-career scientists and engineers who demonstrate excellence in their contribution to public engagement with science activities. In 2022 the awardee was Dr. Kizzmekia Corbett who was honored for her engagement about SARS-CoV-2 vaccinations and for her particular focus on underserved, higher-risk Black communities. She has been described as a model for how scientists, whose research touches upon important and timely social issues, can engage the public in effective and impactful ways. As a postdoctoral researcher at the National Institutes of Health, Dr. Corbett was instrumental in developing the lifesaving Moderna mRNA vaccine against SARS-CoV-2, the infection that leads to COVID-19. She also played a central role in the effort to address vaccine inquisitiveness in communities of color.

In a nationally televised Town Hall meeting during the pandemic, Kizzy Corbett engaged with a young African American man who had not yet been vaccinated. Her interactions with him were respectful and non-judgmental. She re-defined the descriptive terminology that was being used for persons like him—from "vaccine hesitant" to "vaccine questioning." And she was prepared to engage with him until all of his questions were answered. It was later revealed that she met him and his mother at the pharmacy where they received their vaccines. It is likely that this was not all about questions asked and answered, but also about building trust, where scientists, who are members of affected populations, are very visible, are willing to engage on the terms of the audiences, and open to being transparent. These are scientists who are open to accepting and responding to people who are *curious*.

United in our shared curiosity about ourselves and our world, there is an opportunity to support positive interactions between science and society.

References

1. https://www.cnn.com/2020/11/30/us/word-of-the-year-pandemic-dictionary-trnd/index.html
2. Lewenstein B, editor. When science meets the public. American Association for the Advancement of Science; 1992.
3. https://www.scientificamerican.com/article/interview-with-neal-f-lan/
4. https://www.pewresearch.org/science/2015/01/29/public-and-scientists-views-on-science-and-society/
5. https://www.pewresearch.org/fact-tank/2020/02/12/key-findings-about-americans-confidence-in-science-and-their-views-on-scientists-role-in-society/
6. Gopnik A, Meltzoff AN, Kuhl PK. The scientist in the crib: minds, brains, and how children learn. William Morrow & Co; 1999.
7. http://www.project2061.org/publications/earlychild/online/perspect/malcom.htm
8. Kohlstedt SG, Sokal MM, Lewenstein BV. The establishment of science in America: 150 years of the American Association for the Advancement of Science. Rutgers University Press; 1999.
9. https://www.aaas.org/events/annual-meeting/national-meetings-date
10. https://www.aaas.org/govmod?adobe_mc=MCMID%3D40040005947295379212985481204078443008%7CMCORGID%3D242B6472541199F70A4C98A6%2540AdobeOrg%7CTS%3D1672422323

The World in 2050 and Beyond

Martin Rees

Scientists are rotten forecasters—not much better than economists. But there are some trends we can predict with fair confidence even with a cloudy crystal ball. I'll first focus on two of these.

1 Two Predictable Global Trends

First the world in 2050 will be more crowded. Fifty years ago, world population was below 4 billion. It's now about 8 billion. The growth's been mainly in Asia and Africa. But the number of births per year, worldwide, peaked a few years ago and is going down in most countries. Nonetheless, world population is forecast to rise to around 9 billion by 2050. That's mainly because most people in the developing world are young and they will live longer. The age histogram in the developing world will become more like it is in Europe.

Despite doom-laden forecasts in the 1960s, food production has kept pace with rising population; famines still occur, but they're due to conflict or maldistribution, not overall scarcity.

To feed 9 billion in 2050 will require further-improved agriculture—low-till, water-conserving, and GM crops—and maybe dietary innovations: converting insects and maggots, highly nutritious and rich in proteins, into palatable food and making artificial meat. To quote Gandhi—enough for everyone's need but not for everyone's greed.

Population projections beyond 2050 are uncertain. Falling infant mortality, urbanisation and women's education trigger the demographic transition towards lower birthrates—but there could be countervailing cultural influences. If, for whatever reason, families in Africa remain large, then according to the UN that

M. Rees (✉)
Trinity College, Cambridge, UK
e-mail: mjr36@cam.ac.uk

U. A. Betz (ed.), *Curious Future Insight*,
https://doi.org/10.1007/978-3-031-41781-8_15

continent's population could double again by 2100, to 4 billion, thereby raising the global population to 11 billion. Nigeria alone could by then have as big a population as Western Europe and North America combined.

Optimists say that each extra mouth brings two hands and a brain. But it's the geopolitical stresses that are most worrying. As compared to the fatalism of earlier generations, those in poor countries now know, via the Internet, etc., what they're missing. And migration is easier. Wealthy nations, especially those in Europe, should urgently instigate a mega-Marshall plan for Africa, and not just for altruistic reasons.

So the world's getting more crowded. And there's a second firm prediction: it will get warmer. In contrast to population issues, climate change is certainly not under-discussed, though it is under-responded-to. The message of the 6th IPPC is stark: under 'business as usual' scenarios we can't rule out, later in the century, really catastrophic warming, and tipping pints triggering long-term trends like the melting of Greenland's icecap. Politicians focus on immediate threats like Covid-19. But they won't prioritise the global and long-term measures needed to deal with climate change and biodiversity because their worst impact stretches beyond the time-horizon of political and investment decisions—and far beyond their own territory: nations far away from ours that will suffer most from climate change. We're like the proverbial boiling frog—contented in a warming tank until it's too late to save itself.

Moreover, if humanity's collective impact on land use and climate pushes too hard, the resultant 'ecological shock' could irreversibly impoverish our biosphere. Extinction rates are rising—we're destroying the book of life before we've read it. Already, there's more biomass in chickens and turkeys than in all the world's wild birds. And the biomass in humans, cows and domestic animals is 20 times than in wild mammals. Biodiversity is crucial to human wellbeing. But the richness of our biosphere has value in its own right. To quote the great ecologist E O Wilson 'mass extinction is the sin that future generations will least forgive us for'.

Unsurprisingly, it's the young—who expect to live to the end of the century—whose clamour for action is loudest, and surely welcome. Their leverage on voters and the media is amplified by charismatic individuals—especially the disparate quartet of Pope Francis, David Attenborough, Bill Gates, and Greta Thunberg.

But—to insert a bit of good cheer—there's a 'win-win' roadmap. Nations should accelerate R and D into all forms of low-carbon energy generation, and into other technologies where parallel progress is crucial—especially storage (batteries, compressed air, pumped storage, hydrogen, etc.) and smart transcontinental grids.

This should ease Europe and North America's path to sustainability and 'net zero'. But there's something even more important. The faster these 'clean' technologies advance, the sooner will their prices fall, so they become affordable to poor nations in the 'global south'. These nations can't reach acceptable living standards without generating *more* power than they do today. Not only will their per capita energy needs rise, unlike ours—but they will collectively harbour a billion more people by 2050. 'Bending the trajectory' of CO2 emissions from these countries is crucial: they must be enabled to leapfrog more speedily to clean energy rather than building coal-fired power stations—just as they leapfrogged to mobile phones without having landlines first.

> It would be hard to think of a more inspiring challenge for young engineers than devising clean and economical energy systems that can achieve 'net zero' for the world.

2 Avoiding the Downsides of Technology

We should be evangelists for new technology, not luddites—without it the world can't provide food and sustainable energy, for an expanding and more demanding population. But many are anxious that it's advancing so fast that we may not properly cope with it—and that we'll have a bumpy ride through this century.

Nuclear war still looms over us. Nuclear weapons are twentieth-century technology; and threats of nuclear war still loom over us. But this century has brought surges in new technologies: bio, cyber and AI. Advances in microbiology—diagnostics, vaccines and antibiotics—offer prospects of containing natural pandemics. But the same research raises (and it's my number-one fear) the prospect of engineered pandemics.

For instance, in 2012 groups in Wisconsin and in Holland showed that it was surprisingly easy to make the influenza virus both more virulent and more transmissible—to some, this was a scary portent of things to come. Such 'gain of function' experiments can be done for coronaviruses too. And the CRISPR Cas 9 technique for gene-editing is hugely promising, but there are ethical concerns—about experiments on human embryos—and worries about possible runaway consequences of 'gene drive' programmes to wipe out species—as diverse as mosquitos or grey squirrels. Regulation of biotech is needed. But I'd worry that whatever regulations are imposed, on prudential or ethical grounds, can't be enforced worldwide—any more than the drug laws can—or the tax laws. Whatever can be done will be done by someone, somewhere.

And that's a nightmare. Whereas an atomic bomb can't be built without large scale special-purpose facilities, Biotech involves small-scale dual-use equipment. The rising empowerment of tech-savvy groups, by biotech—and by cyber tech as well—will pose an intractable challenge to governments and aggravate the tension between freedom, privacy, and security. The global village will have its village idiots, and their idiocies can cascade globally.

These concerns are relatively near-term—within 10 or 15 years. What about 2050 and beyond? On the bio front we might expect two things: a better understanding of the combination of genes that determine key human characteristics and the ability to synthesis genomes that match these features. Serious attempts at 'human enhancement' are then likely to follow. Do we want biohackers to 'play God on a kitchen table' (as it were)?

And what about another transformative technology: robotics and AI? Already, because of their ever-rising processing speed, computers can cope better than humans with data-rich fast-changing networks, traffic flow, or electric grids. The

Chinese could have an efficient planned economy that Marx could only dream of. And it can help science too: with protein folding, drug development, and perhaps even by one day settling whether string theory can really describe our universe.

The implications for society are already ambivalent. In particular, acquiring 'common sense' won't be so easy for AI. It involves observing actual people in real homes or workplaces. A machine would be sensorily deprived by the slowness of real life—for the machines, it would be as boredom-inducing as watching trees grow is for us.

> [Let me here insert a parenthetic comment.
>
> It's always harder to forecast the speed of technological changes than their direction. Sometimes there's a spell of exponential progress – like the spread of IT and smartphones in the last decade. But then an inflection or even stagnation.
>
> Two examples:
>
> From Alcock and Brown's first transatlantic flight in 1919 to the first jumbo jet Boeing's 747, was 50 years. But 50 years later we still have the jumbo jet. (and Concorde came and went!)
>
> It was only 12 years from Sputnik 1, in 1957, to the moon landings, but 50 years later that's still the high point of human spaceflight. Experts are getting less optimistic about how quickly stage 5 fully driverless cars will become acceptable. And the iPhone 24 may not be too different from the iPhone 14 – its technology may 'plateau', to be superceded by a surging advance in 'metaverse'or holography.]

But, mindful of the uncertainties, let's speculate still further ahead. What if a machine developed a mind of its own? Would it stay docile, or 'go rogue'? Futuristic books portray a 'dark side'—where AI gets out of its box, infiltrates the Internet of things, and pursues goals misaligned with human interest. Some AI pundits take this seriously. But others, like Rodney Brooks (inventor of the Baxter robot), think it will be a long time before artificial intelligence will worry us more than real stupidity.

Be that as it may, it's likely that society will be transformed by autonomous robots, even though the jury's out on whether they'll be 'idiot savants' or display superhuman capabilities—and whether we should worry more about breakdowns and bugs, or about being outsmarted.

The visionary futurologist Ray Kurzweil argued in his book *The Age of Spiritual Machines* that humans will transcend biology by merging their brains with computers. In old-style spiritualist parlance, they would 'go over to the other side'.

But Kurzweil is worried that his nirvana may not happen in his lifetime. So he wants to be preserved until it's reached. A company in Arizona will freeze and store your body so that when immortality's on offer you can be resurrected or your brain downloaded.

I was surprised to find that three academics in England had gone in for this 'cryonics'. Two have paid the full whack; the third has taken the cut-price option of wanting just his head frozen. I was glad they were from Oxford, not from my university. I told them I'd rather end my days in an English churchyard than an American refrigerator.

But of course research on ageing is being seriously prioritized. Will the benefits be incremental? Or is aging a 'disease' that can be cured? Dramatic life-extension would plainly be a real wild card in population projections, with huge social ramifications. But it may happen, along with human enhancement in other forms. Some US billionaires have set up the 'Altos' labs—two in California and one in Cambridge (UK) to focus on this. When they were young, they wanted to be rich. Now they're rich, they want to be young again—not so easy!

It's surely on the cards that human beings—their mentality and their physique—may become malleable through the deployment of genetic modification and, if Kurzweil is right, by cyborg technologies. Moreover, this future evolution—a kind of secular 'intelligent design'—would take only centuries, in contrast to the thousands of centuries needed for Darwinian evolution.

This is a game changer. When we admire the literature and artefacts that have survived from antiquity, we feel an affinity, across a time gulf of thousands of years, with those ancient artists and their civilizations. But we can have zero confidence that the dominant intelligences a few centuries hence will have any emotional resonance with us—even though they may have an algorithmic understanding of how we behaved.

Still further ahead, we then confront the classic philosophical problem of personal identity. Could your brain be downloaded into a machine? If so, in what sense would it still be 'you'? Should you be relaxed about you original body than being destroyed? What would happen if several 'clones' were made of 'you'? These are ancient conundrums for philosophers, but practical ethicists may, one distant day, need to address them.

It's beyond our Earth—in environments hostile to humans—that cyborg and AI technologies have the most spectacular scope where these changes will happen fastest and should worry us less. So now I turn to another technology—space.

3 Our Future in Space

During this century the whole solar system will be explored by swarms of miniaturized probes—far more advanced than the wonderful Cassini, designed in the 1990s, which spent 13 years exploring Saturn and its moons, or the New Horizon probe that transmitted amazing pictures from Pluto, 12,000 times further away than the moon.

Think back to the computers and phones of the 1990s, when these probes were designed, and realize how much better we can do today. The next step will be the deployment in space of robotic fabricators, which can build large structures—for instance, giant telescopes or solar energy collectors.

What about human spaceflight? The practical case gets ever-weaker with each advance in robots and miniaturization. So will it have a resurgence?

> It's 50 years since Harrison Schmitt and Eugene Cernan made the final lunar trip on Apollo 17. Hundreds more have ventured into space subsequently—but, anticlimactically, they have done no more than circle the Earth in low orbit—most in the international space station. Their exploits don't seem glamorous. They make news when something goes wrong: when

the toilet fails, for instance; or when they perform 'stunts', such as the Canadian Chris Hadfield playing his guitar.

Will there ever again be any inspirational 'Apollo'-style projects? There's no denying that NASA's Perseverance robot that's now trundling on the Martian surface will miss startling discoveries that no human geologist could overlook. But machine learning is advancing fast, as is sensor technology. In contrast, the cost gap between human and robotic missions remains huge.

NASA's manned programme, ever since Apollo, has been impeded by public and political pressure into being exceedingly risk-averse. The Space Shuttle failed twice in 135 launches. Astronauts or test pilots would willingly accept this 2 percent level of risk; but the Shuttle had, unwisely, been promoted as safe for civilians. Because of this 'safety culture', NASA will confront political obstacles in achieving any grand goal within a feasible budget. Sending humans to Mars, with 6 months of provisions, and supplies for the return journey, is immensely more expensive and hazardous than going to the Moon. The phrase 'space tourism' should be avoided because it lulls the public into believing that these dangerous ventures are low-risk.

I would argue that all human missions should be private-enterprise ventures. Elon Musk's Space X and Jeff Bezos's Blue Origin could operate a cut-price programme far riskier than western nations could impose on publicly supported civilians. There would still be many volunteers—some perhaps even accepting 'one-way tickets'—driven by the same motives as early explorers, mountaineers, and the like.

It's beyond our Earth—in environments hostile to humans—that cyborg and AI technologies have the most spectacular scope, where these changes will happen fastest and should worry us less.

By 2100 courageous thrill-seekers may have established 'bases' independent from the Earth. Musk himself says he wants to die on Mars—but not on impact. But don't ever expect mass emigration from Earth. Nowhere in our solar system offers an environment even as clement as the Antarctic or the top of Everest. Here I disagree with Musk. It's a dangerous delusion to think that space offers an escape from Earth's problems. Dealing with climate change on Earth is a doddle compared to terraforming Mars. There's no 'Planet B' for ordinary risk-averse people.

Nonetheless, if a few exceptionally brave adventurers have established a permanent base on Mars, we should cheer them on, because they will have a pivotal role in spearheading the post-human future in the twenty-second century and far beyond. This is why. They'll be ill-adapted to their new habitat, so they'll have a more compelling incentive than those of us on Earth to re-design themselves. They'll harness the super-powerful genetic and cyborg technologies that will be developed in coming decades. These techniques will, one hopes, be restrained here on Earth, on prudential and ethical grounds; but 'settlers' on Mars will be beyond the clutches of the regulators. We should surely wish them good luck in modifying their progeny to adapt to alien environments.

So it's these spacefaring pioneers, not those of us comfortably adapted to life on Earth, who will spearhead the post-human era that Kurzweil envisages.

4 An Astronomer's Perspective

I'm often asked, do astronomers bring a special perspective to global issues? Let me explain why I think they do.

We're familiar with timecharts depicting that we and our biosphere are the outcome of nearly 4 billion years of Darwinian evolution. But many somehow think we humans are the culmination—the top of the tree. But no astronomer could believe this. The Sun's less than half way through its life. And the cosmos may have an infinite time ahead. To quote Woody Allen "eternity is very long, especially towards the end". So we may be nearer the beginning than the end of the emergence of ever-greater complexity in the cosmos. So what about the more distant future, where all we can do is make conjectures that are exceedingly tentative?

Human brains have changed little since our ancestors roamed the African savanna. It's surely remarkable that these brains have allowed us to make sense of the quantum and the cosmos—far removed from the 'common sense' everyday world in which we evolved. Nonetheless, some key features of reality may be beyond our conceptual grasp. There may be phenomena, crucial to our long-term destiny, that we are not aware of, any more than a monkey comprehends the nature of stars and galaxies.

I think it's likely that the machines will gain dominance. This is because there are chemical and metabolic limits to the size and processing power of 'wet' organic brains. Maybe we're close to these already. So, by any definition of 'thinking', the amount and intensity that's done by organic human-type brains will be utterly swamped by the cerebrations of AI. Moreover, the Earth's biosphere is far from optimal for AI– interplanetary and interstellar space will be the preferred arena where robotic fabricators will have the grandest scope for construction, and where non-biological 'brains' may develop insights far beyond our imaginings.

But we humans shouldn't feel too humbled. Even though we are surely not the terminal branch of an evolutionary tree, we could be of special cosmic significance for jump-starting the transition to silicon-based (and potentially immortal) entities, spreading their influence far beyond the Earth, and far transcending our limitations.

So, to return to the assertion with which I began: even in the context of a 'concertinered' timeline—extending billions of years into the future, as well as into the past—this century is special. Our creative intelligence could jump-start the transitions from an Earth-based to a space-faring species, and from biological to artificial intelligence—transitions that could inaugurate billions of years of post-human evolution even more marvellous than what's led to us. On the other hand, humans could trigger bio, cyber, or environmental catastrophes that foreclose all such potentialities.

So what should be our message to the younger generation who will live to the twenty-second century? It's surely that there's no scientific impediment to achieving a sustainable world, where all enjoy a lifestyle better than those in the 'west' do today. We live under the shadow of new hazards—but these can be minimized by a culture of 'responsible innovation', especially in fields like biotech and advanced AI, and by reprioritizing the thrust of the world's technological effort.

We can indeed be technological optimists. But the intractable politics and sociology engenders pessimism. The scenarios I've mentioned in this talk—environmental degradation, unchecked climate change, and unintended consequences of advanced technology—could trigger serious, even catastrophic, setbacks to our society, and our world is so interconnected that a collapse, societal or ecological, would be a truly global setback.

Scientists have an obligation to promote beneficial applications of their work and warn against the downsides. Universities should offer their staff's expertise, and their convening power, to assess which scary scenarios—ecothreats, or risks from misapplied technology—can be dismissed as science fiction and how best to avoid the serious ones. We in Cambridge have set up a centre to address just these issues.

And most of the challenges are global. Coping with Covid-19 plainly is. And the threats of potential shortages of food, water, and natural resources—and transitioning to low carbon energy—can't be solved by each nation separately. Nor can the regulation of potentially threatening innovations—especially those spearheaded by globe-spanning commercial conglomerates. Indeed a key issue is whether, in a 'new world order', nations need to give up more sovereignty to new organizations along the lines of the IAEA, WHO, etc.

This 'pale blue dot' in the cosmos that is our habitat is a special place. It may be a unique place. And we're its stewards at a specially crucial era. That's an important message for us all. We need to think globally, we need to think rationally, we need to think long-term—we need to be 'good ancestors' empowered by twenty-first-century technology but guided by values that science alone can't provide.

Save Children's Lives: Climate-Responsible Banking Survival Guide

Frederique Seidel

Abstract

According to evidence provided by UNICEF (The impacts of climate change put almost every child at risk, 2021), the impacts of the climate emergency put almost every child at risk. Children and youth around the world are marching in the streets and taking action, urging adults to take adequate steps towards solving the climate emergency.

In the meantime, global CO_2 emissions from fossil fuels hit a record high in 2022 (CarbonBrief, Analysis: global CO2 emissions from fossil fuels hit record high in 2022, 2022). The latest research reveals that 425 “climate bombs” (Garric A and Mouterde P. Global warming: the 425 ‘carbon bombs’ that could thwart the battle against rising temperatures, 2022), spread across 48 countries, could lead to CO_2 emissions that are twice the global carbon budget, thwarting the objectives of the Paris climate agreement and the fight against climate disruption. These carbon bombs are financed by many of the major banks, asset managers, and pension funds.

Children have no influence on the banking choices that allow for these new carbon bombs to put their future at risk. This chapter presents an initiative based on research supported by the Keeling Curve Prize. The goal of this initiative is to scale up an impactful measure which every adult can undertake to accelerate climate solutions and give hope to children: climate-responsible banking. The lever we use the least turns out to be the most powerful tool we have—engaging with our banks to ensure that our own assets are not contributing to global warming.

F. Seidel (✉)
Programme for Child Rights, World Council of Churches, Sergy, France

U. A. Betz (ed.), *Curious Future Insight*,
https://doi.org/10.1007/978-3-031-41781-8_16

1 Introduction

When I was a teenager in Germany 35 years ago, we distributed sun cream with a protection factor of 100 and bananas "grown in Germany" to alert people in an engaging way about global warming. The curve of greenhouse gas emissions has only continued to go up ever since.[1] Today, 75% of children and youth worldwide are scared about global warming, according to a recent report by *The Lancet* [2].

So, being a mother myself today, why do I have hope for children's future? Because I discovered the massive potential of everyone's banking choices for an effective response to the climate emergency.

Every bank account holder can request that their assets not be invested in sectors known to accelerate global warming but instead be shifted to the development of renewable energies.

This powerful lever for climate solutions is one that none of the children suffering from eco-anxiety have any influence on. Most of them do not yet have a bank account and cannot undertake stakeholder engagement to defend their right to a future. Children worldwide try to do what they can at the expense of sacrificing school and their right to play.

But we adults can influence the investment choices of our banks. As Valerie Rockefeller summarizes it, "The lever we use the least turns out to be the most powerful tool we have—where and how we choose to bank."

2 Climate Destruction: A Crime Against Children

Children are the ones who are most impacted by the climate emergency [3]. They are the least responsible for it and have the least influence on decisions to halt global warming. The facts shared by leading scientists in April 2022 are crystal clear: the Inter-Governmental Panel on Climate Change documents that we must make choices permitting CO2 emissions to peak by 2025, and be halved by 2030, to ensure the world remains inhabitable for future generations [4]. Already today, the education of around 38 million children is disrupted each year by the climate crisis [5]. Nearly 90% of the burden of disease attributable to climate change is borne by children under the age of five [6].

In 2015, the World Council of Churches (WCC)[2] asked me to develop a child rights programme, responding to churches worldwide that witness the suffering of children across the globe. In partnership with UNICEF, we developed the Churches' Commitments to Children programme [7], anchored in consultations with 245 experts from churches, child rights organizations, and children themselves. It

[1] CarbonBrief, "Analysis." [1].

[2] The WCC represents over half a billion Christians worldwide through churches in 120 countries. The organization has 70 years of experience uniting Christians, leaders of other religions, and secular partners behind joint solutions for a better tomorrow.

addresses the question, "How can churches best use their influence to improve the lives of children?" The outcome is an action plan with three commitments: child protection, child participation, and climate justice. It was the children themselves who urged us in 2016 to include climate justice as one of the three pillars of this programme. In this commitment, the programme supports churches in 120 countries with tools and know-how to accelerate behavioural change towards reducing greenhouse gas emissions for and with children and young people.

In 2019, when the Keeling Curve Prize was given to the child rights programme, our focus was to scale up churches' support to children's and youth engagement for climate solutions. The award enabled us to expand and accelerate existing good practices through education and child and youth activities worldwide. We made available a toolkit entitled *Get Informed, Get Inspired, Take Action* [8]. It helps to build capacity through success stories of teenagers who, through their Sunday schools and summer camp initiatives, managed to influence policies and legislation for a reduction of CO_2 emissions.

Examples range from community-based behavioural change measures to court cases by young people suing adults for putting their future at risk through climate inaction. In Nigeria [9], churches engage schools across the country to support children's advocacy for climate solutions. In the United States, pastors [10] help teenagers to file a lawsuit against the government for putting their future at risk through lack of climate action. In Australia, a nun [11] supports teenagers in trying to stop a coal mine.

Throughout the implementation of the programme, we observed an alarming level of eco-anxiety expressed by many of the children engaged in climate solutions. The impact of the climate emergency on children's mental health was confirmed in December 2021 by the leading health journal *The Lancet*.[3] It documents that 75% of young people around the world are scared of the future because of the climate emergency, whether they live in countries that are most or least affected by humanitarian disasters resulting from global warming.

You may ask yourself, what is behind this figure? Many of the children and youth have this anxiety because they know that the most vital actions cannot come from children themselves. Young people are discouraged and scared by the absence of progress in halting global warming. Indeed, as documented by the recent update published by the World Meteorological Organization, the United in Science [12] report shows that the world is heading in the wrong direction. Global CO_2 emissions from fossil fuels hit a record high in 2022.[4] Fossil fuel emission rates are now above pre-pandemic levels after a temporary drop due to lockdowns. The report documents how many of the extreme weather events that we are experiencing have become more likely and more intense due to human-induced climate change.

---

[3] Hickman et al., "Climate Anxiety in Children and Young People" [2].

[4] CarbonBrief, "Analysis" [1].

In light of the growing evidence [13] about the physical and psychological impacts of the climate emergency on children, a new appeal was launched in 2022: Climate Destruction is Child Abuse [14].

In parallel, efforts to develop an international Ecocide Law [15] are progressing. The recognition of ecocide as a crime will also increase awareness that climate inaction, and choices which fail to address the climate crisis, are a crime against children [16].

Once appropriate legal frameworks will be in place, those in decision-making roles who have the choice today between investing in renewable energies or investing into new fossil fuel expansion are likely to be held accountable retroactively for their choices, when the children of today will be facing the consequences of inadequate choices and huge swaths of the globe may become uninhabitable [17].

3 Carbon Bombs Fuelled by Our Financial Service Providers

Teenagers engaged in the WCC child rights programme expressed dismay that their efforts and sacrifices responding to the climate emergency are in vain if, at the same time, the banks pour funding into sectors known to accelerate global warming. The hope that many leaders place in the younger generations is experienced by many children as an unrealistic, far too heavy burden on their shoulders. Indeed, the 2023 *Banking on Climate Chaos Report* [18] shows that fossil fuel financing from the world's 60 largest banks has reached USD $5.5 trillion in the 7 years since the adoption of the Paris Agreement, with $742 billion in fossil fuel financing in 2021 alone. It includes a league table showing which banks are contributing most to the climate emergency by financing fossil fuels.

The flexibility of the Keeling Curve Prize allowed us to react quickly through an initiative that ensures tangible hope for children through adults' engagement with their banks. The prize enabled us to develop *Cooler Earth—Higher Benefits* [19]. This research paper showcases champions among WCC members and partners who led by example, urging their banks to increase their ambitions to align with the Paris Agreement. It documents the power of climate-responsible banking to bend the CO_2 emissions curve and explains how every adult can influence their banks and pension funds. If we considered the greenhouse gas emissions generated by the average assets sitting in our bank account, our carbon footprint would be double what we think it is [20]. While climate-responsible choices are key to addressing various forms of violence against children fuelled by global warming,[5] the measures and opportunities presented in the research cannot be undertaken *by* children. The research thus underlines the urgency for all asset owners to demand transparency on how their assets are invested. Ensuring that assets currently invested in fossil fuels are shifted towards renewable energy investments leads to higher benefits for all.

[5] Seidel, *Facts and Findings* [13].

Asking financial service providers for transparency is essential to strategically defuse carbon bombs. A carbon bomb [21] is a fossil fuel extraction project, such as a coal mine, that can cause over a gigatonne of CO_2 emissions during its lifetime. There are 425 of these carbon bombs worldwide. Collectively, they can unleash over 1000 gigatonnes of CO_2 emissions, which far exceeds the world's carbon budget for staying below 1.5 °C of warming (around 500 gigatonnes in 2017 [22])—the world's agreed target for limiting climate change.

Even though it is now recognized, even by the conservative International Energy Agency (IEA), that to avert catastrophic climate change, no new fossil fuel projects [23] must be built, fossil fuel companies are working on setting off dozens of new carbon bombs [24]. The IEA underlines that t*here is no need for investment in NEW fossil fuel for the net zero pathway* [25], despite supply disruptions from Russia's invasion of Ukraine. New projects would take too long to fill the supply gap; the Paris-based agency instead suggests investments to extract oil and gas more quickly from *existing* fields.

A moratorium on new carbon bombs could avoid about a third of potential emissions from carbon bombs.[6] In the current absence of effective legal frameworks prohibiting new carbon bombs, the most powerful alternative and source of hope for children's future comes from asset owners.

Every individual, institution, and business can help to defuse carbon bombs through their banking. The report "The Carbon Bankroll," [27] through its data, shows how one of the levers we use the least turns out to be one of the most powerful tools we possess to address the climate emergency: where and how we choose to bank and invest. The report documents that corporate cash and investments are a major source of emissions and that for many of the world's largest companies, the carbon footprint generated by their investments and cash held in big banks is a significant source, and sometimes their largest source, of emissions. The new resource "Greening Cash Action Guide" (footnote: https://exponentialroadmap.org/greening-cash-action-guide/) is available for every institution, business and individual to reduce their emissions from cash holdings in the bank by engaging with their financial firms to forge systemic climate progress.

In September 2021, the United Nations Committee on the Rights of the Child (CRC) for the first time expressed concern [28] about investments in fossil fuels made by governments' financial institutions and highlighted the negative impact of climate change on children. In Switzerland, the national bank, with its investments into oil, gas, and coal, currently doubles the CO_2 emissions of the entire country. By shifting into renewables, the bank could halve Switzerland's CO_2 emissions [29]. The CRC recommended that the state party of Switzerland ensure that privately and publicly owned financial institutions take into consideration the implications of their investments on climate change and the resulting harmful impacts for children. They asked for evaluations of financial institutions with regard to these investments and recommended adopting binding rules for these institutions. Through these

[6] Garric and Mouterde, "Global Warming" [26].

recommendations, the CRC helped to increase awareness on the impact of financial choices on child rights. But in the absence of effective legal frameworks, the clients of financial institutions play a key role in demanding the implementation of these recommendations.

4 Climate-Responsible Finance: The Most Urgent Child-Protection Measure of Our Times

The recommendations of *Cooler Earth—Higher Benefits* were turned into action by the appeal "Climate-Responsible Finance: A Moral Imperative towards Children", [30] an initiative developed by the WCC's Churches' Commitments to Children Programme with the United Nations Environment Programme [31] and interfaith partners. It started with an appeal in May 2022, urging our financial service providers to align with the Paris Agreement. The United Nations Secretary-General, Antonio Guterres, supported the launch of the initiative, underlining that "It is now time for financial service providers to accelerate the shift to renewables. They have the power—and the responsibility. The scientific and moral imperative is clear: there must be no new investment in fossil fuel expansion".

As a next step of this initiative, all institutions and individuals who care about children are invited to use the guide "Save Children's Lives: Climate-Responsible Banking Survival Guide" (footnote: https://www.oikoumene.org/resources/documents/save-childrens-lives-climate-responsible-banking-survival-guide) and to help raise awareness about it. Many people in decision-making seats are not yet aware how much their assets contribute to global warming. They do not intend to harm children.

Therefore, the initiative aims to increase capacity building for all on climate-responsible finance as a moral imperative towards children (add new footnote: https://www.oikoumene.org/news/if-your-banks-dont-care-about-climate-change-ask-them-why). Resources that are available [32] include a draft letter [33] for broad dissemination to be sent to banks and pension funds. It is available in English, French, German, and Spanish.

5 Conclusion

By engaging with our banks and pension funds, we can help billions of young people worldwide to lift a far too heavy burden from their shoulders. We need to ensure that our assets are removed from investments into drilling for new fossil fuel projects [34] and redirected into sectors accelerating climate solutions, such as renewable energies. It is amazing to see that so many nature-based climate solutions already exist. Most of the green technologies cannot yet be spread broadly due to the lack of investment in innovation research. In Dubai, a school has become a net zero building by converting all energy produced in gym classes into energy. In Canada, the first airplane entirely fuelled by hydrogen has transported passengers.

A sustainable future is possible if all of us use the powerful lever of engaging with our banks and pension funds, ensuring that our assets do not fuel the climate emergency but are invested into climate solutions and research for related innovation.

An effective remedy against the eco-anxiety of children is to reassure them that we, as adults, are doing everything we can to stop global warming by tackling its root causes. The power of climate-responsible banking, which every asset owner can engage in, has remained one of the best-kept secrets for too long. Now it needs to be popularized.

Let us come together and influence how money is invested: family money, an institution's money, a nation's money. We need everyone to take this step for children.

As scientists, your engagement with financial service providers will be particularly impactful. Thanks again to the Keeling Curve Prize for enabling us to launch this initiative.

The children's future is in our hands.

References

1. CarbonBrief. Analysis: global CO2 emissions from fossil fuels hit record high in 2022. 11 Nov 2022. https://www.carbonbrief.org/analysis-global-co2-emissions-from-fossil-fuels-hit-record-high-in-2022
2. Hickman C, Marks E, Pihkala P, Clayton S, Lewandowski RE, et al. Climate anxiety in children and young people and their beliefs about government responses to climate change: a global survey. Lancet Planet Health. 2021;5(12):e863–73. https://www.thelancet.com/journals/lanplh/article/PIIS2542-5196(21)00278-3/fulltext
3. UNICEF. The impacts of climate change put almost every child at risk. 19 Aug 2021. https://www.unicef.org/stories/impacts-climate-change-put-almost-every-child-risk
4. Intergovernmental Panel on Climate Change. The evidence is clear: The time for action is now. We can Halve emissions by 2030. 4 Apr 2022. https://www.ipcc.ch/2022/04/04/ipcc-ar6-wgiii-pressrelease.
5. UNICEF. The climate crisis is a child rights crisis. Aug 2021. https://www.unicef.org/reports/climate-crisis-child-rights-crisis.
6. UNICEF. Fact sheet: The climate crisis is a child rights crisis. 6 Dec 2019. https://www.unicef.org/press-releases/fact-sheet-climate-crisis-child-rights-crisis.
7. World Council of Churches. WCC's engagement for children. www.oikoumene.org/children.
8. World Council of Churches. Get informed, get inspired, take action: climate justice with and for children and youth in churches. 2020. https://www.oikoumene.org/en/resources/publications/climate-justice-with-children-and-youth
9. Lutheran World Federation. Earth day: Nigerian Youth contribute to climate justice. 21 Apr 2022. https://www.lutheranworld.org/news/earth-day-nigerian-youth-contribute-climate-justice
10. Larkman C. UCC clergy, congregants stand with young plaintiffs in climate justice case. United Church of Christ website. 30 Oct 2018. https://www.ucc.org/ucc_clergy_congregants_stand_with_young_plaintiffs_in_climate_justice_case_10302018.
11. McCubbing G. Kids, Nun take ley to court over coal mine. 7 News website. 26 Oct 2022. https://7news.com.au/news/kids-nun-take-ley-to-court-over-coal-mine-c-1302077.
12. World Meteorological Organization. Press release: United in science: we are heading in the wrong direction. 13 Sept 2022. https://public.wmo.int/en/media/press-release/united-science-

we-are-heading-wrong-direction; World Meteorological Organization. United in science: we are heading in the wrong direction. 2022. https://public.wmo.int/en/resources/united_in_science.
13. Seidel F. Facts and findings: climate solutions as a vital child protection measure. Documentation on violence against children fuelled by the climate emergency. 2022. https://docs.google.com/document/d/14PGIgdfFO3nQdYnri6as1gkdiB3JK82AitTFoIxvIQE/edit.
14. Generations Together. Climate destruction is child abuse. 2022. https://climatechangeischi.wixsite.com/generations-together.
15. Faith for Ecocide Law website. https://www.faithforecocidelaw.earth.
16. Bail J. Ecocide and its influence on children's rights. Humanium website. 15 Feb 2022. https://www.humanium.org/en/ecocide-and-its-influence-on-childrens-rights/#_ftn1.
17. Figueres C, Rivett-Carnac T. 'The only uncertainty is how long we'll last': a worst case scenario for the climate in 2050. The Observer. 15 Feb 2020. https://www.theguardian.com/environment/2020/feb/15/worst-case-scenario-2050-climate-crisis-future-we-choose-christiana-figueres-tom-rivett-carnac.
18. Rainforest Action Network. Banking on climate chaos: fossil fuel finance report 2022. 30 Mar 2022. https://www.ran.org/wp-content/uploads/2022/03/BOCC_2022_vSPREAD-1.pdf.
19. Seidel F, de Martel E, Behaghel E. Cooler Earth – higher benefits: actions by those who care about children, climate and finance. 3rd ed. Geneva: World Council of Churches; 2022. https://www.oikoumene.org/publications/cooler-earth-higher-benefits3
20. Oxfam France. Comment les banques françaises aggravent les changements climatiques. 3 May 2022. https://www.oxfamfrance.org/climat-et-energie/impact-banques-climat; Artisans de la transition website. Agir avec nous. 2020. https://artisansdelatransition.org/agir-avec-nous/desinvestir/rapport-bns#anglais.
21. Carrington D, Taylor M. Revealed: the 'carbon bombs' set to trigger catastrophic climate breakdown. The Guardian. 11 May 2022. https://www.theguardian.com/environment/ng-interactive/2022/may/11/fossil-fuel-carbon-bombs-climate-breakdown-oil-gas.
22. Intergovernmental Panel on Climate Change. Special report: global warming of 1.5 °C. Summary for policymakers (IPCC, 2018). https://www.ipcc.ch/sr15/chapter/spm.
23. International Energy Agency. Pathway to critical and formidable goal of net-zero emissions by 2050 is narrow but brings huge benefits, according to IEA special report. 18 May 2021. https://www.iea.org/news/pathway-to-critical-and-formidable-goal-of-net-zero-emissions-by-2050-is-narrow-but-brings-huge-benefits
24. Cabot C. Multinationals open new gas sites despite climate change warnings. France24. 20 Nov 2022. https://www.france24.com/en/france/20221120-fossil-fuel-projects-on-rise-despite-warnings-from-scientists
25. International Energy Agency. World energy outlook 2022 shows the global energy crisis can be a historic turning point towards a cleaner and more secure future. 27 Oct 2022. https://www.iea.org/news/world-energy-outlook-2022-shows-the-global-energy-crisis-can-be-a-historic-turning-point-towards-a-cleaner-and-more-secure-future.
26. Garric A, Mouterde P. Global warming: the 425 'carbon bombs' that could thwart the battle against rising temperatures. 16 May 2022. https://www.lemonde.fr/en/environment/article/2022/05/16/the-425-carbon-bombs-that-could-thwart-the-battle-against-global-warming_5983635_114.html
27. Vaccaro J, Moinester P. The carbon bankroll: the climate impact and untapped power of corporate cash. The Climate Safe Lending Network, The Outdoor Policy Outfit, and BankFWD. 2022. https://www.carbonbankroll.com.
28. UNHCR and UNCRC. Concluding observations on the combined fifth and sixth periodic reports of Switzerland: impact of climate change on the rights of the child. 20 Sept 2021. https://seafile.ecucenter.org/f/0794403395334b509803/?dl=1.
29. Artisans de la Transition. The Swiss national bank's investments in the fossil fuel industry inflicts heavy losses to Switzerland. Artisans de la Transition. Apr 2018. https://artisansdelatransition.org/assets/images/bns/bns-en-2.pdf.

30. World Council of Churches. Climate-responsible finance: a moral imperative towards children. 9 May 2022. https://www.oikoumene.org/climate-responsible-finance.
31. UN Environment Programme. Religious leaders, UNEP appeal for climate-responsible finance as a moral imperative towards children. 9 May 2022. https://www.unep.org/news-and-stories/press-release/religious-leaders-unep-appeal-climate-responsible-finance-moral.
32. World Council of Churches. Resources and tools available for the implementation of churches' commitments to children: commitment 3: raise churches' voice for intergenerational justice, supporting initiatives for and with children and adolescents. 21 June 2017. https://www.oikoumene.org/resources-children#commitment-3.
33. World Council of Churches. Halting the climate crisis: Letter to financial service providers (Template for customisation). Undated. https://docs.google.com/document/d/1jiLrjssdv59LVXO6iyg0P0HwT1BY3n0Y18ZMLxSC20c/edit#heading=h.d5c6fmwi3skk.
34. Carrington D. UK and US banks among biggest backers of Russian 'Carbon Bombs', data shows. The Guardian. 24 Aug 2022. https://www.theguardian.com/environment/2022/aug/24/uk-and-us-banks-among-biggest-backers-of-russian-carbon-bombs-data-shows.

Fridays for Future Breaking Social Systems with Non-violent Civil Disobedience

Janine O'Keeffe

1 Summary of Speeches Including Scientist Rebellion

I am scared for humanity!
Scientists are scared!
Climate Scientists are scared!
NASA Climate Scientists are scared!
UN IPCC (United Nations Intergovernmental Panel on Climate Change) Climate Scientists are scared! [1]
Johan Rockström, Sweden's top Climate Scientist, is scared!

As an engineer who calculates risks, I am scared.
I would never be given approval to build a bridge with a high risk of collapse,
but I live in a society where we are silently approving a very high risk of collapse.
Top scientists speak openly about civilisation collapse in only a few decades [2].
Adult silence has been approving this risk for decades.
We are the adults giving our final approval with silence in this decade; only 7 years left.

For decades we have seen unprecedented disasters.
We are moving towards breakdown.
On top of this, we are seeing the scariest part of climate breakdown—the accelerating, irreversible tipping points—moving towards us.
In 2018, the earliest tipping points were between 1.5 and 2C degrees.
In 2021, only 3 years later, 5 tipping points will occur earlier than the 1.5C degree limit.
Remember, as Scientist Rebellion says, '1.5 degrees is dead.' This terrifies me.
We currently have a good chance of pulling on the brake to reduce this climate breakdown,
but we are silent at the wheel.

J. O'Keeffe (✉)
Stockholm, Sweden

U. A. Betz (ed.), *Curious Future Insight*,
https://doi.org/10.1007/978-3-031-41781-8_17

If that does not scare you, you have not understood the UN IPCC science.

If that does not scare you, you have not understood that we cannot negotiate with the physics of climate science.

My media is virtually silent. This petrifies me.

The fact that my government shows no sign of fear. This petrifies me.

The fact that fossil fuel companies have empirically known our path since 1959. This revolts me.

It is beyond criminal that the fossil fuel companies' predictions have been breathtakingly accurate.

Did you give your approval for this?

Most of us were not even born.

I have been working with Climate Breakdown for 5 years.

Now with Scientist Rebellion, the first time I wore a white lab coat, we blocked a motorway exit in Germany. I could feel the respect from the police.

This was also true when Swedish scientists recently threw fake blood on the Finance Department's entrance.

Often charges for people in white lab coats are quickly reduced or dropped,

whereas without a coat, charges have been increased.

Sweden, with good democracy, where police are often respectful.

Even Sweden is now stretching interpretation of the law and using the courts to squash climate activism.

I have personally been charged with sabotage.

There are around 100 cases of suspected sabotage.

Just under 20 sabotage convictions.

Even a child has been convicted of sabotage for sitting on a road for a short time.

This is ridiculous and out of line with UN recommendations [3].

I implore you to courageously go against the silence.

Communicate about the tipping points every day.

Be it to your family, your boss, your readers and your friends.

You have power.

Find your voice.

Finding courage to use your voice gives me hope. [4, 5]

2 Top Scientists Are Speaking Up!

Everyone needs to gain a deeper understanding of how serious our planetary multilevel collapses are. Not only climate and ecological collapse but also societal behaviour which accelerates collapse. Examples are media silence, political systems focused on short-term voter cycles, political lobbying, wealth extremes which virtually remove democratic relevance and many business processes and tax subsidies which continue to assume the more fossil fuels the better. All of this seems to ride on a belief in the 'fantasies of unlimited growth on a limited planet' (Greta Thunberg 2019).

Professor Dr. Johan Rockström, Swedish Climate Researcher, Director of Potsdam Institute for Climate Impact Research begged:

> I now really want to address you who are politicians, business leaders, or power holders. It's high time for you, as for all of us, to take responsibility!
>
> The climate and the earth's ecosystems are in a crisis that threatens all of humanity.
>
> You must act Now!
>
> You must act with Power!
>
> If you do, you will get a lot of support. Because as a leader, with power and responsibility for the development of society, it is your responsibility to take the lead.
>
> It is you and I, the adult generation on earth today, who have the responsibility to leave behind us a planet in at least as good a condition as the one we were born on.
>
> The next election result or quarterly report is not more important than the future of our children and grandchildren.
>
> You and I must do 'Everything' in our power to stop the negative spiral the climate is in that threatens our entire earth and all life on the planet.
>
> It can not wait! It must happen now! I trust you! Thanks! [6]

> UN Secretary, Antonio Guterres chimes in with,
>
> '... our planet is fast approaching tipping points that will make climate chaos irreversible. We are on a highway to Climate Hell with our foot still on the accelerator!' [7].

Even after many scientists have stepped up to talk about our dire situation, and there is widespread public acknowledgement of our Climate Emergency, there is no widespread sense of immediacy, no COVID-style weekly press conferences or programmes to continuously inform the public.

Climate-knowledgeable people are dumbstruck at the level of institutional silence effectively creating system stupidity. The people have an expectation that they will be informed of emergencies, and for 65 years democratic institutions have been virtually silent. When shall the silent democratic institutions be held accountable? Isn't it their job to ensure that people are aware of our climate and ecological emergencies, i.e. to continuously ring the climate 'fire' alarm bell?

Most older activists are young enough to be children of Hans Joachim Schellnhuber. Dr. Schellnhuber has been the Director Emeritus of Potsdam Institute and one of the world's most eminent scientists: 'I'm telling you that we are putting our kids onto a global school bus that has a 98% chance of being fatal' [8].

At what point are the risks unacceptable and action is required?

Yet our media and politicians are virtually silent.

'Humanity has a choice: cooperate or perish. It is either a Climate Solidarity Pact—or a Collective Suicide Pact', at COP27, UN Secretary, Antonio Guterres [7].

> 'Those who are in a position to know, have a duty to act!'

3 UN Recommends 'the Only Reliable Motor'

> 'Civil society is to a great extent the only reliable motor for driving institutions to change at the pace required' [9].

It is fascinating to work together with many people to build this 'only reliable motor'. This civil society motor is at its core 'human' connectivity. Human connectivity to ourselves, other humans, to nature and the laws of nature, including planetary boundaries.

4 Fridays for Future

One of the success factors of Fridays for Future is to build and accelerate this civil society 'motor' to create the needed societal change. Fridays for Future is one of many movements accelerating this 'only reliable motor'.

2018-0908 Ramelshovsparken: Starting with a simple call for weekly action from Greta Thunberg, [https://www.youtube.com/watch?v=ATRWnlK1SyA&t=9s].

Every Friday stand in front of your parliaments and local town halls and demand action. 'Everyone is Welcome, Everyone is Needed!' [10] (Fig. 1).

5 Why Record on Our FridaysforFuture Action Map and Lists? [11, 12]

1. Courage is Contagious: Each person 'speaking truth to power' will spark others to dare to do the same. Courage comes from the heart. It contagiously spreads empathy and active hope, while building trust.
2. Acceleration: The Harmony Effect. One voice generates some waves.
   Two is a duet and generates four times as many waves.
   Three is a choir and generates nine times as many waves, etc.
3. Solidarity:
   (a) Map and Statistics enable people to see themselves as part of the greater good. Both before and after every Friday making sure people know they are counted.
   (b) Call in appropriate people and give as much credit as possible.
   (c) Use common hashtags when speaking about each other, which makes it harder for the powerful to look away.

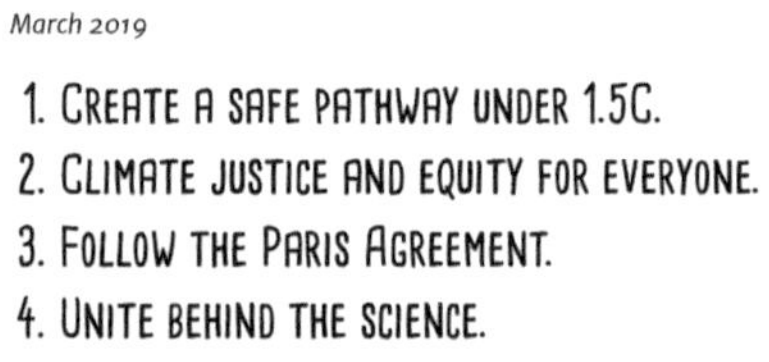

Fig. 1 Fridays for Future Actions, March 2019. Credit: Webpage Fridaysforfuture.org

4. Remove the doubt: near-immediate statistics of the many by the many eliminates doubt about the numbers, something which has plagued many movements for decades and has led to rampant denial.
5. Social scientists use data from the map to develop activism strategies, which grow the entire movement.
6. Journalists use the map to find activists, and contact them for interviews.

6 Introducing Fridays for Future Web, Map and Lists

FridaysforFuture Web, Map and Lists accelerated the many advocacy calls with simple systems:

1. How to create a strike, demonstration or event [13].
2. Register your action with FridaysforFuture Web [14].
3. Announce on the FridaysforFuture Map [11].
4. Show solidarity with the FridaysforFuture Lists [12].
5. Know you are one of many with transparent, regular FridaysforFuture statistics.
6. FridaysforFuture OutReach: use the #Fridaysforfuture, #ClimateStrike, #FFF and #SchoolStrike4Climate hashtags to share and lift many voices.
7. Use the FridaysforFuture Map to share people's work and plans [11].
8. Accept the Vision Proposal often used by FridaysforFuture to join in with Fridays for Future OutReach and lift many voices [15].
9. Start a Fridays for Future Local Group [13].
10. Join here to find work groups to collaborate with many movements [15].

7 Simple Systems

By creating simple systems people can themselves find a way to Step Up their activism. They can create a way to nurture their need to improve their local issues and connect to the global situation. When systems are built with simplicity, inclusiveness, transparency and local autonomy, they can accelerate social feedback loops growing 'the only reliable motor' for societal transformation.

8 Social Feedback Loops

There are many 'Social Feedback Loops' which can be on the individual level, the group level and the system level. The more these three levels can interact, the stronger the social feedback loop. Here are a few examples.

8.1 Social Feedback Loop 1: Simple and Courageous

- Simple, easy to copy: strike.
- A leader needs to have the courage to stand out just like 'The Shirtless Dancing Guy!' [16].
- The courage to even use civil disobedience. Civil disobedience is breaking a minor law to show that a larger one is being broken. In effect, activists are just trying to break the glass so that the climate alarm rings loud, clear and continuously. Our planet, our only home, is on fire, and our media, our politicians and influencers are virtually silent.
- Simple Location: Gather in front of your parliament or town hall.
- Simple to join, inclusive: 'Everyone is Welcome, Everyone is Needed!'
- #ClimateStrike #FridaysforFuture.
- Nothing else needed.
- Individuals can do it on their own.

8.2 Social Feedback Loop 2: Regular, Repetitive, Ritual, Simple to Promote

- Regular, weekly and repetitive, 'Same place, same time'.
- People move their timetables to join you.
- Promote with hashtags: #ClimateStrike #FridaysforFuture.
- FridaysforFuture Register.
- FridaysforFuture Map.
- Promote social media beforehand.
- When more strikers join, teach them the simple promotion loop.
- Public Recognition is created by using the same Sign repetitively.
- Social media Photo, share and ask people to share. Write a flyer and promote to people passing by.
- Record to the FridaysforFuture Lists, FridaysforFuture Map or FridaysforFuture Gamechanger. It is as simple as: #FFFMapRecord, 11, #FridaysforFuture, Germany, Berlin (Fig. 2).

8.3 Social Feedback Loop 3: Recognition, Thanks, Collaboration and Solidarity

- Welcome all followers as equals, just like 'The Shirtless Dancing Guy!'.
- Thank people for joining.
- Build public recognition via social media platforms.
- Share other strikers' work and try to share your social media followers to other strikers, show solidarity, while also respecting the local autonomy needed. Let's learn techniques to spread followers to other strikers.

Fig. 2 Fridays For Future call of action

- Things will go wrong. It may even 'appear' that another striker has done something wrong; reach out and show support and solidarity, even though this is not easy. At least, remain publicly silent.
- Collaborate with other issues and movements. System repression has been clear for centuries with human rights, and decades for climate and ecocide. Reflect each other's hashtags and plans.
- Solidarity to other movements, share followers. Recognize the barriers to civil disobedience, and when something goes wrong, worst case remains silent.
- Building your 'piano keyboard' and playing in harmony. Many activists are part of other movements. If each movement is seen as a single piano key, and each activist having their own piano. Each piano can be played both separately with autonomy or in harmony with other people's piano. Each person has a piano with local autonomy, and only you shall be able to remove a piano key.

8.4 Social Feedback Loop 4: FridaysforFuture Map Activist Public Email Breaks the Editor 'Glass Ceiling'

- Journalist: 'Look there is a strike here'.
- Editor: 'No, that's 'just' local!'
- Map, 2–3 countries, email.
- Interview.
- Journalist: 'Look, I found more than three countries with strikers, the strike is global'.
- Editor: 'That is News!' And the Editor's 'glass ceiling' is broken.
- Article in newspaper!

- This creates more strikers on the FridaysforFuture Map with emails showing! (loop).
- Feeding the Journalists' activist hearts!

With so many emails on the map, many journalists were able to find a way to make climate appear in the media. After writing an article about a local climate event, the editor often rejected the article by saying, 'That's just some local prankster and not news'. Within a short period, the journalist could find the FridaysforFuture Map statistics, contact and get quotes from local strikers in several different countries and continents. They could quickly return with an article and fairly claim 'This is global and many people are involved'.

Journalists are often activists themselves. They became journalists because their voice was often best heard via the written word and photography. They threw their hearts into this work, beautifully breaking the conservative 'ceiling' for climate activism. There were several who came by Stockholm to thank FridaysforFuture Map for the map work. Remember: newspapers follow each other and many more will join once a few start.

8.5 Social Feedback Loop 5: Simple and Safe Social and Technical Platforms

Developing our activist heart with strength together. The weekly nature of Fridays for Future enabled quick community growth and with far less depression and reduced shame. This weekly ritual of community building became a natural form of therapy for many. Many went from being isolated and depressed about climate to publicly requiring action.

This joint process enabled many to express their anxiety and develop a way to speak about this government failure and show anger together.

FridaysforFuture GDPR Declaration of Understanding [17]

Vision Statement often used by FridaysforFuture [15]

Extinction Rebellion arriving made FridaysforFuture able to grow with less state system and denial attacks. Diversity helped.

Once a social feedback 'loop' is described, then it can be built and obstacles removed.

8.6 Social Feedback Loop 6: Fridays for Future Action Outreach

Finding Activists:

The task is to find as many instances of climate activism as possible. 'We need everyone!'

1. Use hashtags to find activists: #FridaysforFuture #ClimateStrike #SchoolStrike #PeopleNotProfit to find climate striker and activist posts.

2. Share the posts in the Fridays for Future weekly thread on social media [18].
3. Show solidarity: Thank each climate activist and comment on their work.
4. Teach them how to have their work appear on the Fridays for Future Map.
5. Utilise the daily spreadsheet created by FridaysforFuture Map workgroup to directly contact new strikers using the FridaysforFuture Weekly Report via contact with email map@fridaysforfuture.org and new strikers social media.

Make it easy for everyone to find us!

1. Create an event.
2. Put out a call beforehand.
3. Use a social media platform thread so people can join in and find us.
4. Thank them, and show appreciation for their courage.

We appreciate them by growing their voice!
We comment and share their work, including our FridaysforFuture Map.

9 Removing Obstacles: Enabling Feedback Loops to Flow

There are often obstacles for activists even within FridaysforFuture and also for other movements. It is important to recognise and remove them. Other people and organisations also create obstacles for climate activism. The UN recognises state systematic repression of climate justice activists and recommends that states take special care to stop any 'chilling effect' and ensure this 'only reliable motor' is able to stop climate breakdown [3]. Another well-documented and acknowledged obstacle is the money and effort poured into the Climate Change Denial machine, which creates mass 'silence' about our climate breakdown. This silence and culture of entitlement accelerates the 'endless' useless task of seeking the perfect activist 'method'.

For every person who has said to Greta, 'You should go to school and become a scientist and change the system from within'. They are effectively saying, 'I believe in 'your' cause and your right to express yourself, but not your method!'

Firstly, climate is not one person's political view or 'cause', it is scientific fact. Climate breakdown has 9800 scientists who have written peer-reviewed articles for decades and gradually discovered how humans, mainly in the Global North, together undermine the earth to produce what wealthy humans think they 'need'.

Climate is better agreed upon than gravity. Remember Galileo was placed on trial in Rome. He was accused of heresy and disobedience. Eventually due to repression and pressure, he renounced his theories. He officially accepted that the Sun revolved around the Earth.

Secondly, to disagree with an activist's 'method', while not doing any work to find a better method, is to accelerate the silence. Every climate activist is actively seeking a better 'method'. To give feedback and actively test a method is important and useful. But to just say, 'Stop, your method is bad!', and to do no more is to join the chorus of climate change deniers. This is accelerating the act of silencing a climate activist.

10 Silencing Climate Activists

To create an obstacle or silence a climate activist is to aid and abet the fossil fuel companies in their decades of criminal and planned climate change denial. As the UN Secretary-General, Antonia Guterres said, 28 February, 2022, 'The abdication of leadership is criminal. The world's biggest polluters are guilty of arson of our only home!' Every literate adult has power, and a 'voice'. They have a duty to listen to the science. Those that have the privilege to know have a duty to act. If an adult instead uses their power to silence or create an obstacle to the voice of another climate activist, they are supporting this criminal denial. As in many previous cases of genocide, even silence and standing on the sideline is likely to be seen as criminal.

If not guilty of arson, then homicide?
While Guterres has called the world's biggest polluters 'guilty of arson', many lawyers are taking the next step and seeing the deadly decades of climate change denial by the fossil fuel companies as homicide.

'Fossil fuel companies learned decades ago that what they produced, marketed, and sold would generate "globally catastrophic" climate change. Rather than alert the public and curtail their operations, they worked to deceive the public about these harms and to prevent regulation of their lethal conduct. They funded efforts to call sound science into doubt and to confuse their shareholders, consumers, and regulators. And they poured money into political campaigns to elect or install judges, legislators, and executive officials hostile to any litigation, regulation, or competition that might limit their profits. Today, the climate change that they forecast has already killed thousands of people in the United States, and it is expected to become increasingly lethal for the foreseeable future' [19].

Meanwhile, well over 100 UK lawyers, including prosecutors and seven King's Councils, have signed the Lawyers Are Responsible Declaration to not prosecute climate justice activists [20].

If only your 'method' was better!
Often the criticism of the climate activism 'method' is because people fear that some methods encourage people to be against climate activism. It seems unrealistic to not see that climate activists are not only scared for their own lives but also have empathy for the lives of so many, for example, for the decades of climate breakdown in the global south. Everyone will have trouble agreeing on how to break the glass in front of the climate 'fire' alarm bell. And yet climate scientists have repeatedly said, humans cannot negotiate with the natural climate laws of physics.

Civil disobedience was a successful part of societal change for democracy, women's suffrage, civil rights, gay rights, Berlin wall, the Baltic way, and many more, even the 1856 '888' hour movement. Almost everything everyone takes for granted in our lifestyle, e.g. the weekend, sick pay, salaries, all of this has been fought for using civil disobedience. It was a key method to create and develop democracy and has regularly been used to sound the alarm when our society has been silent and unjust.

School striking is a form of civil disobedience and has shown to be very successful.

How do the adults step up? The unions have resisted joining, often arguing that 'We use democratic methods!'

'We use democratic methods!'

Many of the workers' striking methods have wilted away and unions would find it difficult to get a large body of workers on board without a campaign of deeper climate education. Even the comment 'We use democratic methods', is a denial of union history, which normally involved strikes, civil disobedience and jail terms for leaders. Only afterwards does it become possible and attractive to 'photoshop' their past and sit silently on the sideline. As Guterres says, the abdication of leadership is criminal, especially the silence of our 'democratic' institutions. There are many who cannot afford to go to jail, there are others who will find it difficult to strike, but our democratic institutions and those who are in a position of power have the duty to listen to the science and act or step down from power. Those who have the privilege to know have the duty to act!

The duty to act!

Many climate justice activists from every group respond to this 'duty to act' by stepping into the more recent waves of activism, e.g. Last Generation in Germany and Austria, Just Stop Oil in the UK, Återställ Våtmarker in Sweden, Stop Fossil Fuel Subsidies in Australia and Scientist Rebellion. The media finds it harder to dismiss these people as merely 'young unemployed'. Calling climate activists 'young unemployed' is a classic dog whistle that activists are living off others' handouts, and is perhaps even a form of 'age cleansing'. Our states ('countries') try to repress and silence these waves of activism with jail terms and charges of sabotage, treason and sedition and 'Climate Trials'.

Have our states understood that every hour everyone delays climate transformation costs lives and 'the equivalent of millions of Swedish Krona?' [21].

Have our states understood that women's suffrage, civil rights and many other movements after decades of asking nicely were finally moved by civil disobedience and even by the threat of violence?

Climate Justice Activists know that love and peace are far better ways.

No one envies the example of South African darkness and violence [22].

Spiralling chaotic and brutal climate and societal breakdown is collective suicide and not to be seen as a method of choice! Collective Action would be far better!

We are already too late!

No one disagrees with our states (countries) being very late.

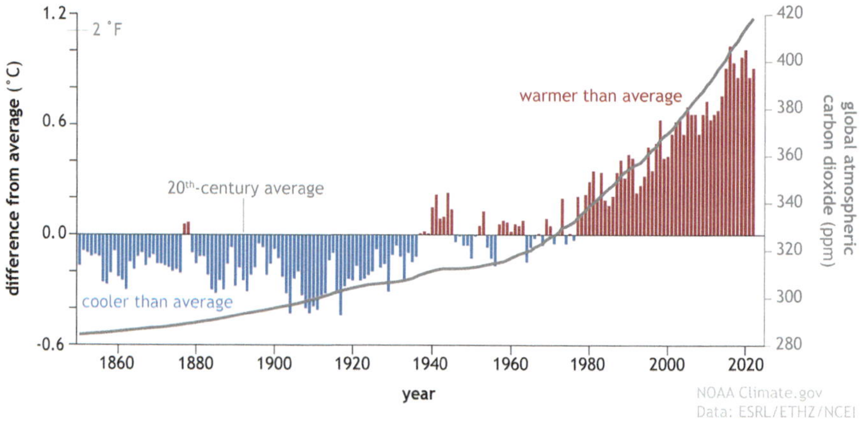

Fig. 3 Yearly global surface temperature and atmospheric carbon dioxide (1850–2022). Source: https://www.climate.gov/media/13840 [25]

Peter Kalmus, 'It will never be "too late" we will simply lose more and more the longer we fail to act!' [23].

Society and people in power knew about climate breakdown well before most of us were born. In 1856, Eunice Foote's experiments first proved the warming we see today. Scientists are saying that the coming years are going to be worse than the past. The tipping point of behaving justly regarding the future has been passed.

British Petroleum, BP, would like each of us to count our Carbon Footprint. While it is very useful to reduce CO2 emissions, individual efforts alone cannot solve the climate crisis. And blaming the individual for their emissions often distracts people from looking for the real criminals.

Adaptation was never and can never be enough. Humans are moving towards situations that are beyond adaptation. Unfortunately, at present, most states put most of their efforts into adaptation:

- There is no way humans can adapt their way out of this. We either reduce, redraw and repair now, or it keeps getting harder the longer we delay.

So, what is 'too' late? Surely we shall try to save as much as possible.

Clearly, until our states have turned off the money tap to fossil fuels, e.g. subsidies, and other methods which push carbon consumption upon us all and invest seriously in every process to Reduce, Redraw and Repair, our states are not even really trying:

- Reduce (CO2 emissions at emergency speed).
- Redraw (CO2 atmosphere excess probably even as far down as 280–300 ppm) [24].
- Repair (also called Refreeze) (use direct cooling and other measures to protect us all till safe atmospheric concentrations of CO2 are restored) (Fig. 3).

Until our states and everyone starts really trying, how can anyone know 'that it is too late'? As long as our states continue to pay out millions in subsidies to fossil fuel companies, the powerful are not even trying. And with millions, even trillions/year in subsidies, why would a Fossil Fuel industry listen to the IPCC science.

11 Collective Action or Collective Suicide

Using collective action is really our only option and until the silence is broken thoroughly, similar to what occurred from the first few weeks of COVID, then a very powerful tool is to keep using civil disobedience to break the glass in front of the climate 'fire' alarm bell. Seeing all the courage needed to break this silence gives enormous hope!

Appendix 1: Sweden Sabotage Statistics

In May, 2023 the number of Sabotage cases approximately were:

Sabotage Climate Justice Cases charged: more than 100
Sabotage Climate Justice Trials in Swedish County Courts (including one child): 27
Sabotage Climate Justice Trials awaiting decision from Swedish County Courts: 7
Sabotage Climate Justice convictions in Swedish County Court (including one child): 19
Sabotage Climate Justice prosecution appeal to Appeals Court: 1
Sabotage Climate Justice conviction appeal to Appeals Court: 19
Sabotage Climate Justice appeal denied by Appeals Court: 2Sabotage appeal overturned by Appeals Court (child): 1
Swedish High Court Climate Justice conviction appeals: 2

About the Author
Janine O'Keeffe—School strike 4 Climate adult supporter, FridaysforFuture original international organiser, Klimataktion Board member, member of Klimatriksdagen, active in Klimatsverige, Scientist Rebellion, Extinction Rebellion, Last Generation, and Återställ Våtmarker, Bachelor of Engineering in Electronics, Master of Business Administration and author of this text.

Background Invitation and Disclaimer
After the speech each speaker was invited to compose a chapter of a book to give more background to the speech and work. This is an attempt to document some of my recollections and thoughts. A lot is missing, especially mentioning the beautiful work of and showing gratitude to so many people who have worked diligently to build the Fridays for Future social infrastructure and many other wonderful climate and environmental advocacy movements. All lack of mentions and recollection faults are mine. This is purely some recollections and ideas and not an attempt to

represent School Strike For Climate, Fridays For Future, Extinction Rebellion Sweden, Extinction Rebellion, Last Generation, Återställ Våtmarker, Scientist Rebellion or any movement.

With Gratitude to

With more than eight million people on the street in September 2019, it is impossible to name each person individually. But it would be neglectful not to mention at least some who I am aware of and did a lot of work to get us all there. Any missing names are Janine's fault.

Brice Montagne, Edit Liedberg, Minna Dahlin, Eira Frohling, Eva Lenke, Jens, Jan, Martin Lindblad, Benj, Maxime Senza, Anna Bokström, Alfred Westh, Alexandra Wagner, Dirk Jan Luiting, Ola, Patrick, Julia, Sam, Peter Forsberg, Lena Hammarback, Bengt Sundbaum, Sibylla Jamting, Henrik Garbergs, Lisa Kjellberg, Maj-Lis Lindholm, Anne O'Hara, Kevin Hecht, Bhavreen Kandhari Jonathan Olwenyi, Mo Markham, Caterina Lindman, Mattes Liebsch, Abdul Wahab Watan Dost, Finlay Pringle, Geoff Pringle, Rachael Pringle, Ella Pringle, Aruba Faruque, Arrhenius Müntzing family, Adebola, Lena Schiller, Sabine Thie-Olliges, Anthony C Gleeson, Michael Gravenor, Anders Bäcklund, Anita Maria Eng, Fred E, Patrick Gallagher, Stefan Sommer, Auro Ricking, Anders Dahlner, Ken Batts, Fidele Uwihoreye, Yolanda Alfonso, Jürgen Voskuhl, Felix Werker, Ushindi Kashali, Horst Günther Bulander, Chance Bahozi, Solveig Forsthoff, Patrick O. Ehrmann, Nicole Rostock, René Becker, Constanze Küppers, Jonas Ohlsson, Farzana Faruk Jhumu and many more.

Organisations:

Klimataktion, KlimatSverige and many more climate and environmental advocacy movements, both well-known and less known.

References

1. https://news.un.org/en/story/2022/11/1130247
2. https://twitter.com/MrMatthewTodd/status/1648402404050714625
3. https://actionnetwork.org/groups/un-recommends-disruption-of-traffic/
4. Link to speech, 2022-07. https://mdigital.sharepoint.com/:f:/s/ExternalShareCuriousFutureInsight/ElURt6CQo0lNlOpv-ZvdLuMBsHTuSHfkzdPpTA0UWttm_g
5. 2023-05-15 Scientist Rebellion, 'The Science is Clear!' speech, from 40 minutes. https://drive.google.com/file/d/1GHa7yJCu1xSOJEZlw1s02qtiJZbxR7Hx/view
6. 2021-08-06 TV4 Sweden (translation Janine O'Keeffe). https://www.facebook.com/janine.okeeffe/posts/pfbid0T2LrXbweJFdTtCQteDYY3DCq6G2GxVberj4H4Y5ssVMYQNyyhGkAQWYAhu9c5w4Xl
7. 2022–11 COP27. https://news.un.org/en/story/2022/11/1130247
8. https://twitter.com/visevic/status/1431012057143463937
9. https://www.ohchr.org/en/documents/thematic-reports/a76222-exercise-rights-freedom-peaceful-assembly-and-association
10. https://fridaysforfuture.org/

11. https://map.fridaysforfuture.org/
12. https://map.fridaysforfuture.org/lists
13. https://actionnetwork.org/groups/getting-started/
14. https://fridaysforfuture.org/register
15. https://actionnetwork.org/forms/vision/
16. https://www.youtube.com/watch?v=V74AxCqOTvg
17. https://actionnetwork.org/forms/gdpr-agreement-form/
18. https://www.facebook.com/groups/FridaysForFuture.org/posts/2952657131533493/
19. https://papers.ssrn.com/sol3/papers.cfm?abstract_id=4335779
20. https://www.lar.earth/
21. https://www.gp.se/debatt/s%C3%A5-mycket-kostar-det-att-inte-agera-p%C3%A5-klimatkrisen-1.85030005
22. https://www.youtube.com/watch?v=ofwx-kyxHq4
23. https://twitter.com/ClimateHuman/status/1509256737416638465
24. https://www.climate.gov/news-features/understanding-climate/climate-change-atmospheric-carbon-dioxide
25. https://www.climate.gov/media/13840

The Keeling Curve Prize: How a Climate Action Prize Gets It Done

Jacquelyn Francis, Lindsay DeTroia, and Gautam Barua

Abstract

When implementing solutions to an incredibly complex problem, one does not immediately think of offering prizes, but when it comes to highlighting solutions to climate action, prizes appear to be effective. The Global Warming Mitigation Project's flagship program, the Keeling Curve Prize (KCP), has a track record of an impressive team of experts and a robust process for attracting applications that includes screening and scoring global organizations while incorporating a coordinating supporting system. The prize has demonstrated measurable results. Climate action needs planetary transformational systems changes to happen quickly, and prizes are proving to be one of the effective vehicles to rapid progress.

1 Historically

The Industrial Revolution sparked unprecedented economic growth and improvements to standards of living worldwide. Billions of people have ready access to food, electricity, transportation, and consumer goods. However, much of the transformation has been powered by burning fossil fuels, and their emissions

J. Francis (✉) · L. DeTroia
Global Warming Mitigation Project, Aspen, CO, USA
e-mail: director@kcurveprize.org; l.detroia@kcurveprize.org

G. Barua
Aclaria Partners, San Francisco, CA, USA
e-mail: gautam@kcurveprize.org

U. A. Betz (ed.), *Curious Future Insight*,
https://doi.org/10.1007/978-3-031-41781-8_18

represent nearly 75% of heat-trapping greenhouse gases released into the atmosphere,[1] raising global temperatures by nearly 2 °F since the nineteenth century.[2]

Much has been studied and written about carbon dioxide (CO_2) levels in the atmosphere and oceans. On Mauna Loa the record of carbon dioxide accumulation has been directly and precisely measured since 1958.[3] In 1961, Charles David Keeling published the first study showing the steady increase of carbon dioxide in the atmosphere. Dr. Keeling's paper included the "Keeling Curve"—a graph displaying increasing concentrations of atmospheric CO2. Today, the Keeling Curve is the most recognizable scientific depiction of rising greenhouse gas emissions.

Named in honor of Dr. Keeling and launched in 2018, the Keeling Curve Prize (KCP) sustains the belief that the symbolic power of the Keeling Curve is rooted in scientific objectivity and integrity of data. It annually awards a monetary prize and recognition to ten organizations whose project or program displays the greatest impact and potential for increasing the uptake of CO2, or reducing emissions of CO2 and other heat-trapping gases. This prize is designed to inspire change-makers and honor those in communities of all sizes, whose efforts lead to changing the direction of the Keeling Curve by creating and optimizing new pathways to a sustainable future. The Prize draws attention to leaders, advocates, and innovators—giving them global visibility and leverage for their work, as well as a financial boost to further their mission and vision.

2 Currently

As heat-trapping gases accumulate in our world, the subsequent weather-related chaos is the mounting body of evidence currently being experienced across our planet. Escalating local, national, and global challenges have been dominating the news during the last few years, with tragic climate catastrophes becoming more frequent. We are aware of specific weather events defying predictability, but the reality that our planetary climate is changing can no longer be ignored. Many weather events are becoming increasingly devastating as global warming intensifies them. A deadly heat wave in the Pacific Northwest melted power lines, buckled roadways, and led to the incineration of an entire town in Canada. Torrential rains and subsequent flooding in Europe washed away Bavarian villages, leaving behind a wake of destruction and hundreds of casualties. Ongoing drought is crippling places as widely varied as Madagascar, South America, Australia, much of Africa, and the western United States, just to name a few of the most devastated regions.

Even with stories of climate chaos everywhere and exploding awareness of climate change, humanity continues to add more carbon dioxide, methane, and

[1] Our World in Data, 9/18/20 [1].

[2] Energy Information Agency, 12/21/22 [2].

[3] Carbon Cycle Greenhouse Gases, 12/30/22 [3].

other greenhouse gases to the fragile atmospheric and oceanic natural balance. In fact, 2022 was on track to see a record high in fossil fuel-related greenhouse gas emissions.[4]

What was once thought of as a *future problem* is now a present and constant roar in our ears, with the term "existential" understating the intensity and complexity of the climate challenge. A one-size-fits-all solution to current emissions and the accumulation of past emissions is unlikely to materialize. Instead, a mosaic of approaches will be needed.

Growing interest in quickly and broadly reducing greenhouse gases is appearing on multiple fronts, from the Inflation Reduction Act in the USA, which allocates nearly $400 billion to clean energy and emissions reductions,[5] to record levels of venture capital investment in "climate tech"—that now represent 25% of VC funding across all sectors.[6]

3 The Prize

The climate challenge is multifaceted, requiring insights on topics ranging from market and industry dynamics to science and engineering. The Keeling Curve Prize uses robust, evidence-based analyses to identify technological, financial, societal, and natural innovations that can pragmatically address the climate crisis.

The Keeling Curve Prize has so far been awarded to 60 programs and projects working around the world. Examples include a globally accessible micro-hydropower technology developer, a solar company electrifying hubs in remote regions of Africa, a company applying innovative chemistries to produce low carbon products (including jet fuel), a proprietary chemical-free energy storage system using carbon captured from the air, and a startup harnessing natural solutions through seaweed farming and mangrove restoration. Winning the prize has not only elevated these organizations to levels of success beyond conventional expectations, but has also shown the world that climate action is thriving everywhere.

The existence of society as we know it is reliant on the shift away from our collective dependence on fossil fuel energy. Towards that end, the continued success of the Keeling Curve Prize is demonstrating how this is possible, practical, and profitable, with KCP Laureates well on their way to becoming leaders of climate action through their global energy and mobility transition efforts.

A team of 12 analysts review KCP applications (currently limited to 500 per year). The analysts share a high level of climate knowledge but have different backgrounds, different geographic perspectives, and a variety of scientific

[4] World Economic Forum, 11/11/22 [4].

[5] McKinsey & Company, 10/24/22 [5].

[6] DGB Group, 12/13/22 [6].

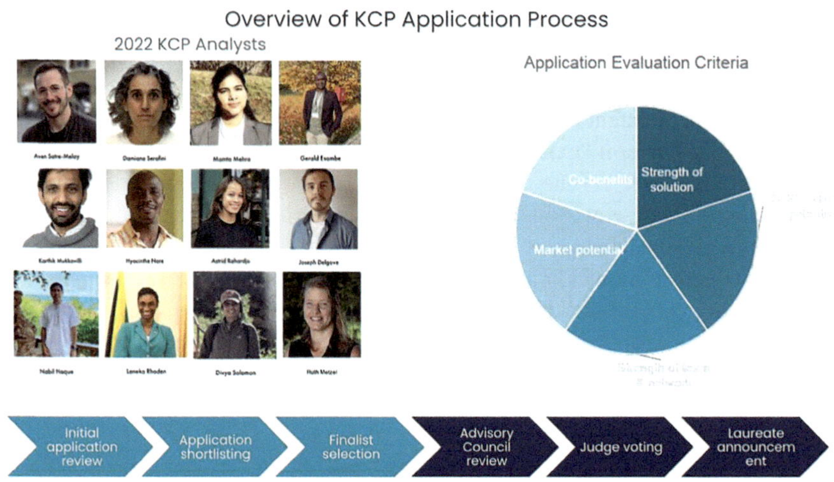

Fig. 1 Keeling Curve Prize application and selection process

expertises. Their areas of expertise include technology, artificial intelligence, forestry, finance, chemistry, physics, and culture.

An open application period runs for several months, during which hundreds of applications from around the globe are accepted. Each application is evaluated, ranked, and scored by at least two analysts. The entire analyst team then reviews and evaluates approximately 50 of the highest-scoring applications. The process becomes both quantitative and qualitative, with discussions aiming to narrow the field through information sharing and sometimes voting. The scoring matrix, the normalization metrics, and the team of experts with their diversity of knowledge all make this process highly competitive and rigorous (Fig. 1).

4 Prize Categories

The five prize categories are wide-ranging to reflect the need for climate solutions from across a vast range of sectors and areas of activity.

5 Carbon Sinks (Natural and Engineered)

Projects in this category activate and accelerate natural and/or engineered systems while also devising strategies for capturing and/or utilizing heat-trapping gases from the air and oceans, with a particular emphasis on long-term sequestration. Additionally, they bring their strategies to the forefront of the marketplace while providing methods for verification. Examples: land-use practices, agricultural methods, efforts to reduce deforestation, reforestation efforts, technology for GHG capture systems, advances in verification for natural sequestration, markets for captured gases, and verification systems. Approaches may include afforestation, soil enhancement, gas recycling, waste-to-energy, and improved cement products.

6 Energy

Projects in this category decarbonize energy, support zero-carbon energy innovations, and lead the way in the supply, distribution, access, infrastructure, and improvements in low or zero-emissions energy systems worldwide. Approaches can be regional, national, or international, and should consider economic accessibility, efficacy, scalability, and strategic competitiveness with fossil fuels. Of particular interest are projects focused on clean energy for equitable energy access, including innovations in smart utility-scale grids with low end-user costs, as well as mini-grids and off-grid solutions to reach "last mile" customers in underserved areas worldwide. These should improve energy access or reduce emissions from existing energy systems, focusing on less-developed communities—especially those vulnerable to increased coal and fossil fuel use—with demonstrable scalability.

7 Finance

Projects in this category facilitate the economic viability of greenhouse gas reduction or removal, counterbalancing market failures and shortfalls. Illustrative approaches include financing products for lower or zero emissions businesses, mechanisms that factor in ecological and human sustainability, empowerment of climate solution programs, investment strategies, eliminating financial barriers, carbon accounting, and market-shaping policy. Successful applicants will emphasize widespread participation by marginalized communities and economic valuation of thriving ecosystems where positive impact has been independently verified.

8 Social and Cultural Pathways

Projects in this category are changing the way people consider, understand, and engage with humanity's impact on the livability of Planet Earth. They seek to answer the question of what it takes—socially and culturally—to move beyond fossil fuels, such as closing gaps in human behavior and implementation. These projects are influencing consumption, awareness, communication, mobilization, activism, policy, international relations, and cultural norms. They encourage individuals, communities, businesses, and governments to make progress toward emission reductions and carbon drawdown, now and in the future. Projects should ideally demonstrate a path to efficacy and scalability.

9 Transport and Mobility

Projects in this category should be reimagining and reinventing all types of vehicles, fuels, and mobility options for both people and products. These projects confront the carbon footprint of the vehicles themselves and routes traveled. Creative solutions might include travel avoidance, innovative manufacturing, roadways or other systems over which transportation operates, waste-reduction in supply chain distribution, improvements in fleet management, mass transit scheduling and efficiency, rural access to zero-emissions transportation, power sources and charging for electric vehicles, and more. The KCP strongly encourages applications for projects that accelerate power source and modal shifts, address weak and wasteful transport systems, or challenge the viability of the internal combustion engine.

10 Impact

In its first 5 years, the Keeling Curve Prize selection panel has consistently picked the most promising innovators in the field. The impacts and trajectories of KCP Laureates reflect this promise.

Chosen from among 1700 applicants, KCP Laureates are projected to collectively reduce greenhouse gasses by 100 gigatons—roughly 5–10% of forecasted global emissions through 2050. The market has rewarded them accordingly, with laureates demonstrating increased traction against their targets. They have raised 4x as much capital in the years they received the KCP and beyond, in contrast to their years prior to winning the KCP (Fig. 2).

The impact of the KCP on laureate organizations themselves is amply demonstrated by their leaders' consistently appreciative statements about the prize. Some example quotes:

> The Keeling Curve Prize was the first true validation that what we were doing made sense. It was super-helpful in bringing us into the light.—*Jennifer Holmgren, CEO, LanzaTech*

> Mechanisms such as the Keeling Curve Prize are essential for trying out new ideas and seeing what can scale and what sticks. The Keeling Curve Prize was essential early support for our work.—*Nate Aden, Finance Sector Lead, Science-Based Targets Initiative*

> Winning the Keeling Curve Prize was an incredible boost. The recognition and validation from the prize inspired other funders to support our work.—*Nicole Crescimanno, Our Climate*

Category	Laureate	Year of prize	Total funding raised before year prize awarded ($M)	Total funding raised since and including year prize awarded ($M)
Carbon Sinks (Natural and Engineered)	Air Company	2022	$8.8	$30.0
	Symbrosia	2022	1.7	7.0
	Carbfix	2020	0.0	117.0
	twelve	2019	1.4	198.0
	Bolder Industries	2018	0.0	80.0
	Lanzatech	2018	204.3	606.1
	Takachar	2018	0.1	0.3
Energy	Nitricity	2022	6.1	20.9
	Jaza Energy	2020	2.4	1.7
	Oorja Development Solutio	2020	0.0	1.3
	African Clean Energy	2019	0.0	1.0
	Natel Energy	2018	19.2	46.6
	WattTime	2018	0.0	1.7
Finance	Odyssey Energy Solutions	2020	0.0	12.6
Transport & Mobility	BasiGo	2022	1.0	3.4
	EnZinc	2022	1.3	6.3
	CLIP	2021	0.4	1.6
	EVmatch	2020	0.0	0.7
	Three Wheels United	2019	0.0	9.9
		Median	**$0.4**	**$7.0**
		Total	**$246.5**	**$1,146.1**

Fig. 2 The impacts and trajectories of Keeling Curve Prize Laureates

11 Conclusion

Anthropogenic global warming is an urgent and complex issue that requires a multipronged approach to innovation. It is also critical to not lose sight of the importance of equitable outcomes for lower-income communities in industrialized nations, along with 80% of the global population in nations with emerging market economies.[7]

The Keeling Curve Prize represents a unique approach to incentivizing for-profit and nonprofit entrepreneurship in terms of climate impact and social outcomes. While the planned trajectory of the prize program is to expand further into including regional focuses and an oceans track, it also serves as an inspiring model for creatively spurring action towards positive outcomes for our climate and life across the planet.

[7] Corporate Finance Institute, 12/15/22 [7].

References

1. https://ourworldindata.org/ghg-emissions-by-sector
2. https://www.eia.gov/energyexplained/energy-and-the-environment/greenhouse-gases-and-the-climate.php
3. https://gml.noaa.gov/ccgg/trends/
4. https://www.weforum.org/agenda/2022/11/global-co2-emissions-fossil-fuels-hit-record-2022/
5. https://www.mckinsey.com/industries/public-and-social-sector/our-insights/the-inflation-reduction-act-heres-whats-in-it
6. https://www.green.earth/blog/climate-tech-investment-is-expected-to-reach-79b-in-2022
7. https://corporatefinanceinstitute.com/resources/economics/emerging-market-economy/

Challenging the Scientific Ethic in the Age of Biotechnology

George P. Smith, II

Abstract

This paper analyzes the extent to which there is an a priori or a situational scientific ethic in play during scientific investigations and studies the extent to which such an ethic would mandate an explanation of the nature and the consequences of these investigations; and, furthermore, whether a co-ordinate responsibility of self-restraint among scientists exists when consequences and foreseeable dangers outweigh the positive values of an investigation. This inquiry is tested within the context of the science and biotechnology of genetics and the great promise they hold for advancing a better, healthier world order.

The thoughts of two great intellectual giants in science and in philosophy serve as the predicates to this paper.

Max Planck lived from 1858 to 1947 [1]. He received a Noble Prize in 1918 for his work in advancing the science of physics; more specifically, for his original authorship of the quantum theory, acknowledged as the theoretical basis of modern physics which explains the nature and behavior of matter and energy on the atomic and subatomic processes.[1] This theory is said to have revolutionized both theoretical physics and scientific philosophy [2].

A paper presented at CURIOUS 2021, Future Insight Conference, Merck International, Darmstadt, Germany on July 14, 2022.

[1] One photon of light carries exactly one quantum of energy. Planck is considered the father of the

G. P. Smith, II (✉)
The Catholic University of America, Washington, DC, USA

The Institute for Advanced Study, Indiana University, Bloomington, IN, USA
e-mail: smithg@cua.edu

U. A. Betz (ed.), *Curious Future Insight*,
https://doi.org/10.1007/978-3-031-41781-8_19

Planck's book, *The Philosophy of Physics*, established his pre-eminence as both a scholar and a scientist. As well, this book set forth his philosophy of science which seeks to understand what is known rather than "discover" new things.[2] He maintains that ". . . the laws governing the growth and effect of scientific ideas can never be reduced to an exact formula valid for the future." Indeed, any new idea is but the work of an author's imagination.[3] For Planck, the significance of a scientific idea depends upon its value, rather than its truth.[4] Whether an idea has a definite meaning is relatively unimportant. Rather, "what does matter is that it shall give rise to useful work."[5] Work, as applied to science, is determinative of success.[6]

Acknowledging the validity of Planck's scientific ethic, the Association of German Engineers adopted a motto in the 1930s stating clearly that, "what is needed is investigation."[7]

Martin Heidegger lived from 1889 to 1976 and is recognized as "one of the most original and important"—as well as controversial—philosophers of the twentieth century [4]. Regarded as his Magnum Opus, *Being and Time*, was published in 1927.[8]

A considerable amount of Heidegger's scholarship seeks to present and promote an epistemological understanding of the place and challenges which technology has in contemporary life [9] and, further, to study its very foundation or essence[10]; for simply, technology is a way to understand the world.[11] For Heidegger, the fundamental analytical question is always, "How," and never, "Why."[12]

Inasmuch as technology can never be overcome, it is best to accept it as a part of a whole way of life, and thus see it as a means to an end.[13] The manner and form of manipulation is, then, determinative of humankind's relationship to technology—as measured and forthcoming, or, as fractions and deceitful[14]; for, within modern technology, is a "possibility of a deeper relationship between man and being (i.e.,

---

Quantum Theory. According to Planck: E = h[latex]/nu[/latex]. where h is Planck's constant (6.62606957(29) × 10^{-34} J s), v. is the frequency, and E is energy of an electromagnetic wave.

[2] Max Planck (1936) [3].

[3] *Id.* at 96.

[4] *Id.* at 103.

[5] *Id.* at 106.

[6] *Id.* at 106–107.

[7] *Id.* at 107.

[8] 1927.

[9] *See* Martin Heidegger (1977) [5, 6].

[10] Mark Blitz [7].

[11] Lovitt, *supra* note 12 at 3–35.

[12] Blitz, *supra* note 13.

[13] *See* Lovitt, *supra* note 12 at 5.

[14] Richard Walton (2000) [8].

truth, "unconcealment" or what it means to be) and hence between man and all that is and has ever been."[15]

Modernly, both science and technology are used interchangeably, but they are different fundamentally. Put succinctly, science is knowing and technology is doing.[16] Science seeks the pursuit of knowledge for its own sake while the goal of technology is to create products that solve problems and improve human life.[17] Biotechnology is, quite simply, technology based on biology. Its goal is to harness cellular and bio-molecular processes which endeavor to not only improve the quality of lives but the health of the world.[18] Additionally, biotechnology studies ways to reduce rates of infectious disease worldwide, creates more precise tools for the detection of disease and anti-viral therapeutics, as well as enhances agriculture productivity and genetically modified food.[19]

Regrettably, contemporary biological science and biotechnology have become political issues because of one principle fact: they put in focus the extent to which the government can restrict private medical research undertakings—either in the name of safety, morality, or the public good.[20]

The complex ethical, philosophical, socio-legal, and medical issues of this Age of the New Biotechnology are often said to be "biopolitic" in that many of the issues have become "embryocentric"—this, simply because of limitations on federal funding for human embroyonic stem cell research.[21]

1 Containing Technology

The European Union (EU) has ambitious plans to see that the technology industry is kept in check by seeking to fill the void seen as a "political paralysis," which the United States now finds itself in.[22] Even though the EU has been contending with serious issues of sluggish growth, political turmoil seen dramatically with BREXIT challenges, together with Asian influence, the Union is setting rules for the world economy.[23] EU rules are being transformed into global standards through market

[15] Lovitt, *supra* note 12 at 3–35.
See Krell, *supra* note 12 at 245, 349 (XXVII).

[16] *See* XIV, Oxford English Dictionary at 648–649 (2nd ed. 1999).

[17] *Id.*
See XVII, Oxford English Dictionary at 705–706.
See generally, Liebe F. Cavalieri (1882) [9].

[18] *See* II, Oxford English Dictionary at 210 (2d ed. 1999).

[19] *See* Doron Weber (2020) [10].
See generally, George P. Smith, II (1989) [11].

[20] *See* George P, Smith, II (2002) [12].

[21] *Id.*
See George P. Smith, II (2008) [13].

[22] Trade Regulation (2020), at p. 63 [14].

[23] *Id.*
See Anu Bradford (2019) [15].

mechanisms.[24] Indeed, the direct consequence of what is termed, The Brussels Effect, is the Europeanization of many aspects of global commerce—not only through business practices but public policies especially data privacy, consumer health and safety,[25] environmental protection,[26] antitrust, and online hate speech.[27]

In 1996, the Nobelist, Robert F. Curl of Rice University, opined that while the twentieth century was "the century of physics and chemistry," the twenty-first century must be acknowledged as "the century of Biology."[28]

The success of scientific contributions to an understanding of the biology of life is undeniable. It is a statement of fact that support for basic scientific work is crucial to the continued advancement of social order. Yet, care must be taken to foreswear any effort to recognize science as an exclusive sovereign in determining "where we come from, who we are and where are we going."[29] Justifiable concern is given properly to the extent to which biologists are "steering human progress" and have been allowed to re-define not only "what life is for" but—as well—to first reconceive and define subsequently "what life is."[30]

Establishing the scientific good to be derived from any experimentation invariably involves evaluating the risks vs. benefits which will accrue to societal interests if a scientific achievement is achieved.[31] Inasmuch as "science is not a stand-alone enterprise," social involvement in this decision-making process is essential—for it remains for society to approve or disapprove any given scientific advancement—to accept the achievement, restrict its operation, or forbid its use.[32] As will be shown, a largely complacent, uninformed society refuses to accept shared decision-making responsibility with the scientific community.[33] Consequently, science is allowed to assert sovereignty over the on-going scientific revolution.[34]

---

[24] *Id.*

[25] *Id.*

See generally, Barbara Osimani (2013) [16].

[26] Cass R. Sunstein (1997) [17].

[27] The Economist, *supra* note 25; BRADFORD, *supra* note 26.

[28] Robert F. Curl [18].

But see, Naomi Oreskes et al. (2019) [19].

See generally, A Whole New World, The Economist, April 6, 2019 at 3.

[29] Doran Weber (2019) [20].

See Sheila Jasanoff (2018) [21].

See generally, A Whole New World, The Economist, April 6, 2019 at 3.

[30] *See* Jeremy Rifkin (1998) [22].

See generally, George P. Smith, II (1993) [23].

[31] *Supra* notes 31, 32.

[32] *Id.*

[33] William J. Lederer (1961) [24].

See Sheila Jasanoff, *supra* note 32.

[34] Jasanoff, *id.*

For ethicists, the foundational issue which confronts the application and use of synthetic biology is the extent to which this science may be misused and result in "biological terrorism of warfare." As well, the means by which this new biological knowledge is processed must be assessed together with careful evaluation, and of the types of technology which will ultimately be developed and disseminated from this original basis of scientific knowledge.[35]

In other words, the basis ethical dilemma confronting science is whether it should be totally utilitarian—providing the greatest good to the greatest number—even if its result compromises the rights of some.[36] Perhaps the most equitable approach to resolving this dilemma would be to utilize a situation—as opposed to a prior—ethic in testing or measuring the scientific efficacy of each scientific experimentation before it is undertaken.[37]

2 Good or Evil Consequences?

Most scientists maintain that the positive or the negative consequences of pure scientific research are of little or no relevance,[38] for it is only at the technological or applied level of scientific work where "good or evil consequence" is to be evaluated.[39] A stronger and preferred position acknowledges that "the whole of research whether pure or applied"[40] should be tied to an ethical and moral ethic of collective responsibility.[41] This duty, then, requires scientists to not only explain the focus and application of scientific discovery, but, similarly, "the perils which they see may arise from use or abuse of new knowledge."[42] As new "word-menacing problems" arise, there should be an evolving scientific ethic to resolve these problems.[43] Yet, conduct which is likely to produce "limit-situations," for mankind as a whole—or, in other words, those activities "likely to produce dangers of cataclysmic physical or psychological proportions,"[44] should be subject to a duty of restraint.[45]

[35] Thomas Douglas and Julian Savulescu (2010) [25].

[36] George P. Smith, II (1976, 1981) [26, 27].

[37] *See* George P. Smith, II (1992) [28].

[38] Julius Stone (1972) [29].

[39] *Id.*

[40] *Id.*

[41] *Id.*

See George P. Smith, II, *supra* note 39.

[42] Julius Stone, *supra* note 41 at 236–238.

[43] *Id.* at 241.

[44] *Id.* at 240.

[45] *Id.* at 244.

See generally, Neil B. Cohen (1985) [30].

It has been suggested, quite simply, that the commitment to knowledge made by scientists expresses but a basic drive for the enlargement of human powers—a *libido dominandi*.[46] This will to power or desire to dominate is said to be a part of the cultural shift in modern society where this very drive supersedes the search for meaning.[47] Scientists demand the total freedom to experiment and admit that they are guided "only by the very demands of science" and by their "own conscious."[48] Any restrictions placed on this freedom are honored if there are "pressing reasons for public policy."[49]

This position is wholly consistent with the philosophy of Andre Gide, which holds that rather than conform to external standards, personal internal standards are the only valid source of restraint.[50] Thus, one needs only be true to themselves.[51]

There has been a progressive application of scientific knowledge ever since the discovery of fire.[52] Over the years, this discovery spurred one of the most significant transformations seen in human history for the past 500 years—namely, the beginning of the scientific revolution.[53] Today, the individual sense of self and of society is changing just as it did when the early Renaissance spirit swept over medieval Europe.[54]

Both science and democracy encourage not only unconventional opinions and vigorous debate but—as well—demand adequate reasoning, coherent argumentation and vigorous standards for honesty and evidence.[55] Indeed, it has been positioned that humanity's future is inextricably aligned with the future of science.[56] Put simply, science is seen correctly as the best force to satisfy the fundamental quest for knowledge.[57] How successful new technological advances can be utilized—safely and ethically—in harnessing the vast potentialities that derive from it, will determine the extent to which humanity is secured or is limited.[58] A central

---

[46] Julius Stone, *supra* note 41.

See generally, E. Michael Jones (2005) [31].

[47] *Id.*

[48] Stephen L. Carter (1984) [32].

[49] *Id.*

[50] Joel Finebert and Jules Coleman (2004) [33].

[51] *Id.*

[52] *See* Isaac Asimov (1989) [34].

For the past 420 million years, fire has been a part of the story of earth. The first stage of human intersection with fire was recorded 1.5 million years ago. Concrete evidence of the use of flints to start fires was determined to have occurred some 40,000 years ago. The use and control of fire was found to have occurred only 7001 years ago. *See* times.com/5295907/discover-fire/.

[53] *Id.*

[54] Jeremy Rifkin, *supra* note 33.

See generally, George P. Smith, II, *supra* note 33.

[55] Carl Sagan (2003) [35].

[56] Martin Rees (2018) [36].

[57] *See generally*, John D. Bernard (1939) [37].

[58] Martin Rees, *supra* note 59.

challenge to any effort taken here is acceptance of the hard reality that whatever regulations are, prudently, set to constrain science, will simply never be enforced worldwide—because, no effective transnational process exists for supervising.[59]

3 Whole Science

Today, "science claims a monopoly over the steering of human progress;" a process secured and then implemented "through the kinds of engineered solutions that a biology armed with awesome" powers can only be imagined.[60] Science "arrogates, to itself, the right to determine what life is for, along with the capacity to discover and redesign what life is."[61] The fundamental question of the twenty-first century emerging from today's scientific dialogue with law, the then, is how one lives—how "human needs, expectations and desires" are responsive to the hopes, and—indeed—promises, of the New Biology and to the Era of Biotechnology.[62]

4 Junk Science

Inasmuch as few can recognize, clearly, what makes a scientific study good or bad, this uncertainty becomes the basis for questioning the validity of scientific evidence upon which federal-state regulatory programs are structured.[63] Scientific research has many built-in uncertainties—this is because scientists must extrapolate from studies, specific evidence which allows them to recommend proactive measures.[64] "Absolute certainty is rarely an option."[65] Even using the best evidence available for scientific investigations does not protect the work-product from being challenged by corporations on the grounds of insufficiency.[66] Major industries, such as tobacco, chemical, asbestos, lead, and platinum, routinely seek to "manufacture uncertainty" in scientific reports done principally for federal administrative agencies under congressional authority to set rules and regulations governing these industries and their products.[67] These manufactured uncertainties are termed junk science and are

[59] *Id.*

[60] Sheila Jasanoff, *supra* note 32 at 168.
See Isaac Asimov (1960) [38].

[61] *Id.*

[62] *Id.* at 165.
See generally, George P. Smith, II (1998, 1989) [11, 39].

[63] David Michaels and Celeste Monforton (2011) [40].
See Sound Science for Endangered Species in Science and Technology in Congress at 1 (Sept. 2002).

[64] Michaels and Monforton, *id.*

[65] *Id.*

[66] *Id.*

[67] *Id.*

described as "faulty scientific data and analysis used to further a special agenda."[68] Aided by US federal legislation in the Data Quality Act, formal challenges may be made to admit administrative agencies on the grounds that they are of insufficient quality, objectivity, utility, or integrity.[69]

Compounding efforts to maintain research integrity is the underlying realization that much of the scientific information used to formulate regulation comes directly from groups and industries that the government is regulating.[70] Limited federal research funding is the reason for this state of affairs.[71] Greater transparency within the scientific community—itself—is also of considerable concern to the very integrity of scientific investigations.[72] Termed an epidemic of fraud, often scientists fail to report conflicting data in their investigations. Indeed, often researchers[73] use misleading analytical methods of research.[74] And other researchers may succumb to self-deception and proceed to over emphasize only the portions of evidence which support a preferred conclusion[75] which gives rise to falsification of experiments.[76] Although enforcement actions related to misconduct in research and development are focused primarily on the actions of individual researchers, the impact of falsified research extends to affected companies, industries, and the public at large.[77]

5 The Coronavirus Pandemic and Junk Science

Throughout the global coronavirus pandemic, junk science—so to speak—became codified as a work product of many members of the scientific community.[78] Misrepresented personal "scientific" data was routinely presented to the public by pharmaceutical commercial interests[79] and by media outlets such as Facebook,

[68] *Id.*

See Don Agin, Junk Science (2006).

[69] Consolidated Appropriations Act of 2001, Sec. 515, P.L. 106–554.

[70] David Michaels and Wendy Wagner (2003) [41].

[71] *Id.*

[72] Nicolas Chevassus-au-Louis, Fraud in the Lab (2019).

[73] *Id.*

[74] *Id.*

See Sally Satel (2019), at A13 [42].

[75] *See* George J. Annas (1996) [43].

See Richard H. Girgenti, *passim*, (2016) [44].

[76] Nicholas Chevassus, supra note 75.

But see, Betsy McKay and Katie Camero (2020), at 3 [45].

[77] George P. Smith, II (1999) [46].

See generally, *Liberation Theology: The Future*, The Economist at 11, April 6, 2019 (discussing the engineering of living organisms and the changes that will come with).

[78] *See generally*, Marc Zimmer (2020) [47].

See Rammya Matthew (2020) [48].

[79] Sheera Frenkel (2021), at A1 [49].

Instagram, Twitter, and Clubhouse.[80] The White House proceeded to create a "political atmosphere" by pushing the pace of clinical trials for a number of vaccines in order to establish their efficacy with the hope that a "vaccine could be secured before the November election."[81] Moderna, a "Big Pharma," received nearly $2.5 billion to develop, manufacture, and sell to the federal government an efficacious coronavirus vaccine.[82] Interestingly, Pfizer, another major pharmaceutical, chose to keep an "arms length" distance from government assistance and thus declined research and development monies.[83]

Evidence-based science was replaced by countless pseudoscientific assertions and claims made by people with questionable scientific qualifications, which actually complimented "conspiracy theories," which in turn gave rise to medical scams, and promoted specious scientific work.[84] Literally, anything could be published as scientific fact in unaccredited "pay for play" type journals as well as online journals without quality review and any verification of the provenance for sourced material.[85] Unvetted data was published "haphazardly" resulting in a flow of disinformation which only led to public suspicion and confusion.[86]

6 Genetic Modifications

Sir Isaac Newton's third law of physics applies as much to scientific development as it does to other aspects of life itself.[87] Accordingly, for every action there is an equal and opposite direction. Thus, for every new and daring bio-technological advancement, a new medico-legal challenge is presented—a challenge rooted in complex socio-political, religious, moral, and ethical vectors of force. So it is with the issue of

[80] *Id.*

[81] Sharon LaFranier et al. (2020), at A1 [50].
See Paul D. Thacker (2020) [51].

[82] *Id.*

[83] *Id.*

[84] Walter Scheirer (2020) [52].

[85] *Id.*

[86] *Id.*
See Manas Sharma et al. [53].
Interestingly, even with the serious mis-steps seen in the roll-out of an efficacious vaccine to combat the coronavirus, as a consequence of the pandemic, the whole field of science—and more specifically, the pharmaceuticals—has a new burnished image. "Big Science" is no longer thought by the American to be as "money grubbing." Rather, the notable success of Pfizer (with Biotech of Germany), which is the speedy development of a variance against COVID-19, has now placed "Big Pharma" in "a seat firmly at the table" which allows "a chance to be 'good' again." *Reformulated: The Future of Drugmaking*, The Economist, April 10, 2021, at p. 62.

[87] *See* George P. Smith, II (1985) [54].
See also, George P. Smith, II (2005) [55]; George P. Smith, II, *supra* note 39.

germline editing where vast global interest and development is being shown today.[88]

Although genetic research has expanded in recent years, the motivating force behind the New Biology has been basic to human society. Since the time of Plato, people have attempted to improve the human race and research and experimentation in genetics have followed this tradition, seeking to relieve or totally alleviate human suffering that is genetically determined. These research efforts reflect the belief that society as a whole would prosper from methods to make humans more fit because it would be populated by the best physical specimens who, in turn, would begat superior offspring. Some individuals, over the course of history, have been motivated to undertake genetic experiments by the power of possible scientific creation and manipulation.[89]

Essentially, gene therapy—by which alterations are made of genes—may be performed either in germ cells (e.g., sperm or egg cells), or in somatic cells (cells comprising other body tissues).[90] Use of germ-line therapy produces alternations which would be inherited by future generations.[91] Contrariwise, somatic cell therapy affects only the treated individual.[92] Apprehension over the users of gene therapy are focused on the possibility that these interactions will alter the genetic composition of human beings permanently and thereby gradually erode concept of humanity and personhood.[93] Indeed, intervention into the reproductive process creates deep concerns—if not fears—that "biological knowledge" will give rise to "biological reductivism" and be used to denigrate the rights of personhood—rights which all individuals enjoy to autonomy, dignity, and personal integrity; and rights justified traditionally as protected civil liberties ensured by the United States Constitution.[94]

There are latent fears that the state could well use gene therapy to not only modify human behavior, but to engineer new breeds of humans, possibly through cross-species transfer of genes or even by cloning existing individuals.[95] The net effect of the use of these scientific techniques is that the very diversity of the whole gene pool would be compromised.[96] Presently, the most rational line of defense against such "what-if" scenarios is gene therapy treatments which will most likely be feasible

---

[88] Adolf Hitler, through his program termed, *Lebensborn*, or Fountain of Life, undertook genetic experimentation designed to create a "Master Race" in Germany by promotion of a positive eugenics which for him required the ruthless extermination of those seen as carrying inferior genes (e.g., Jews, homosexuals). Smith, *supra* note 44 at 699 *passim*.

[89] Sheila Jasanoff (1990) [56].

[90] *Id.*

[91] *Id.*

[92] *Id.*

See Barbara Pfiffer Billauer (2020) [57].

[93] *Id.*

See Thomas Douglas and Julian Sevulescu, *supra* note 38.

[94] *Id.*

[95] *Id.*

[96] *Id.* at 276.

only in the treatment of a limited group of disorders caused by single genetic defects and not multi-factorial conditions such as schizophrenic apprehension.[97]

As with medical treatments which carry positive benefits as well as serious risks, should gene therapy become—over the course of time—a common technique, it will present a wide range of legal issues with constitutional significance.[98] For example, questions of religious freedom may well arise in connection with parental refusals to allow gene therapy treatment for minors.[99] Mandated treatment of genetic disorder, as a precondition to receiving a marriage license, would surely raise issues of due process and equal protection.[100] So long as the judiciary analyzes such policies within a traditional public health framework, state action could be validated.[101] Efforts to control genetic disorders could be analogized to compulsory vaccination, which the United States Supreme Court upheld in the case of Jacobson v. Massachusetts in 1905 was a legitimate state policy designed to prevent the spread of communicable diseases.[102] Accordingly, it could be argued that mandatory gene therapy would similarly prevent the vertical transmissions of disease from one generation to the next.[103]

6.1 Gene Editing

In the spring of 2014, the White House Office of Science and Technology held firm to the policy that embryonic gene-editing, at least for the present, should not be undertaken. In other words, no alteration of the human germline for clinical purposes could be funded with federal research funds.[104] This was policy buttressed by reports from the National Academics of Science, Engineering, and Medicine in 2016–2017,[105] and by a specific 1996 Congressional prohibition, in the Dickey-Wicker Amendment, for federal research monies to be used on embryonic assisted reproduction.[106]

Human embryos were first edited in 2017 and, then, in 2018, a Chinese scientist reported that he had gene-edited twin girls born in 2018.[107] Both of these scientific

[97] Smith, *supra* note 39 at 705 *passim*.
[98] Jasanoff, *supra* note 92 at 276–277.
[99] *Id.*
[100] *Id.*
[101] 197 U.S. 11 (1905).
[102] Jasanoff, *supra* note 92 at 276–277.
[103] *See* Paige Winfield Cunningham (2015) [58].
[104] Raymond C. O'Brien (2019) [59].
See Myrisha S. Lewis (2021) [60].
[105] *Id.* at 426, 451.
[106] *Id.* at 449 *passim*.
[107] *Id.*

achievements were achieved "independent of governmental regulations or recommendations" and established clearly once again the sovereignty of science.[108]

In 2019, some 18 international scientists and ethicists urged that a voluntary global moratorium, of indefinite length, on all clinical uses of human germline editing be imposed and self-regulated. This moratorium would not apply to germline editing for research purposes. One distinguished scientist at the University of California-Berkeley faulted this proposal specifically because "no pathway toward possible responsible use" was set out in the moratorium, itself.[109]

7 The Precautionary Principle: An Ethic of Restraint or Precaution?

Throughout global society, efforts are made on a daily basis to regulate risk—even when risks of harm are remote.[110] Protecting health and the environment is of paramount importance for maintaining a good society.[111] Stated succinctly, "the [precautionary] principle imposes a burden of proof on those who create potential risks . . . and requires regulation of activities even if it cannot be shown that those activities are likely to produce significant harms."[112]

It has been asserted that the precautionary principle is fast becoming—if not already established in fact—as a binding part of customary law.[113] Whether it is "logically coherent, internally consistent and intellectually appealing" is an open question and largely dependent upon fact-sensitive applications.[114] Indeed, finding a Cartesian gloss of clarity, distinction, and objectivity within the principle of precaution is problematic.[115] Even though incorporated into the laws of the European Union through the Treaty on the Functioning of the European Union (TFEU), the Principle defies a uniform interpretation.[116] The European Commission adopted the principle and the implementing guidelines in 2000.[117] The Principle applies to all EU actions in all areas of health and safety.[118] In the United States, the Congress has—principally through the National Environmental Policy Act[119] and the Clean

---

[108] Eli Adashi and I. Glenn Cohen [61].

[109] Smith, *supra* note 39 at 719–720

[110] *Id.*

[111] *Id.*

[112] *Id.* at 1003

[113] *Id.* at 1005

[114] Patrick Jiang (2014) [62].

[115] *See* Albert R. Jonsen (1990) [63].

[116] Cass Sunstein (1997) [64]; Jiang, *supra* note 117.

[117] Sunstein, *id.* at 1007
See Jonathan B. Weiner and Michael Rogers (2002) [65].

[118] Jiang, *supra* note 117 at 495.

[119] N.E.P.A., 42 U.S. Code Sec. 102 et seq. (1969).

Air Act[120]—brought into clear focus a notion of precaution and protection of the environment.[121]

The first international recognition of the Principle of Precaution was seen in the 1982 United Nations World Charter for Nature where it was suggested that in circumstances were "potential adverse effects are not fully understood, the activities should not be proceed."[122]

Today, there are several differing viewpoints regarding implementation of the precautionary principle,[123] with one understanding being that "a lack of decisive evidence of harm should not be grounds for refusing to regulate."[124] The Commission on the Precautionary Principle is agreed that measures based on the Principle, "should not be blindly precautionary" but—rather—be "set within a structural approach to the analysis of risk" and through assessments of the risk, itself, and from "risk management and risk communication."[125] Others suggest that the Principle requires that a "margin of safety" should be evident in all decision-making made under it.[126]

7.1 Judicial Recognition of the Precautionary Principle

Within the law of the European Community, the case law of the Court of First Instance, now, the Court of General Jurisdiction, dealing with the Precautionary

See Richard Lazarus (2012) [66].

Environmental impact statements are essentially cost-benefit models. A number of states have adopted state Environmental Protection Acts which parallel the Federal Act and require state environmental impact statement of undertakings which could adversely affect state environments. *See* ballotpolicies.org/state_environmental_ policy_acts.

See Sarah Langberg, *Environmental Impacts on NEPA EIS: The Case for Addressing the Importance of Substantive Regimes*, 124 Yale L. J. 576 (2014–2015).

[120] The Clean Air Act, 42 U.S.C. Sec. 7409(b) 1 (2000).

In setting national primary ambient air quality standards, the Act requires the Environmental Protection Agency to maintain "adequate margins of safety to protect the public health in the national standards." Sec. 109.

[121] Sunstein, *supra* note 117 at 1006–1007 at n. 17, p. 1031.

No doubt drawing upon the powerful mandate of NEPA to safeguard the environment, in 1993, President William J. Clinton issued Ex. Order No. 12866, requiring all federal agencies to consider the degree and nature of risks posed by their activities and reduce all risks to public health, safety and the environment. *See* 58 FED. REG. 51, 735 (Oct. 4, 1993). In 2007, the OMB updated the Principles of Risk Analysis. *National Research Council, National Academy of Sciences, Scientific Review of the Proposed Risk Assessment Bulletin from the Office of Management and Budget* (2007).

[122] World Charter for Nature, G. A. Res. 7, U.N. GAOR. 37th Sess., Annex, Agenda Item 21, at 5 U. N. Doc. A/RES/37/7 (1982).

[123] Sunstein, *supra* note 119 at 1014.

[124] *Id.* at 1011.

[125] *Id.* at 1017.

[126] *Id.* at 1013.

Principle, shows a judicial temperament which has found a point of balancing in its decision-making, while respecting the European Communities' legal order and conceding that the Community's institutions have "a certain right of appreciation in this field," yet stating clearly that the judicial review of decisions made under the present system "is thorough enough to prevent abusive reliance on the precautionary principle."[127]

The Council of Europe represents the governments of individual member countries, while the European Commission represents the interests of the European Union as a whole.[128] The Commission has adopted working "doctrines" which govern the precautionary principle, [129] and the European Court of Justice has chosen to give wide deference to the Commission's decisions.[130] The Commission has determined that, in addition to following the general principles of EU law-making, the Principle of Precaution should "be informed, reasoned, and not arbitrary,"[131] as well as show respect for the common principles of proportionality, non-discrimination and legal certainty.[132] Further, "all available scientific evidence" should be considered in decision-making in order to attain full knowledge of known facts, and, further, that consideration should be "holistic" which in turn requires a cost-benefit analysis of economic and non-economic factors[133] in both the short and the long term be undertaken.[134]

Interestingly, even with agreement upon "ground rules" for use of the precautionary principle by the courts, in practice while "policymakers ostensibly pay due deference to scientific opinion, the final assessment of risk and application of the precaution principle will be policy-driven rather than based on science."[135] When doubt arises, a preference may be to eliminate risk by imposing a ban, rather than a cost-benefit analysis that includes the damage caused by banning a potentially useful product.

---

[127] Patrick Jiang, supra note 117 at 515, 516.
See Olivier Segnana (2002) [67].

[128] *See* Veerle Hegvaert (2006) [68].
See also, Segnana, *id.*

[129] *Id.*

[130] *Id.*
See generally, Sunstein, *supra* note 119.

[131] Jiang, *supra* note 117 at 491, 496.

[132] *Id.*

[133] *Id.* at 497.

[134] *Id.*

[135] Forrester and J. C. Hanekemp (2007) [69] [this conclusion is drawn from the decision of the Court of First Instance in the Pfizer judgment, T-13/99 at http://curia.eu.int. with reference to the Alpharma judgment, T-7011 also at http://curia.edu.int.].
See Segnana, *supra* note 130.

8 Legal Weaknesses

The sentiments that Federal Judge David L. Bazelon expressed in 1977 regarding the extent to which the judiciary is challenged by the emerging new technologies of the day are still pertinent today.[136] Then, as now, judges were seen as "technically illiterate," with little knowledge or training allowing them to assess competing scientific arguments. So, today—it may be asserted the same situation prevails.[137] The central role of judges is to "scrutinize and monitor the decision-making process to make sure that it is thorough, complete, and rational; that all relevant information as possible, those who will be affected by a decision have had an opportunity to participate in it."[138] When judges are required to consider highly technical and scientific evidence,[139] fulfilling this central role is exceedingly problematic in order to attain a level of "scientific consciousness."[140]

At the very core of this complex issue of scientific judicial review is the need for broader public participation in the administrative process.[141] Legislatures traditionally make value choices in reviewing legislative proposals and proceeding to enactment into legislation.[142] Today, this level of legislative scrutiny is delegated to administrative agencies.[143] In order to "manage" judicial review of regulatory actions by administrative agencies, citizens must be informed and "activated" at the basic or very first level of problematic issues. Ballot referenda and initiatives are—without doubt—the best way for the public to express their view and preferences to actual legislative proposals.[144] Absent voter participation and choice over problematic issues in legislative proposals, when real legal issues arise, these issues are ideally presented at the regulatory stage for proper hearings. It is at this stage that the judiciary is asked to referee the issues and resolve them.[145]

The "technical illiteracy" of the courts in a distinct way tracks with the passive ignorance and indifference shown by many young people in public and industrialized countries—an indifference only shaken when socio-political affairs

---

[136] David L. Bazelon (1942) [70].

See Smith, *supra* note 80.

[137] *Id.*

See Sheila Jasanoff, *supra* note 32.

[138] Leon R. Yankwich (1975) [71].

A more succinct judicial philosophy is to be found with Justice Brett Kavanaugh's notion that the duty of judges is to be impartial and merely follow the law and not re-make it. Brett Kavanaugh (2016) [72].

[139] Bazelon, *supra* note 139 at 823.

[140] *Id.* at 826, 828.

See Smith, *supra* note 80.

[141] Bazelon, *supra* note 139 at 829.

[142] *Id.*

[143] *Id.* at 829, 830.

[144] Caroline J. Tolbert and David A. Smith (2005) [73].

[145] Bazelon, *supra* note 139 at 282–830.

have an impact on their individual and immediate well-being.[146] Compounding this situation with young people is (a) realization that a considerable fraction of the whole populations of developed countries have had no education in science.[147] A revealing public opinion survey conducted by Eurobarometer in 2005, found disturbing data.[148] The survey revealed that on average, only half of the Europeans surveyed knew that electrons are smaller than atoms; almost a third of the sample believed that the sun goes around the earth and, nearly a quarter of them affirmed that the earliest humans co-existed with dinosaurs.[149]

"Law lag," as a term of art, captures nicely that law is reactionary to science and technology and not in a close partnership with it.[150] Indeed, encoded within this phrase is a tacit recognition of a hierarchical relationship which exists presently between science and law as it "regards the protection of life."[151]

Science promotes progress, while the law seeks to "extricate itself from outdated principles" and maintain its relevance by updating social values—all in order to keep pace with new scientific knowledge and maintain social order.[152] Throughout the 1980s, the courts largely accepted the "promises"—often tantalizing—made by "scientific optimism."[153] This optimism, originating as it did during the European Enlightenment, emphasized "reason and individuality rather that corporate tradition."[154] Man considered himself a part of nature, if not a member.[155] Fear of science was replaced by an attitude of positivism and participation.[156] The judicial attitude which evolved during the 1980s was to embrace science and not restrict its "claims of technological progress."[157] Today, although there are some small signs of a scientific confluence between law and science, the predominant mantra remains "scientific sovereignty," with the judiciary still retreating.[158]

---

146 Valenti Rull (2014) [74].

See David Bazelon, *supra* note 139 at 90.

147 Europeans, Science and Technology (2005) [75].

148 *Id.*

149 Sheila Jasanoff, *supra* note 32 at 68 *passim*.

150 *Id.*

151 *Id.*

152 *Id.*

153 *Id.* at 79.

154 O'Brien, *supra* note 107 at 477–478.

155 *Id.*

156 *Id.*

157 Sheila Jasanoff, *supra* note 32 at 79.

158 *Id.* at 79.

9 Conclusions

Fundamental to recognizing a philosophy of science is to understand, and then to accept, as Max Planck did, that science—in and of itself—is a positive value to society when its "products" are useful and practical and, as a matter of course, provide work opportunities.[159] Thus, science is a source for good. For Planck, in order to shape and utilize a philosophy of science which serves as a frame for rational investigations of principles of being, knowledge, or conduct, it must be understood fully before discoveries should be pursued.[160]

Martin Heidegger's philosophy of science seeks to study and then utilize the technology of science as a tool for viewing the whole of life, and as a construct for pursuing a good life, a healthy life. Put simply, science is a means to an end.[161]

In contemporary society, science advances applications of scientific knowledge which are then used to satisfy basic human needs and maintain proper healthful living standards which, in turn, secure social order.[162] Most of the tools of technology—specifically biotechnology—are by-products of scientific efforts.[163] Scientific research is said to satisfy the human thirst for knowledge and, thus, enhance human cultural heritage which is knowledge-based. The vexing question in this, the Age of Modernity, is whether scientific research should be dedicated to the service of human needs and social order[164] or, whether scientific research should be unshackled and allowed to follow the paths that scientists wish to pursue for the advancement of knowledge. Accepting this second alternative assures the sovereignty of science.[165]

In order to maintain social order, laws need to coalesce with values which undergird the whole of society.[166] Today's contemporary values should be understood as being shaped not only by cultural norms, but by science, medicine and biotechnology, all which in turn guide and, indeed, establish news order of conduct.[167] In order to attain a point of equilibrium in the modern state, the law needs to not only oversee, but direct and regulate the courses of scientific conduct which safeguard life, liberty, and the pursuit of happiness.[168] In poplar government, the powers of common sense weigh heavily against "the powers of paradox" or, in other words, the "treasury of scientific knowledge."[169] But, if the standard of living and,

---

[159] *Supra* notes 2–9.

[160] *Id.*

[161] *Supra* notes 10–17.

[162] *Id.*

[163] See Rull, *supra* note 149.

[164] *See* John D. Bernal, *supra* note 74.

[165] *See* Sheila Jasanoff, *supra* note 32 at 67 *passim.*

[166] *Id.*

[167] *Id.*

[168] *Id.*

[169] Daniel J. Boorstein (2011) [76].

indeed, survival—is to have an enduring significance, a democratic society must use common sense in allowing scientific progress.[170]

No doubt, the central weakness to the quest for an ideal state of equilibrium is that the law's responsibilities are simply not being met.[171] The courts continue to struggle to understand complex scientific cases, the regulatory agencies charged with executing legislative mandates are regularly entangled by uninformed and mis-directed mandates for rule-making and scientific certainties, while society remains uninformed, lethargic, and unwilling to accept any responsibilities for participation in a deliberative democracy.[172]

Put simply, scientists need to explain to lawyers and the general society more fundamentally and in understandable terms the work of science.[173] And, at the same time, lawyers need to listen more intently and participate to the scientific dialogue.[174] Or, as Kenneth C. Frazier, the CEO of Merck said in May 2013, at the 90th Anniversary Dinner of The American Law Institute, members of the legal profession need to adopt a "grander vision" of what lawyering is about and understand that problem-solving requires "creative thinking" drawn from sources of knowledge found in cross disciplines.[175]

References

1. https://www.google.com/search?q=The+Philosophy+of+Max+Planck&rlz=1C1GCEU_enUS825&oq= The+Philosophy+of+Max+Planck&aqs=chr.
2. Stein J. Planck's constant: the number that rules technology, reality and life. https://wwwpbs.org/wbgh/nova/article/plancks-constant/.
3. Planck M. The philosophy of physics. W. H. Johnson, trans. London: George Allen and Unwin, Ltd; 1936.
4. https://www.iep.utm.eduheidegge/; https://www.lep.utm.edu/heidegger/InternetEncylopediaofPhilosophy.
5. Heidegger M. The question concerning technology and other essays. William Lovitt, trans. New York: Garland; 1977.
6. Heidegger M. Basic writings., David Farrell Krell, ed. New York: Garland; 1977.
7. Blitz M. Understanding Heidegger on technology. https://thenewatlantic.com.publications/understanding-heidegger-on-technology.
8. Walton R. Heidegger in the hands-on science and technology center: philosophical reflections on learning in informal settings. J Technol Educ. 2000;12:50.
9. Cavalieri LF. Ch. 9: Science as technology. In: Smith II GP, editor. Ethical, legal and social challenges to the brave new world. Port Washington: Associated Faculty Press; 1882.
10. Weber D. Reining in the hubris of science and scientists. Washington Post. 8 Feb 2020.
11. Smith GP II. The new biology: law, ethics and biotechnology. New York: Plenum Press; 1989.

[170] *Id.*

[171] *See* George P. Smith, II (1988) [77].

[172] *See* Smith, *supra* note 80.

[173] Lewis Thomas (1985) [78].

[174] *Id.*

[175] Kenneth Frazier (2013) [79].

12. Smith GP II. Distributive justice and health care. J Contemp Health L & Pol'y. 2002;18:421.
13. George P. Smith, II, Policy making and the new medicine: managing a magnificent obsession, 3 J Health Biomed Law 3, 303 (2008).
14. Trade Regulation. The Brussels effect. The Economist. 22 Feb 2020, p. 63.
15. Bradford A. The Brussels effect: how the European Union rules the world. Oxford: Oxford University Press; 2019.
16. Osimani B. The precautionary principle in the pharmaceutical domain: a philosophical enquiry into probabilities reasoning and risk aversion. Health Risk Soc. 2013;15:123.
17. Cass R. Sunstein, Beyond the precautionary principle. Pa. L Rev. 1997;151:1002.
18. Curl RF. Banquet Speech, The Nobel Prizes. http://www.nobleprize.org/prizes/chemistry/1996/curl.speech/.
19. Oreskes N, Lindee MS, Edenhofer O. Why trust science. Princeton: Princeton University Press; 2019.
20. Weber D. Reining in the hubris of science and scientists. Washington Post. 8 Feb 2019
21. Jasanoff S. Can science make sense of life? Cambridge: Polity Press; 2018.
22. Rifkin J. The biotech century: harming the gene and remaking the world. New York: Jeremy P. Tarcher/Putnam; 1998.
23. Smith GP II. De Lege Latta, De Lege Ferenda. J Contemp Health Law Policy. 1993;9:233.
24. Lederer WJ. A nation of sheep. New York: Norton; 1961.
25. Douglas T, Savulescu J. The ethics of knowledge. J Med Ethics. 2010;36, 687
26. Smith GP II. Manipulating the genetic code: jurisprudential coneendrums. Georgetown Law J. 1976;64:697–725.
27. George P. Smith, II, Genetics, ethics and the law (1981). Associated Faculty Press, Gaithersburg
28. Smith GP II. Toward an international standard inquiry. Health Matrix. 1992;2:167.
29. Stone J. Knowledge, survival and the duties of science. Am U L Rev. 1972;28:231–6.
30. Cohen NB. Confidence in probability: burdens of persuasion in a world of imperfect knowledge. NY U L Rev. 1985;60:385.
31. Michael Jones E. Libido Dominandi: sexual liberties and political control. South Bend: St. Augustine's Press; 2005.
32. Carter SL. The Bellman, The Snark and The Biohazard Debate. Yale Law Policy Rev. 1984;3:358–66.
33. Finebert J, Coleman J. Philosophy of law. 7th ed. Belmont: Thomson/Wadsworth; 2004. p. 479.
34. Asimov I. Chronology of science and discovery. New York: Harper & Row; 1989.
35. Sagan C. Describing the world as it is, not as it would be. In: Arana M, editor. The writing life: writers on how they think and work. New York: Public Affairs; 2003. p. 309.
36. Rees M. On the future: prospects for humanity. Princeton: Princeton University Press; 2018.
37. Bernard JD. The social functioning of science. New York: Macmillan; 1939.
38. Asimov I. The intelligent man's guide to science. New York: Washington Square Press; 1960.
39. Smith GP II. Family values and the new society: dilemmas of the 21st century. Westport: Praeger; 1998.
40. Michaels D, Monforton C. Manufacturing uncertainty: contested science and the protection of the public's health and environment. Am J Pub Health. 2011;95:S39.
41. Michaels D, Wagner W. Disclosures in regulating science. Science. 2003;302:2073.
42. Satel S. Fraud in the lab: experiments in doubt. Wall St J. 2019:A13.
43. Annas GJ. Questing for grails: duplicity, betrayal & self-deception in past medical research. J Contemp Health L Pol'y. 1996;12:297.
44. Girgenti RH. The new era of regulatory enforcement. New York: McGraw Hill Education; 2016. p. 247.
45. McKay B, Camero K. Sharing data faster to fight an epidemic. Wall St J. 2020;22:3.
46. Smith GP II. Judicial decision-making in the age of biotechnology, Notre Dame. J Ethics & Pub Pol'y. 1999;13:93–4.
47. Zimmer M. The state of science: what the future holds and the scientists making it happen. Guilford: Prometheus Books; 2020.

48. Matthew R. We must not be guided by bad science on COVID-19. Brit Med J. 2020; https://doi.org/10.1136/bmj.m2241.
49. Frenkel S. Misinformation deepens gap in vaccinations. N Y Times. 11 Mar 2021.
50. LaFranier S, Thomas K, Weiland N, et al. Politics, science and the remarkable race for a viable vaccine. N.Y. Times. 22 Nov 2020.
51. Thacker PD. The U.S. politicization of the pandemic: Raul Grijalva on masks, BAME, and Covid-19. BMJ. 2020;370:m3430. https://doi.org/10.1136/bmj.m3430.
52. Scheirer W. A pandemic of bad science. Bull At Sci. 2020;76:175.
53. Sharma M, Scarr S, Kelland K. Speed science: the risk of swiftly spreading Coronavirus research. https://www.reuters.com/article/us-china-health-research-analysis/speed-science-the-risks-of-swiftly-spreading-coronavirus-research-idUSKBN20D21S.
54. George P. Smith, II, Genetics, eugenics and public policy, 1985 South Ill Law Rev 435.
55. Smith GP II. The Christian religion: a search for principled decision making. Dordrecht: Springer; 2005.
56. Jasanoff S. Biology and the bill of rights: can science reframe the constitution? Am J L Med. 1990;13:249–75.
57. Billauer BP. Wrongful life in the age of CRISPR-CAS: using the legal fiction of the concepted being to redress wrongful gamete manipulation. Penn St L Rev. 2020;123:435.
58. Cunningham PW. Bipartisan worries over designer babies. Washington Examiner. 2015;12:13.
59. O'Brien RC. The immediacy of genome editing and mitochondrial replacement. Wake Forest J L Pol'y. 2019;9:419., 442–444, 471, 479–480.
60. Lewis MS. Is germline gene editing exceptional? Seton Hall L Rev. 2021;51:735.
61. Adashi E, Cohen IG. JAMA forum: is a moratorium needed?. https://newsatjama.com/2019/06/03/jama-forum-heritable-genome-editing-is-a-moratorium-needed/.
62. Jiang P. A uniform precautionary principle under EU law. Peking University Transnational Rev. 2014;2:492.
63. Albert R. Jonsen, The new medicine and the old ethics 1 (1990), Harvard University Press, Cambridge
64. Sunstein C. Beyond the precautionary principle. Pa L Rev. 1997;151:1003–7.
65. Weiner JB, Rogers M. Comparing precaution in the United States and Europe. J. Risk Res. 2002;5:317.
66. Lazarus R. The National Environmental Act in the United States Supreme Court: A re-appraisal and a peek behind curtains. Geo L J. 2012;100:1507.
67. Segnana O. The precautionary principle: new development in the case law of the court of first inovance. German L J. 2002;3. http://www.germanlawjournal.com/index.php?pageID=198
68. Hegvaert V. Facing the consequences of the precautionary principle of European Community Law. European Law Rev. 2006;31:185.
69. Forrester I, Hanekemp JC. Precaution, science and jurisprudence: a test case. J Risk Res. 2007;9:297.
70. Bazelon DL. Copying with technology through the legal process. Cornell L Rev. 1942;62:817.
71. Yankwich LR. The art of being a judge. In: Winters GR, editor. Handbook for judges. Princeton: Princeton University Press; 1975. p. 12.
72. Kavanaugh B. The judge as umpire: the principles. Cath U L Rev. 2016;65:681.
73. Tolbert CJ, Smith DA. The educative effects of ballot initiatives on voter turnout. Am Politics Res. 2005;33:283.
74. Rull V. The most important application of science. EMBO J. 2014;15:919.
75. Europeans, Science and Technology. 2005. http://ed.europa.eu/public_opinion/archives/eles_224_report_en.pdf.

76. Boorstein DJ. Cleopatra's nose: essays on the unexpected. New York: Vintage eBooks; 2011. p. 177.
77. Smith GP II. Biotechnology and the law: social responsibility or freedom of scientific inquiry? Mercer L Rev. 1988;39:437–8.
78. Thomas L. Overview: regulating the biohazard debate. Yale L Pol'y Rev. 1985;3:309–14.
79. Frazier K. Remarks. Washington, DC: American Law Institute; 2013.

Epilogue: 'Did You Know?' (that you are likely to be awarded the Nobel Prize)

Robert Huber

This is probably the most frequently asked question for Laureates and a frequent answer is: **'No, it was a great surprise'**

My answer is different, as I was quite sure that it would happen in 1988 for the Nobel Prize in Chemistry.

The story is as follows:

I had a large research group at the Max-Planck Institute of Biochemistry in Martinsried near München which was engaged in several research projects, in parallel, focussing on the structure and function of proteins, one was the photosynthetic reaction centre, in a collaboration with Hartmut Michel and Johann Deisenhofer at the same institute (work that led to the Nobel Prize)

One of these projects was a plant ascorbate oxidase, a member of the blue copper oxidase family. A pioneer of this protein family was Professor Bo. G. Malmström, who was chairman of the Nobel Committee for Chemistry from 1977 to 1988. He had focussed on the closely related fungal laccase, which he studied with biophysical, spectroscopic methods and identified different types of copper, but, not unexpected, could not define their spatial arrangement and ligation in the protein.

Albrecht Messerschmidt in my research group together with Italian colleagues was engaged in the crystal structure determination of the large protein assembly of ascorbate oxidase and we were working on the manuscript for publication by the end of 1987, which described the structure of the copper complex in full atomic detail. This achievement did not remain secret in the blue oxidase research community and I was approached by Bo. Malmström asking for visit in Martinsried in February 1988 to see 'his' molecule in all its beauty. I was honoured and happy to receive him for two days to discuss structure and function

R. Huber (✉)
Max-Planck-Institut für Biochemie, Martinsried, Germany
e-mail: huber@biochem.mpg.de

U. A. Betz (ed.), *Curious Future Insight*,
https://doi.org/10.1007/978-3-031-41781-8_20

of ascorbate oxidase. Before he left, I accompanied him to the airport, he asked me to find for him passport photographs of Johann Deisenhofer, Hartmut Michel, and myself. This I did. For me, a most likely reason for the collection of photos of the three people involved in the reaction centre study was Nobel. Indeed months later in December 1988, these photos were reproduced on the Nobel Poster. I did not talk to anybody, not to my family, not to my two colleagues about Malmstöm's request and my conjecture, but in October I answered the call from Sweden and may say now: **'I did know'**

Robert Huber,
10 8 2022.

Correction to: Curious Future Insight

Ulrich A.K. Betz

Correction to:
U. A. Betz (ed.), *Curious Future Insight*,
https://doi.org/10.1007/978-3-031-41781-8

Due to an unfortunate error, the chapters "From a 350th Anniversary to a Global Movement: United by Science for a Better Tomorrow" Written by Ulrich A. K. Betz and "Advancing Human Progress as a Twenty-First-Century Science and Technology Pioneer" written by Belén Garijo, Matthias Heinzel, Peter Guenter, Kai Beckmann, and Marcus Kuhnert were originally published electronically on the publisher's internet portal without open access. The copyright of these chapters changed on 23rd February 2023 to © The Author(s), 2024, and these chapters are forthwith distributed under a Creative Commons Attribution 4.0 International Licence (http://creativecommons.org/licenses/by/4.0/), which permits use, sharing, adaptation, distribution, and reproduction in any medium or format, as long as you give appropriate credit to the original author(s) and the source, provide a link to the Creative Commons licence, and indicate if changes were made. The images or other third-party material in this chapter are included in the chapter's Creative Commons Licence, unless indicated otherwise in a credit line to the material. If material is not included in the chapter's Creative Commons Licence and your intended use is not permitted by statutory regulation or exceeds the permitted use, you will need to obtain permission directly from the copyright holder.

The updated versions of the chapters can be found at
https://doi.org/10.1007/978-3-031-41781-8_1
https://doi.org/10.1007/978-3-031-41781-8_2

U. A. Betz (ed.), *Curious Future Insight*,
https://doi.org/10.1007/978-3-031-41781-8_21

Keynotes with Abstracts

Peter C. Doherty
University of Melbourne
Challenges, opportunities, and removing roadblocks

The discussion will focus on issues that threaten the future of humanity, including our apparent inability to take meaningful action in the face of obvious dangers. It may also emphasize a Southern perspective.

Sally Rockey
The Foundation for Food and Agriculture Research
Changing the World Through Agriculture Innovation

Growing populations, climate change, depleted natural resources threaten humanity. Agriculture must nourish the planet with healthy food while using less land and resources, remain resilient to shifting weather patterns, reduce its environmental impact, all while providing economic opportunity for farmers. Investing in agriculture innovation today ensures we are a healthy and vibrant society. The rapidly increasing pace of science has proven to be a game changer for agriculture and food production as new technologies may be applied to this sector quicker than any other. How the new innovations apply to the biggest challenges in agriculture is the topic of the talk.

Ada Yonath
Weizmann Institute of Science
From origin of life to next-generation therapeutics

In ribosomes, the site for peptide bond formation, the PTC, is located within a highly conserved internal pocket made exclusively of rRNA. The high conservation implies its existence irrespective of environmental conditions and indicates that it may represent a prebiotic RNA machine, which could be the kernel around which life originated. Lab constructs imitating this pocket possess capabilities for peptide bond formations, thus indicating that a molecular prebiotic bonding entity still exists and functions within ribosomes of all living cells. In contrast, specific structural

U. A. Betz (ed.), *Curious Future Insight*,
https://doi.org/10.1007/978-3-031-41781-8

features of ribosomes related to genetic diseases, or in antibiotics resistant pathogens are being used as bases for next-generation therapeutics.

Sandrine Dixson-Declève
Club of Rome

From The Limits to Growth to Wellbeing economics: a new approach for human and planetary health

In 1972 The Limits to Growth was the first report to model our planet's interconnected systems and make clear that if growth trends in population, industrialization, resource use, and pollution continued unchanged, we would reach and then overshoot the carrying capacity of the Earth at some point in the next one hundred years. 50 years on, we are experiencing the real impact of the encroachment of humanity on these limits through COVID-19, climate change and conflict and more than ever we need to stop pursuing growth at all costs and instead take a holistic approach and move toward wellbeing economics.

Brian Kobilka
Stanford University School of Medicine0

G Protein-Coupled Receptors: Challenges in Drug Discovery

G protein-coupled receptors (GPCRs) conduct the majority of transmembrane responses to hormones and neurotransmitters and represent the largest class of drug targets for the pharmaceutical industry.

Özlem Türeci
BioNTech

How a novel approach against cancer helps combat the COVID-19 pandemic

With almost six million suspected deaths worldwide and counting, the ongoing COVID-19 pandemic has been one of the deadliest disasters in human history. Once its global spread and destructive force became clear at the beginning of 2020, scientists around the world raced to develop a vaccine to stop the novel virus in its tracks and save millions of lives. Prof. Özlem Türeci, Co-Founder and Chief Medical Officer of BioNTech, is a physician, immunologist, cancer researcher, and company founder. With her husband Ugur Sahin, she has pioneered a number of fields including cancer antigen discovery, individualized on demand customized mRNA cancer vaccine approaches and other types of immunotherapies which are currently in clinical development. Along with her team at BioNTech, she developed one of the first mRNA vaccines for effectively preventing COVID-19. Türeci will share her perspective on how to turn innovations into medicines by showcasing the development of the mRNA vaccine technology.

Joachim Frank
Columbia University
Introducing the fourth dimension to cryo-EM

Two techniques will be highlighted that are going beyond the current single-particle cryo-EM of macromolecules and provide information on their dynamic features. One is time-resolved cryo-EM of short-lived states, the other is the mapping of states in a continuum from information hidden in large datasets of projection data.

Jacquelyn Francis, Frederique Seidel, Lukas Mörchen
Global Warming Mitigation Project
Keeling Curve Prize: Winners Announcement

The climate crisis is dire, and our expert team of climate scientists and analysts understand its root cause—and how to solve it. We invite conference attendees searching for the answers to "What can I do to help?" to attend our presentation and recognize the case for climate ACTION. You'll hear an inspiring and thought-provoking discussion facilitated by Jacquelyn Francis, featuring some of the organizations that have won the Keeling Curve Prize in previous years. Jacquelyn will announce the 2022 winners of the prize, that are all paving the way to a livable future.

Jürgen Schmidhuber
NNAISENSE, IDSIA, USI &amp
Past, Present, Future, and Far Future of Artificial Intelligence

Significant historic events since the Big Bang appear to be occurring more frequently as time goes on. Interestingly, it seems like subsequent intervals between these events are shrinking exponentially by a factor of four. This process looks like it should converge around the year 2040. The last of these major events can be said to have occurred around 1990 when the cold war ended, the WWW was born, mobile phones became mainstream, the first self-driving cars appeared, and modern AI with very deep artificial neural networks came into being. In this talk, I'll focus on the latter, in particular, the "miraculous year of deep learning 1990–91" when our team laid foundations of the most cited neural networks. I'll discuss how in the 2000s this has begun to impact billions of human lives, how the timeline predicts the next big event to be around 2030, what the final decade until convergence might hold, and what will happen in the subsequent 40 billion years. Take all of this with a grain of salt though.

Paul Workman
The Institute of Cancer Research, London
Probing and drugging the cancer genome to overcome tumor evolution and therapy resistance

One in two people will be diagnosed with cancer in their lifetime. We have seen progressively reduced cancer mortality with >50% survival at ten years for cancer patients overall. But many cancers continue to show poor outcomes. We have seen the strong impact of genome sequencing (including now incorporating liquid biopsy) leading to the exciting era of personalised medicine based on predictive biomarkers. But only around 9% of patients with metastatic cancer have actionable molecular targets allowing treatment with precision medicine and only around 5% receive such treatments. Many cancer driver proteins are technically hard to drug—notably numerous transcription factors—and so far only 5% of the cancer genome has been drugged. Immmunotherapy with T cell checkpoint inhibitors is highly effective in some patients but we lack biomarkers to predict patients who will benefit. Tumour heterogeneity and the ability to adapt, evolve and become drug resistant is the major challenge we face. I will describe progress towards extending the drugged cancer genome, including tackling hard-to-drug targets. I will stress the importance of more robust target validation. I will emphasise the use of high-quality chemical probes for functional annotation of the genome and target validation, alongside powerful orthogonal technologies such as RNA interference and CRISPR/Cas9 knockout/editing. I will illustrate the importance of selectivity of chemical probes and drugs (on-target versus off-target effects) and how we can control or exploit polypharmacology. I will illustrate the above points using examples from the 20 drug candidates we have discovered since 2005, with twelve of these entering clinical trial, and will describe how these approaches will allow us to overcome or limit cancer adaption, evolution and drug resistance.

Noam Slonim
IBM
Project Debater: How persuasive can a computer be?

Project Debater is the first AI system that can meaningfully debate a human opponent. The system, an IBM Grand Challenge, is designed to build coherent, convincing speeches on its own, as well as provide rebuttals to the opponent's main arguments. In February 2019, Project Debater competed against Harish Natarajan, who holds the world record for most debate victories, in an event held in San Francisco that was broadcasted live worldwide. In this talk I will tell the story of Project Debater, from conception to a climatic final event, describe its underlying technology, and discuss how it can be leveraged for advancing decision-making and critical thinking.

Katja Becker
German Research Foundation (DFG)
Providing Means for Action

Funding research arising from independent curiosity lays the foundations for rapid yet reliable results, not only in the field of vaccine development but also in relation to other research requirements of the future. The pandemic clearly demonstrates that the best way to prepare for new and unforeseeable societal challenges—be they biomedical or other—is to generate knowledge repositories that are not yet related to specific problems but rather to open questions. Funding such research proves to provide the best means possible for political action in an ever more complex world.

Andrew Dzurak
UNSW Sydney

Quantum computing and near-term quantum technologies + Quantum computing: A new silicon revolution

Quantum computing is experiencing explosive growth, with both research activity and investment at record levels. However, large-scale quantum computers may still be some time off. This talk will cover the important milestones reached and challenges that lie ahead on the path to building a quantum computer. I will also discuss near-term quantum technologies and opportunities that have arisen from quantum computing research, specifically in the area of measurement and spectroscopy.

In the 1950s the invention of the silicon integrated circuit opened the path toward ever increasing computing power that has changed the way we live, work, and communicate. We are now entering a new age of quantum computing that opens horizons for solving global challenges related to human health and climate change. Although the concepts underpinning quantum computing are radically different from those of existing computing, it could be that the silicon microchip, now reimagined, could again provide the platform for this new technology of enormous global significance.

Harald zur Hausen
German Cancer Research Center

Recorded Keynote: Novel infectious agents originally derived from specific bacterial plasmid linked to specific cancers and chronic diseases

By analyzing milk and sera of European dairy cattle, we identified a large number of small single-stranded circular DNAs. Some of them have been identified in common human cancers and also in chronic neurological diseases. Diseases linked to these infections have been labeled as plasmidosis.

Jennifer Chan
Icahn School of Medicine at Mount Sinai

Reproductive tract extracellular vesicles are sufficient to transmit intergenerational stress and program neurodevelopment

Extracellular vesicles (EVs) are a unique mode of intercellular communication capable of incredible specificity in transmitting signals involved in cellular function, including germ cell maturation. Spermatogenesis occurs in the testes, behind a protective barrier to ensure safeguarding of germline DNA from insults in the environment. Outside the testes and following DNA compaction, further sperm cell maturation occurs in the epididymis. Here, we report the novel ability of reproductive tract EVs to transmit information regarding stress in the paternal environment to sperm, ultimately altering fetal development. Using the artificial reproduction technique, intracytoplasmic sperm injection, we found that sperm incubated with EVs collected from stress-treated epididymal epithelial cells produced offspring with significant changes in neurodevelopment and adult stress reactivity. Proteomic and transcriptomic assessment of these secreted EVs showed dramatic changes in protein and miRNA content long after stress treatment had ended, supporting a lasting cellular programmatic change in response to chronic stress. Thus, EVs are a normal part of sperm maturation and also perform additional roles in intergenerational transmission of paternal environmental experience.

George Church
Harvard Medical School

Technologies for Reading and Writing Genomes, Organs & Ecosystems

Technologies for reading and writing short pieces of DNA are (thanks to molecular multiplexing (far beyond parallelization)) exponentially improving in cost (and slightly more slowly length and quality). Consequently thereby creating revolutions downstream in genome-scale engineering, gene and cell therapies, organ transplants, vaccines, aging reversal, and ecosystem restoration.

Maria Leptin
European Research Council

The case for investing in bottom-up, frontier research

Maria Leptin will talk about the importance of investing in bottom-up, frontier research, the relationship between science and technology, the importance of international cooperation in science, how we can assess excellent research and her experiences during her first six months as President of the European Research Council.

Robert Huber
Max Planck Institute of Biochemistry

The century of vision: Protein structures for drug design and development: The proteasome and other cage-forming proteases

Methods and instruments to visualize atoms and molecules were discovered in the past century, undergo further rapid development, and revolutionize the life sciences and medicine. They are an essential, often founding step in the discovery and refinement of novel therapies and medicines. Cage-forming proteases, the proteasome, DPP8/9, and DegP are very significant drug targets and will serve for illustration.

Moritz Helmstaedter
Max Planck Institute for Brain Research
The interactions between natural and artificial intelligence

Modern AI was inspired by brain research more than 60 years ago. Neuroscience has moved forward since, now allowing the mapping of neuronal circuit architecture at ever increasing scale and pace. While this progress requires modern AI to succeed, there is the justified hope that connectomes from cognitively capable animals will inform modern approaches to artificial intelligence that may overcome the limitations of energy and label inefficiency in concurrent AI.

Ron Naaman
Weizmann Institute of Science
The molecular symmetry and the electrons' spin: How are they related and how can we utilize them?

Spin-based properties, applications, and devices are commonly related to magnetic effects and to magnetic materials. However, we found that chiral organic molecules act as spin filters for photoelectrons transmission, in electron transfer, and in electron transport. The new effect, termed Chiral-Induced Spin Selectivity (CISS), was found, among others, in biomolecules and in bio-systems. It has interesting implications for the production of new types of spintronics devices and on electron transfer in biological systems. Our findings shed new light on enantio-specific interactions and it opens the possibility to construct novel methods for enantio-separation.

Sara Seager
MIT
The Search for Life Beyond Earth

For thousands of years people have wondered, "Are there planets like Earth?" "Are such planets common?" "Do any have signs of life?" Today astronomers are poised to answer these ancient questions, having recently found thousands of planets that orbit nearby Sun-like stars, called "exoplanets." Professor Sara Seager, one of the world's leading experts on this search for Earth-like planets and life beyond Earth, will share the latest advances in this revolutionary field.

Martin Rees
University of Cambridge
The world in 2050

The lecture will address the challenges posed by the world's growing and more demanding population. The challenges are of two kinds: those stemming from the growing pressure we're collectively imposing on the biosphere.

Jean-Marie Lehn
University of Strasbourg
Toward Adaptive Chemistry

The implementation of constitutional dynamics in chemical entities points to the emergence of adaptive and evolutive chemistry, toward systems of increasing complexity.

Phil Baran
The Scripps Research Institute
Translational Chemistry

There can be no more noble undertaking than the invention of medicines. Chemists that make up the engine of drug discovery are facing incredible pressure to do more with less in a highly restrictive and regulated process that is destined for failure more than 95% of the time. How can academic chemists working on natural products help these heroes of drug discovery—those in the pharmaceutical industry? With selected examples from our lab and others, this talk will focus on that question highlighting interesting findings in fundamental chemistry and new approaches to scalable chemical synthesis.

Thomas C. Südhof
Stanford University
Understanding synapses: Toward rational treatments of brain disorders

Synapses are the fundamental computational units of the brain and the basic building blocks of neural circuits. In many brain disorders synaptic dysfunction drives pathogenesis, ranging from neuropsychiatric diseases such as autism and schizophrenia to neurodegenerative disorders such as Alzheimer's and Parkinson's disease. My laboratory studies the molecular mechanisms that enable synaptic function and their impairments in disease. We are focused in particular on how synapses are constructed and reconstructed throughout life, resulting in the enormous plasticity of information processing by neural circuits. I will discuss some principal mechanisms involved, in particular those that guide the actual formation of synapses, and their impairments in brain disorders.

Christina Smolke
Stanford University
Unlocking the complex power of natural products with synthetic biology platforms

Although natural products comprise 60% of the small molecule therapeutics market, numerous challenges arise in conventional natural products-based drug discovery and production. Recent breakthroughs in the field of synthetic biology have demonstrated how to use microorganisms such as brewer's yeast to make plant-derived medicines that match the most complicated known biosynthetic pathways found in nature. The ability to engineer self-replicating, precision chemical factories that build natural product-inspired scaffolds from the atoms up will accelerate the discovery of new medicines while scaling bioproduction of existing essential medicines.

William E. Moerner
Stanford University
What Can You Learn with Single Molecules and Light?

A single molecule is ridiculously tiny, about 1 nanometer across, maybe 100,000 times smaller than the diameter of a human hair. Yet, individual molecules rule the nanoscale activity and structure in our cells. Thirty years ago, single molecules were first detected optically, but how do we really detect a single molecule today, and what good is this?

Shirley M. Malcom
AAAS
When Science Meets the Public

Even when scientists are excited about their work and public audiences are curious about their science, there are often barriers to connecting them. For the scientists this may relate to weak communications skills, lack of understanding of their audiences, few incentives for undertaking this work, failure to understand the importance of engagement or a lack of opportunity. It is critical to consider strategies for overcoming these challenges and providing access to science for all.

Abstracts for Short "Ignite" Talks

1. Eleonore Eisath, Beworm, Germany, ignite talk, winner best ignite presentation

Beworm tackles the plastic waste problem through a biocatalytic recycling process inspired by nature

Beworm develops a biocatalytic recycling process for polyethylene, the world's most used plastic material. We isolated plastics attacking bacteria and are analyzing the degradation mechanism to find its key enzymes. These enzymes can be optimized to split down low density polyethylene, the flexible version of PE, into virgin-like raw materials for the petrochemical industry. Compared to mechanical recycling methods, our solution creates a fully circular supply chain as the output is reusable with no downcycling of the plastic quality. The developed process fits within current recycling facilities as a complement, enabling the recyclers to increase the output reusable material from the input sorted plastics. The process will have plastics put in reactors under certain environmental conditions, in contact with the enzyme substrate. The enzymatic reaction will break the chain of polymers into states usable to manufacture new plastics.

2. Cécile Echalier, Institute of Biomolecules Max Mousseron, France, ignite talk

From Silylated Biomolecules to Artificial Extracellular Matrices

Compatible with fragile biological cargos and live cells, the sol-gel process is a versatile cross-linking chemistry to develop customized biomimetic matrices for regenerative medicine. Silylated precursors, obtained by introduction of alkoxysilanes onto (bio)molecules, can be combined and polymerized via the sol-gel process to yield covalent bio-inorganic tridimensional matrices. Hydrogels based on silylated synthetic polymers (PEG), peptides (collagen-inspired), polysaccharides (HPMC, hyaluronic acid, chitosan) and, more recently, proteins (gelatin) were prepared. Covalent functionalization of the network with bioactive peptides or fluorophores during the polymerization process provides additional functionality to the materials such as cell-adhesive or antibacterial properties. Live cells, in particular stem cells, can be encapsulated without affecting their viability

U. A. Betz (ed.), *Curious Future Insight*,
https://doi.org/10.1007/978-3-031-41781-8

and function. Cell-laden solutions can be casted or used as bioinks to 3D print scaffolds for tissue engineering.

3. Dragana Savic, University of Oxford, UK ignite talk,

Hyperpolarized Magnetic Resonance Imaging (MRI); Accelerating MRI, Visualizing Metabolism

In the history of medical science, no technology has gone from idea to first-in-human as fast as hyperpolarized MRI. We now can image metabolism in real time in humans, with 10,000-fold better sensitivity than other methods and without the use of radiation or toxic contrast agents. Hyperpolarized MRI makes it possible to map the metabolic activity within a tissue based on the enzymatic conversion of the injected 13C-labeled compound. Unlike Positron Emission Tomography(PET), which uses radioactive compounds, hyperpolarized compounds are safe, non-radioactive and naturally occurring. The first human study showed that early detection of prostate tumors could be detected by imaging the conversion of pyruvate to lactate (Warburg effect) before any physical mass was visible on the MRI. Patient trials are growing weekly in the fields of cancer, diabetes, and autoimmune diseases. This technology could significantly advance breakthroughs in the treatment of cancers and autoimmune diseases.

4. Fähzan Ahmad, Klar2O GmbH, Germany, ignite talk,

Regenerative biochemical microplastics filtration

About 325 different plastic particles are in one liter of drinking water, that were partly absorbed by our bodies and accumulate in the liver, adipose tissue, bloodstream, lungs, kidney, the placenta, and even our brain and can cause serious health damage in humans. We developed a special biochemical adsorbing coating, which is able to bind microplastics of all kinds by molecular interactions from water. Our patented sustainable filter technology is recyclable and regenerable and was recently approved by independent analysis laboratories. Since we do not use a conventional membrane, our filter has no filtration limit due to the pore size. This allows us to filter efficiently even in the nanometer range. Our innovative biochemical coating can bind microplastics pressure-free and is economical and smaller than most filters with higher efficiency. With our technology, we want to renaturalize water, our life-building block, and eliminate the danger of microplastics for a healthy life.

5. Fred Phillips, TANDO, Inc., ignite talk

Re-conceiving the long waves: Implications for innovation and economic policy

The long (Kondratieff) economic waves are named for jumps in human ability to harness greater sources of energy, and for discoveries of new materials that permit new kinds of energy release, finer machining, or stronger structures. This exploratory talk presents evidence that the K-waves should be attributed not just to advances in energetics, but co-equally to advances in economic coordination. It is not only the latest K-wave that's driven by ICT! Though the new equal emphasis on machines and coordination improves economic forecasting only arguably, we indicate how the Leapfrog Theory of Scientific Advance—the constant leapfrogging of theory, methodology, data, and scientific problems to society's forefront—may sharpen forecasting as this thread of research continues. Preliminary evidence shows economic upswings primarily follow the methodology frog's forward leaps, these leaps bringing us new techniques, procedures, instruments, and inventions. Some policy implications emerge.

6. Iuliia Myrgorodska, Astrazeneca, ignite talk

Self-preservation behavior in a community of synthetic protocells

The development of programmable microscale materials with cell-like functions, dynamics, and collective behavior is an important milestone in systems chemistry, soft matter bioengineering and synthetic protobiology. In living systems, feedback is an essential control element used to maintain homeostasis and adapt to environment. Engineering feedback loops on the micrometer scale can further drive the design of biomimetic soft materials with collective properties. In this work a population of artificial cell-like structures has been shown to display self-preservation behavior achieved by careful design of negative feedback loop. Furthermore, I will address some challenges and potential application of synthetic protocells.

7. Marek Checinski, CreativeQuantum GmbH, Germany, ignite talk

Reinventing the MeOH synthesis by advanced quantum mechanical simulations

Methanol is one of the most important chemicals in industry. It is produced at 100 Mt/a scale. Nowadays, the industry standard catalyst is working at 240–260 °C and 50–100 bar. Heating and compressing of 100 Mt of substrates under these conditions makes it a very energy intense and expensive process with a high CO2 footprint. Decreasing process conditions would have a big impact on the environment on CAPEX/OPEX.

This motivation leads us to think about new approaches from scratch, which were evaluated by intense quantum mechanical simulations within a few weeks. An investigation of >100 catalysts by virtual high throughput screenings within two months revealed new promising catalysts. These candidates were tested within 1.5 months in the laboratory, and it was proven that the predicted good catalysts synthesized MeOH under very mild conditions while the bad ones did not. The first optimized catalyst system produced MeOH at 120–150 °C and 10 bar CO pressure with a TON of >3000.

8. Siegfried Schindler, Justus-Liebig University, Germany, ignite talk

Research and collaboration in the context of mixed reality (VR/AR) and AI

Virtual reality and augmented reality (better described now as mixed reality) in combination with AI will completely change our future work environment, cooperation with other people and research itself. The presentation will show what we already can do right now and furthermore will show the next developments coming up.

9. Xin Lai, Friedrich-Alexander-Universität Erlangen-Nürnberg, Germany, ignite talk

Cooperative MicroRNA Therapeutics in Cancer

MicroRNAs (miRNAs) are short, non-coding RNAs that regulate gene expression at the post-transcriptional level. An miRNA can potentially bind many mRNAs, thereby affecting cancer genes and the activity of relevant pathways. miRNA therapeutics can harness this evolutionarily conserved mechanism for the coordinated gene regulation and thus restoring normal phenotypes in cancer cells. However, the promiscuous binding of miRNAs can provoke off-target effects caused by high-dose treatments. Hence, it is desirable to develop miRNA therapeutics with increased specificity and efficacy. To achieve that, we propose the concept of miRNA cooperativity to exert synergistic repression on target genes, thus reducing the undesired effects of miRNAs. We demonstrate the effectiveness of cooperating miRNAs in reducing the chemoresistance of tumors and improving the immunogenic potency of dendritic cells for immunotherapy. Our work shows the potential of miRNAs as monotherapy or adjuvant therapy in cancer.

Abstracts of Presented Posters

Gertrud Morlock, A. Ziltener, S. Geyer, A. Mehl, T. Schreiner, T. Kamel, T., J. Tersteegen, F. Brümmer, Justus Liebig Universität, Germany, joint poster prize,

Dolphins' beauty secrets

A person with a skin rash might go to the doctor and get an ointment to apply. Indo-Pacific bottlenose dolphins also get skin problems, but they get help by rubbing against certain corals and sponges. We analyzed 48 samples from the gorgonian coral Rumphella aggregata, the leather coral Sarcophyton sp. and the sponge Ircinia sp., and found 17 biologically active compounds with antimicrobial, antioxidant, hormonal and toxic properties [1]. The discovery of these biologically active compounds led us to hypothesize that the mucus of the corals and sponges serves to regulate the dolphin skin microbiome. Repeated rubbing brings the biologically active compounds in the corals and sponges into contact with the dolphins' skin. These metabolites could support the homeostasis of their skin and be useful for prophylaxis or adjunctive treatment against microbial infections.

[1] Morlock, G.E., Ziltener, A., Geyer, S., Mehl, A., Schreiner, T., Kamel, T., Tersteegen, J., Brümmer, F. Indo-Pacific bottlenose dolphins self-medicate with invertebrates in coral reefs. iScience (2022) 104271

Natasa Stojanovic Guzvic, Florian Lüke, Giancarlo Feliciello, Christian Werno, Bernhard Polzer, Christoph Klein, Fraunhofer ITEM, Germany, joint poster prize

Exploiting Liquid Biopsy for Personalized Cancer Therapy

Despite the major advances in cancer treatment, identification of most effective therapies and monitoring therapy response remain main challenges. Using comprehensive molecular analysis of biomarkers from non-invasively obtained liquid biopsy (LB) of body fluids, we can monitor the evolving make-up of therapeutic targets over time. Additionally, we established patient-derived in vitro and in vivo models from circulating cancer cells isolated from metastatic cancer patients. Multi-omics analysis and drug screens identified potential intrinsic resistance mechanisms and vulnerabilities to novel drugs. Our approach also enables generation of complex

U. A. Betz (ed.), *Curious Future Insight*,
https://doi.org/10.1007/978-3-031-41781-8

tissue models derived directly from patients and investigation of multiple therapeutic approaches with special relevance in currently expanding immune-oncology field. Cellular LB is particularly useful and should be systematically explored for inoperable, rare, and hard-to-treat cancers with great need for personalized treatment options.

David Adedia, Atinuke O. Adebanji, Simon K. Appiah, University of Health and Allied Sciences, Ho, and Kwame Nkrumah University of Science and Technology, Ghana

Comparative analysis of some structural equation model estimation methods with application to Coronary Heart Disease Risk

This study compared a ridge maximum likelihood estimator, a published ridge maximum likelihood, maximum likelihood, unweighted least squares, generalized least squares and asymptotic distribution free estimators in fitting six models that show relationships in some non-communicable diseases. Uncontrolled hypertension has been shown to be a leading cause of coronary heart disease, kidney dysfunction, and other negative health outcomes. It poses equal danger when asymptomatic and undetected. Research has also shown that it tends to coexist with diabetes mellitus (DM), with the presence of DM doubling the risk of hypertension. The study assessed the effect of obesity, type II diabetes and hypertension on coronary risk, and also the existence of converse relationship with structural equation modeling.

Akinranti S. Ajibola, Juliana B. Agberotimi, Aanouluwapo I. Obanubi, University of Ibadan, Ibadan, Nigeria

Analysis and ecological risk of fluoroquinolone antibiotics in landfill leachates in Ibadan, Nigeria

A quick, easy, cheap, effective, rugged and safe (QuEChERS) extraction method was optimized to investigate the occurrence and ecological risk of three fluoroquinolone antibiotics (ciprofloxacin, norfloxacin and ofloxacin) in leachates from three unsanitary landfills (Lapite, Ajakanga and Aba-Eku) in Ibadan, Nigeria. Recoveries ranged from 102–128%. Method LOQs were 0.35 µg L-1, 1.91 ?g L-1, and 11.69 ?g L-1 for norfloxacin, ciprofloxacin and ofloxacin, respectively. Inter-day precision (% RSD) ranged from 7–22%. High levels of target fluoroquinolones were measured, up to 796 ?g L-1 in leachates from Aba-Eku landfill. Norfloxacin presented low risk to algae, daphnia and fish. Ciprofloxacin and ofloxacin posed medium risk to algae in leachate from Aba-Eku landfill. First report on the occurrence and ecotoxicological risk of target fluoroquinolones in leachates from Nigerian landfills is presented. Sound management of wastes containing pharmaceuticals in Nigeria is recommended.

Anahid Amiri, Christian Dietz, Robert W. Stark. Technische Universität Darmstadt, Germany

The cyto-linker and scaffolding protein "Plectin" disarray leads to softening of cancer cells

Accurate diagnosis of cancer stage is inevitable for the following prognosis in patients struggling with these lesions to promote patient's health and survival rate. Previous studies on survival rate statistics show in some cases failure in cancer stage surveys in which metastasis or recurrence of the disease was not accurately prognosed. Here we show that carcinomas with an occult propensity of metastasis depict a number of poorly differentiated cells with decreased amount of cytoskeleton components in a near well-differentiated population. Force spectroscopy in conjunction with fluorescence microscopy of lung cancer, liver hepatoma and melanoma provided a general view of cells architecture leading to the conclusion that the scarce abnormal-shaped cells with low formation of structural filaments conveys the high risk of metastatic potential of the tumor. The results demonstrate that force spectroscopy complements conventional diagnostic approaches and can improve the following prognosis.

Carmen Andreina Olivares Moreno, Instituto Universitario Politécnico Santiago Mariño, Venezuela

Sustainable polymer poly(dihydroferulic acid) (PHFA) from coconut husk

Most commercially available polymers are oil-based and present low degradation rates, which lead to considerable environmental issues. In addition, these materials are widely used in the manufacture of single-use products. Thus, they remain for centuries in the environment making the development of sustainable alternatives essentially important for reducing their environmental impact.

This study was aimed to obtain the sustainable polymer poly(dihydroferulic acid) PHFA from the mesocarp fibers of Cocos nucifera fruit, i.e. coconut husk. These fibers are a waste product from the coconut industry and currently unexploited. However, given their considerable lignin content, coconut husk is a promising starting material for obtaining aromatic-aliphatic polymers, e.g. PHFA, which is a sustainable alternative for the widely used poly(ethylene terephthalate) PET. Based on this innovative approach, the sustainable polymer PHFA was successfully synthesized from coconut husk via polycondensation.

Kingsley Badu, Stephen Opoku Afriyie, Thomas Kwame Addison, Yilekal Gebre, Abdul-Hakim Mutala, Kwasi Baako Antwi, Dawood Ackom Abbas, Kofi Agyapong Addo, Austine Tweneboah, Cristian Koepfli. Kwame Nkrumah University of Science and Technology, Ghana

varATS RTqPCR a highly sensitive nucleic acid assay detects missed clinical malaria infections: evidence from two hospitals in the Ashanti region of Ghana

The World Health Organization recommends parasitological confirmation of all suspected malaria cases by microscopy or rapid diagnostic tests (RDTs) before treatment. These conventional tools are widely used for point-of-care diagnosis in spite of their poor sensitivity at low parasite density. Here, we present for the first time in Ghana, the accuracy of diagnosis of malaria using highly sensitive varATS qPCR as reference in a clinical setting.

1,040 febrile patients were recruited from two primary health care centers in Ghana and tested for malaria by microscopy and RDT. The sensitivity and specificity were assessed using varATS qPCR as gold standard. Using varATS qPCR as the standard, RDT was more sensitive (55.7% vs 39.3%), marginally less specific (98.2% vs 98.3%), and had higher positive (95.7% vs 94.5%) and negative predictive values (75.3% vs 69.0%) than microscopy. RDT and microscopy missed over 40% of infections that were detected by varATS qPCR. Such novel tools are needed.

Zetty Norhana Balia Yusof, Universiti Putra, Malaysia

Microalgal Vaccines: For Fish Oral Vaccination Today, For Human Oral Vaccination in the Future

Infectious diseases are affecting one of the most reliable food sources which is aquaculture. Vaccination has been proven to be effective, but most vaccines are injected into fish, which is labor intensive, costly, and stressful for the animals. Our work manipulated a green microalga, Nannochloropsis sp. as a potential vaccine carrier for oral vaccination due to its wide usage in the aquaculture industry. Transformation vectors harboring a gene coding for an antigenic protein were constructed via homologous recombination (HR) and clustered interspaced short palindromic repeats (CRISPR) systems. Nannochloropsis sp. was transformed and validated at the genomic, transcriptional and translational levels. Overall, the development of transgenic Nannochloropsis sp. for fish oral vaccination has been proven successful. This could be the gateway to construction of transgenic microalgae as vaccine delivery system to not only aquatic organisms, but hopefully for humans in the future.

Mario Bernardi, A. Carretta, L. Pesce, F. Cardarelli, NEST Laboratory Scuola Normale Superiore, Italy

Probing the synthetic identity of nano- encapsulated drugs by fluorescence lifetime analysis

The FDA often stresses a lack of analytical techniques to thoroughly characterize physicochemical properties of nanoparticle drug delivery systems. This scenario lies behind the financial troubles of next-generation nanotechnology companies.

We rely on intrinsic luminescence signals to characterize encapsulated drugs and dynamically determine the fractions of each coexisting physical state of the active principle. Our approach allows non-invasive study of metabolism and uptake patterns in vitro or ex-vivo, stability, and batch-to-batch analysis suitable for touchless quality control procedures. We are currently testing FDA-approved drugs (i.e., Doxil, Onivyde) to unveil the elusive mechanism of action of next-generation drugs and address state-of-the-art challenges in the field of drug delivery.

Karen Cloete, Giday Gebregziabher, Miche Hess, Nandipha Botha, Ademola Adetunji, Givemore Makonya, Juliet Sackey, Remy Bucher, Jill Farrant, M. Maaza. University of South Africa, Nanosciences African Network and University of Cape Town, South Africa

Nano-enabled agriculture for sustainable agro-production

Climate change and the Ukrainian war is posing a significant indirect threat to sustainable agriculture related to nutrient use efficiency and fertilizer supply. To find solutions to the current crisis within the agro-production sector, nanofertilizers have been suggested as an innovative technology to address sustainable agro-production.-Why? Nanofertilizers can be designed for smart nutrient release, while their unique physicochemical parameters present a more efficient approach to optimizing nutrient use efficiency. We aim to discuss the following questions: How can nanofertilizers be synthesized using safe and green methods? What is the most effective route of application? How can science diplomacy promote the research and development and safety screening of nanofertilizers? Can nanofertilizers ultimately present as one of the answers to the current agricultural crisis triggered by climate change and fertilizer availability?

Arindam K. Dey, Adrien Nougarede, Flora Clément, Carole Fournier, Evelyne Jouvin-Marche, Marie Escudé, Dorothée Jary, Fabrice P. Navarro, Patrice N Marche. Univ. Grenoble Alpes, INSERM, France

Nucleic acid-mediated tuning of the charge of cationic nanocarriers to dampen inflammatory responses

The unique characteristics of nanomaterials (NMs) might give unforeseen toxicity that could adversely affect the immune system.

We investigated non-activated and IL-4 or LPS-activated primary bone marrow-derived macrophages (BMDMs) in response to cationic lipid carriers (cNLCs) (45.18 nm; +45.8 mV). The impact of cNLCs was assessed by measuring proinflammatory molecules and metabolism (glycolytic activity and oxidative phosphorylation). Furthermore, we combined cNLCs with negatively charged siRNA at different N/P ratios (N = nitrogen, P= phosphate groups of nucleic acid). Our results demonstrate that cNLCs significantly increase the secretions of proinflammatory molecules (IL-6, TNF-?, MCP-1, NO) and cellular metabolism in both unactivated and activated BMDMs but reversing the surface charge of cNLCs reverse the effect. We conclude that reversing the surface charge with biomaterial, such as nucleic acid, can dampen the inflammatory activities of the cationic nanocarrier.

Ben Glasspoole, Jasmine Gardner, Marko Hermsen, Philipp Harbach, Robert Grady, Merck KGaA, Darmstadt, Germany

Chemistry Innovation through Collaboration: Digital Tools for Drug Discovery

Much has been made in the last decade about the reproducibility of life science research, and how applicable would-be breakthroughs actually are in drug discovery.

But irreproducibility does not always stem from poorly (or nefariously) reported science—more often, it is simply the lack of access to the reagents or tools used that limits the adoption of a new discovery. Simply put, without commercial availability, most new publications or scientific disclosures are destined to remain interesting—but not reproduced—academic curiosities.

By collaborating with academic and industrial partners (and leveraging our own R&D), our group has helped bridge the gap between inventors and users to make a steady drumbeat of MedChem and Early Discovery tools available to chemists around the world. A few such examples will be discussed, focusing on new developments in Chemical Biology, MedChem, and instrumentation.

Igor Goryanin, Sergey Vesnin, Okinawa Institute Science and Technology, Japan and Medical Microwave Radiometry Ltd. UK

Passive Microwave Radiometry for early diagnostics and drug R&D

Temperature is the first indicator of various disorders in the human body. Thermal changes, as a rule, precede structural changes and arise at the earliest stage of the pathological process, and in many cases characterize the risk of the disease.

We are using first in the world commercially available MWR-2020 (former RTM-01-RES) device which can simultaneously visualize internal (microwave, 60 mm deep) and skin (infrared, 1mm deep) temperature emissions. The unique features of the device are passive sensing, lack of radiation, no side effects, high sensitivity and specificity, detection of pathologies at early stage, and very low examination costs.

In collaboration with University of Edinburgh, we have developed and trained different Neural Networks on MWR data collected during many years. Only using our AI MWR system we have shown that breast cancer could be detected (>90% accuracy), venous diseases (>80% sensitivity), and COVID-19-induced pneumonia (>79% sensitivity).

Nicole Hall, M. Frances Vest, Bo Jarrett Wood, Kariann Lamon, Kevin Sean Murnane. Louisiana State University Health Sciences Center, USA

Revolutionizing our view of addiction: Elucidating neurovascular damage as a critical mechanism in methamphetamine-Induced dopamine dysregulation

Methamphetamine (MA) use is a growing worldwide phenomenon that is devastating millions of lives throughout the world. Decades of research has focused on modulating the mesolimbic dopamine pathway as a target for methamphetamine medications development, yet no current treatment is available. We are conducting studies using animal models of binge exposure and compulsive drug taking, and parallel experiments in patients recently abstinent from methamphetamine to elucidate the critical neurobiological mechanisms that drive the progression of methamphetamine addiction, and to identify new targets for disease-modifying therapeutics. Our main hypotheses are that 1) methamphetamine-related neurovascular damage leads to dopamine dysregulation that impairs neurocognitive control and the capacity to remain abstinent and 2) that anti-inflammatory and neuroprotective drugs can promote resilience and abstinence. I will present data that address these hypotheses.

Pooja Hegde, Michael Howe, Matthew Zimmerman, Anthony Baughn, Veronique Dartois, Courtney Aldrich, University of Minnesota and Hackensack Meridian Health, USA

Modified para-Aminosalicylic Acid (PAS) Analogs as Improved Anti-TB Agents

para-Aminosalicylic acid (PAS), an important second-line agent for treating drug-resistant Mycobacterium tuberculosis, has moderate bioavailability and rapid clearance that necessitate high daily doses, which in turn causes severe gastrointestinal disturbances presumably by disruption of gut microbiota and host epithelial cells. We first synthesized a series of ester prodrugs to increase the oral bioavailability and thereby prevent intestinal accumulation as well as undesirable bioactivation by the gut microbiome to non-natural folate species that exhibit cytotoxicity. We next modified the PAS scaffold to reduce N-acetyltransferase-1 catalyzed inactivation by introduction of groups to sterically block N-acetylation and fluorination of the aryl ring of PAS to attenuate N-acetylation by electronically deactivating the para-amino group. The pivoxyl prodrug and fluorination at the 5-position of PAS address the primary limitations of PAS and have the potential to revitalize this TB drug.

Hungharla Hungyo, Vibha Tandon Jawaharlal Nehru University, India

Identification of KRAS variant in the 3'UTR and Prochlorperazine response in head and neck cancer

SNPs within the 3'UTR of the KRAS gene has been associated with drug resistance/sensitivity and poor patient outcome in various cancers. KRAS variant carrying HNSCC patients do not respond to the standard treatment of cisplatin along with radiotherapy in the presence of cetuximab, suggesting that EGFR is no longer considered as a biomarker for a drug target. Therefore, identification of new biomarkers to treat HNSCC patients targeting the KRAS variant is needed. Moreover, the development of effective radiosensitizing agents will be necessary for further improvement in the radiation therapy for HNSCC. Toward this effort, evaluation of the radiosensitization profile of prochlorperazine in patient-derived KRAS variant cells showed a positive response and suggested decreased cell viability. In different tumor xenograft mice models, it was observed that the tumor volume was significantly reduced with the PCZ treatment in the KRAS variant model, suggesting that PCZ is a radiosensitizer among different tumor xenograft model in vivo with reduced toxicity in mice. In addition, KRAS mutation led to globally altered gene expression profiles.

Fatima Irfan, Dr Panjwani Center of Molecular Medicine and Drug Research, Pakistan

Role of Medicinal Plant Metabolites In Enhancing the Potential of Stem Cells for Wound Healing

The study is divided into two parts, in vivo and in vitro. In the in vivo part of the study cold burn wound model was developed and hUMSCs transplanted. 20μM concentration for both quercetin and rutin is selected based on MTT results. In vitro scratch assays showed a reduction of wound area in preconditioned MSCs groups as compared to the untreated group. In vivo macroscopic evaluation showed a decrease in wound healing time in preconditioned MSCs groups. Histology examination revealed enhanced skin regeneration in treated groups based on reformation of skin adnexa, fibroblast connectivity, and overall tissue integrity. Gene expression profile shows decreased inflammatory cytokine IL-1beta, IL-6, increased anti-inflammatory cytokine IL-4, IL-5, and upregulation of antioxidative markers GPX7, PRDX, and TXNRD2.

Florian Kabinger, Florian Kabinger, Carina Stiller, Jana Schmitzová, Christian Dienemann, Goran Kokic, Hauke S. Hillen, Claudia Höbartner and Patrick Cramer Universität Würzburg, University Medical Center Göttingen and Max Planck Institute for Biophysical Chemistry, Germany

Mechanism of molnupiravir-induced SARS-CoV-2 mutagenesis

Molnupiravir is an orally available antiviral drug candidate currently in phase III/IV trials for the treatment of patients with COVID-19. Molnupiravir increases the frequency of viral RNA mutations and impairs SARS-CoV-2 replication in animal models and in humans. We establish the molecular mechanisms underlying molnupiravir-induced RNA mutagenesis by the viral RNA-dependent RNA polymerase (RdRp). Biochemical assays show that the RdRp uses the active form of molnupiravir, β-D-N4-hydroxycytidine (NHC) triphosphate, as a substrate instead of cytidine triphosphate or uridine triphosphate. When the RdRp uses the resulting RNA as a template, NHC directs incorporation of either G or A, leading to mutated RNA products. Structural analysis of RdRp–RNA complexes that contain mutagenesis products shows that NHC can form stable base pairs with either G or A in the RdRp active center, explaining how the polymerase escapes proofreading and synthesizes mutated RNA. This two-step mutagenesis mechanism probably applies to various viral polymerases and can explain the broad-spectrum antiviral activity of molnupiravir.

Amaranta Kahn, Sandra A. C. Figueiredo, João P. Reis, Ana Vieira, Marine Cuau, Kathleen Abt and Pedro N. Leão Interdisciplinary Centre of Marine and Environmental Research (CIIMAR/CIIMAR) and ICBAS Abel Salazar Institute of Biomedical Sciences, University of Porto, Portugal, and Poitiers University, France

Exploring A New Pathway of Exogenous Fatty Acid Incorporation in Cyanobacteria

Fatty acids (FAs) are involved in multiple biological processes and are key metabolites in living organisms. Due to the high energetic cost of FAs biosynthesis, most organisms have evolved mechanisms to incorporate exogenous FAs (eFAs). By doing so, organisms reduce the energetic burden of FAs synthesis and increase their plasticity as to FAs source. Until recently, all characterized FA incorporation mechanisms were reported to involve activation of the FAs by the acyl-acyl carrier protein (Aas). However, our group recently reported a new cyanobacterial enzyme—BrtB—that directly esterifies free FAs with alkyl halide moieties found in the bartolosides. We hypothesize that bartoloside esters represent a new system of eFAs scavenging and/or storage. This study will open doors to new cyanobacterial synthetic biology tools, potential antibiotics, and biofuel engineering.

Kübra Kaygisiz, Kübra Kaygisiz, Arghya Dutta, Lena Rauch-Wirth, Tristan Bereau, Christopher V. Synatschke, Jan Münch and Tanja Weil. Max Planck Institute for Polymer Research, and Ulm University Medical Center, Germany

Data mining unveils property-activity correlation and predicts infectivity enhancing peptides

The therapeutical application of retroviral vectors is still limited by moderate transduction efficiencies.

Here, we show that data mining is a powerful tool to study and predict novel peptides for the discovery of new sequences providing superior transduction efficiency. A peptide library consisting of more than a hundred derivatives of an active peptide EF-C was created and amyloid fibrils with a positive charge and a strong tendency to aggregate were identified to highly correlate with efficient viral transduction. The universal importance of these properties for peptide–virus–cell interaction was confirmed by screening pathogenic amyloidal peptides. Further, we applied a machine-learning approach for in silico discovery of novel active peptides based on this library. The sequences were converted to vectors and activity was modeled with a supervised machine-learning regression. To our delight, 16 novel sequences were probed for infectivity enhancement with a hit rate of 20%.

Amanda Johnson, Jack Korleski, Sophie Salls, John Laterra, Hernando Lopez-Bertoni, Hugo W. Moser Research Institute at Kennedy Krieger and Johns Hopkins Institute, USA

Glioma stem cells mimic regulatory T-cell function to suppress the immune response and promote tumor propagation

The cell-intrinsic mechanisms of tumor cell immune escape are considered fundamental to clinical GBM growth and recurrence. Single-cell RNA sequencing analysis of GBM neurospheres revealed a previously unrecognized Oct4/Sox2high/FOXP3-cell subpopulation with high expression of TGF-beta1, CD39, CD73, PD-L1, and Galectin-1, a gene expression fingerprint typically associated with Treg cells and their immune suppressive functions. Bioinformatics analysis shows that the above-mentioned genes are enriched in the mesenchymal GBM subtype and highly correlate TGF-beta1 type II receptor (TGFBR2) expression in both clinical GBM and primary GSCs. Importantly, pharmacological inhibition of TGFBR2 depleted CD44+ GSCs and reduced immunosuppression in tumor cell/T-cell co-cultures. Pilot experiments show that TGFBR2 pharmacological inhibitor is well tolerated in mice and reduces tumor growth capacity in sub-cutaneous tumor models.

Jan Luprich, iris.ai, Norway
Iris.ai accelerates CRO's clinical data processing

In light of the Green Deal and other industry challenges of recent months, AI has proven its potential to radically accelerate the innovation needed for future success across the industries. But all R&D starts with scientific knowledge processing and reading through scientific papers, whereas finding and extracting relevant data is a tedious, time-consuming task. This need not be that way, provided that we leverage the latest advances in Natural Language Processing technology.

At Iris.ai we have spent the last 6 years building an award-winning AI engine for scientific text understanding, based on the valuable feedback of R&D professionals around the world. With AI-powered tools, researchers can not only automate the search for highly relevant papers, but also summarize the findings and extract important information from research papers or patents within minutes, saving up to 75% time to drive real innovation.

With the Iris.ai Researcher Workspace, change can happen now.

Irina Meln, European Vaccines Initiative, Germany
Pan-European collaborations to foster digital innovations for vaccine development

Vaccines effectively prevent infection, foster healthy living, and reduce antimicrobial resistance (AMR). Yet, vaccine R&D and production are long, costly, and often fail due to challenges posed by biologically complex pathogens (malaria, tuberculosis), incomplete understanding of host–pathogen interactions, multiple diverse steps in vaccine development, and others. EVI's innovative portfolio includes 1) artificial intelligence (AI)-based approaches for effective vaccine design; 2) mucosal organ-on-a-chip and improved controlled human infection models (CHIMs) for rapid vaccine testing; 3) in silico platform for optimized vaccine biomanufacturing; 4) in silico decision-making tool to evaluate the value of vaccines for boosting investments in vaccine R&D. EVI leverages public–private collaborations, developing much-needed innovative methods to accelerate and de-risk all stages of next-generation vaccine development from antigen design to optimized biomanufacturing.

Volker Morath, Katja Fritschle, Luisa Krumwiede, Milica Zivanic, Markus Anneser, Stephanie Robu, Sarah Dötsch, Linda Warmuth, Tarik Bozoglu, Susanne Kossatz, Christian Kupatt, Katja Steiger, Markus Schwaiger, Dirk Busch, Arne Skerra, Wolfgang Weber Technische Universität München, Germany

Imaging of cell and gene therapies using a novel PET-reporter gene system

Advanced medical treatments necessitate a reliable diagnostic method to image transgenes and quantitatively monitor their localization over time. We developed a novel reporter gene system which is suitable for both monitoring of chimeric antigen receptor (CAR) T-cells and quantification of the in vivo transduction by adeno-associated viral (AAV) vectors.

The reporter protein DTPA-R comprises an extracellular Anticalin binding protein, which binds metal•DTPA complexes, the V5-tag, and a membrane anchor domain. In AAV9 studies, we could detect viral transduction of tiny anatomic structures such as adrenal glands. Furthermore, we were able to monitor migration of human CAR T-cells in mice bearing a CD19-positive lymphoma over a 1-month treatment course.

This novel & proprietary reporter gene provides a promising tool to elevate the understanding of cell and gene therapies and support the development of precision medicine.

Willis Muganda, International Younger Chemists Network and Gesellschaft Deutscher Chemiker, Germany

Younger chemists making a positive change toward a sustainable and peaceful world

The International Younger Chemists Network (IYCN) was launched in 2017 as an associated organization of IUPAC, aimed at creating a virtual platform for young chemists to connect and share ideas. Our members are chemists under the age of 35, or those who are within 5 non-continuous years from their terminal degree. Our membership base currently spreads across six continents and our online presence is growing daily. The primary focus for us is to establish a network that fosters communication, mentorship, and collaboration between our members.

Our initiatives are aimed at developing tools that can be used to foster an interest in chemistry amongst the public, one of our projects works to do this through the publication of experiments, designed to be performed primarily with non-specialized equipment and translated into multiple languages. With this model we are working to remove all barriers between chemistry and the general public, making our passion universally accessible. In addition, we are actively involved in promoting innovation and sustainable chemistry by collaborating with other industry partners and global scientific institution. Our portfolio is ever expanding and we are working to offer more online webinars and to plan broader student exchanges between countries in different parts of the world. This presentation will focus on our achievements to date and the future objectives of IYCN.

Subhashini Pandey, Institute for Stem Cell Biology and Regenerative Medicine, India

Intermittent scavenging storage lesion from stored red blood cells by nanofibrous sheets enhances the quality and shelf-life of stored RBCs

The crisis of stored RBC units has always been a critical problem in India due to the deterioration in their quality throughout 42 days(D). The degrading rate corresponds to Damage-Associated Molecular Patterns (DAMPs) formation in the blood bags during storage. The primary focus of this study is to improve the quality and shelf-life of stored blood by removing these storage lesions on a specific day during the storage window (42 days). This particular removal led to an overall low accumulation of DAMPs at the end of the 42nd D. The primary focus of this study is to improve the quality and shelf-life of stored blood by removing these storage lesions on a specific day during the storage window (42 days). This particular removal led to an overall low accumulation of DAMPs at the end of the 42nd D. Thus, the proposed novel blood cleaning technology that scavenges DAMPs carry immense potential in the arena of blood transfusion.

Ernesto Prado Montes de Oca, Personalized Medicine Laboratory, CIATEJ, Mexico and Clinical Pharmacogenetics Implementation Consortium, NIH, USA

Retinoblastoma pathway regulates in silico DNA methyltransferase DNMT1 and DNMT3B in Lupus

Lupus/SLE is an incurable and potentially lethal autoimmune disease with altered lysosomal autophagy. To discover both novel biomarkers and therapeutic targets, we analyzed 13 genes associated with lupus susceptibility: DNMT1, RASGRP1, NLRP3, CAMP, DNMAP1, FOXP3, DNMT3A, DNMT3B, IFI44L, TLR9, IL1A, IL1B and ITGAM. Using the mRNA of these genes as targets, we searched for miRNAs (TargetScan v.8.0, MirtarBase v.9.0); assessed differential methylation of DNA (patients vs. controls; DiseaseMeth v.3.0); searched for transcription factors (TFs) involved in gene transcription (AnnoMiner); predicted protein–protein interaction (STRING v.11.5); and performed epitranscriptomic predictions (Whistle). We found 16 (14 of them novel) high-confidence (p-values below 0.001) miRNAs that regulate DNMT1. We found differential methylation in DNMT1, DNMT3A, IFI44L, and FOXP3. We predicted five N6-methyladenosine (m6A) modifications in DNMT1 mRNA including one altering lysosomal processes (m6A_766, $p=0.001$).

Giada Quintieri, Andre Gröschel. WWU Münster, Germany

Multicompartment Cubosomes for Catalysis Application

ABC triblock terpolymers are often employed for mimicking the compartmentalized character of cells and organelles. Here, we investigate the formation of multicompartment cubosomes (MCCs) out of ABC triblock terpolymers, through redispersion of polymeric bulk films in selective solvents. In particular, we employ polystyrene-block-polybutadiene-block-poly(tert butyl methacrylate) (SBT) triblock terpolymers in acetone/isopropanol solvent mixtures of varying compositions. Below a critical corona length (7 wt%), we were able to identify the first example of MCC with a lamella membrane. We modified the MCCs for catalytic applications by loading the outer wall of the membrane with catalysts (e.g., Pt NPs). Hydrolysis of the corona block further converts the MCCs into water-soluble and porous catalyst scaffolds, while the inner walls maintain structural stability. We will further employ the MCCs as platform for catalytical reduction of nitrobenzaldehyde dimethyl acetal to aminobenzaldehyde.

Stephanie Scherer, Ethan Patterson, Stacey Ward, Hongyi Zhang, Lisa Rois, Ashley Quiggle, Merck KGaA, Darmstadt, Germany
Custom engineered cell lines for early drug discovery, pre-clinical research, and biomanufacturing

Cell Design Studio® engineers custom, genetically modified cell models specific for your research, pre-clinical, and/or biomanufacturing needs. We partner with our customers to create workflows that address the unique requirements for each project and we prioritize open communication with our customers throughout a project. Using state-of-the-art genome editing technologies (CRISPR/Cas9, ZFNs, lentivirus, shRNA), we have edited over 200 genes in over 300 cell lines, including immortalized and iPS cells. We aim to solve the toughest problems in the industry by collaborating with the global scientific community.

Dieter Schroth, Merck KGaA, Darmstadt, Germany
Licrion: Enabling next-generation data communication

Liquid crystals for microwave components are sought by established material and display industries, along with companies in the satellite and telecommunications fields, driven by the need for new data communication technologies. We detail promising applications that could go mainstream, including tunable microwave filters, beam-steering antenna systems, and millimeter-wave communication [2]. licriOn™ is the liquid crystal material technology , developed at Merck providing the core functionality for realizing a new generation of affordable electronically steerable antennas.

[2] Wittek, C, Fritsch, C, Schroth, D, 2021. Employing Liquid Crystal-Based Smart Antennas for Satellite and Terrestrial Communication. Information Display, 37: 1–52

Julia Schulmeyer, LMU Munich School of Management, Germany
Privacy in the Metaverse: Are we thinking about our data?

The increasing availability and connectivity of digital devices have led to a growing amount of available data. Data as "the oil" of the digital economy is an important asset class and driver of innovation. However, recent data scandals and increased transparency have sharpened users' awareness about organizational data practices and induced concerns about privacy. Privacy assurances (PA) describe organizational measures that provide users with assurances that their privacy is protected. When companies implement corresponding measures, they can address users' privacy concerns which leads to higher willingness to disclose personal data. Privacy-enhancing technologies, as a novel category of PA, are coherent systems of ICT measures that protects privacy by eliminating or reducing personal data (e.g., homomorphic encryption). How can these technologies be used by companies to motivate users to more data disclosure? Are they threat or gain for users? How can companies use their potential?

Karimunnisa Sameer Shaikh, Manoj Madhukar Lokhande. Progressive Education Society's Modern College of Pharmacy, India

Development and evaluation of a Pickering emulsion for colon targeted delivery of resveratrol

Pickering emulsion was investigated as a novel formulation for the colon targeted delivery of resveratrol, a potential biomaterial in treatment of colitis. Poor oral bioavailability of resveratrol is due to its poor water solubility, short half-life, faster metabolism, and excretion. Hence, Pickering emulsion, res-PE, was investigated as a new oral controlled release delivery system to improve water solubility, bioavailability, and colon targeted delivery of resveratrol. The nanoparticulate form of chitosan, PEG-6000, and liquid paraffin were used as emulsifier, hydrophilic polymer, and oil phase, respectively. A 32 factorial design was employed to arrive at an optimized Pickering formulation. The emulsions were characterized for solubility, TEM, DSC, particle size, entrapment efficiency, and % in-vitro cumulative drug release. The solubility of resveratrol, dissolution rate, and in-vivo anti-inflammatory activity were enhanced due to Pickering emulsion formation.

Radhika Sharma, Punjabi University, India

Development and evaluation of a polymer film-forming gel hybridized with clotrimazole-loaded liquid

Dermatophytes, a group of fungal species causing superficial fungal infections possess a major health problem in a wide group of people. Treatment of skin infection by dermatophytes is still limited, and the application of conventional topical formulations (ointments, creams, etc.) causes patient discomfort due to repeated administration and low efficacy. Clotrimazole (CTZ), a broad-spectrum imidazole antifungal agent, is widely used to treat fungal infections. Conventional topical formulations of clotrimazole are intended to treat infections by effective penetration of drugs into the stratum corneum. However, drawbacks such as poor dermal bioavailability, poor penetration and retention, frequent dose application, and variable drug levels limit the efficiency. This study describes the polymer film-forming gel (PFFG) hybridized with clotrimazole-loaded liquid crystalline nanoparticles (CTZ-LCNP) for enhanced antifungal activity against dermatophytes.

Parvesh Singh, Lalitha Gummidi, Md. Shahidul Islam University of Kwazulu-Natal, South Africa

Antidiabetic evaluation of 1,3,4-thiadiazole-thiazolidinone molecular hybrids

A new library of 4-thiazolidinedione molecular hybrids was developed via a one-pot multicomponent reaction between 5-substituted phenyl-1,3,4-thiadiazol-2-amines, aromatic aldehydes, and 2-mercaptoacetic acid [3]. The full structure elucidation of all synthesized compounds was established using 1D and 2D NMR and other spectroscopic techniques. The in vitro antidiabetic evaluation of compounds via **α**-glucosidase and **α**-amylase inhibition subsequently disclosed several potent compounds as exemplified by the representative compound (4e) that exhibited 1.49- and 13.7-fold superior activity as compared to the standard inhibitor acarbose against **α**-glucosidase and **α**-amylase, respectively. Structure-activity relationship (SAR) analysis further revealed that the nature and position of substituents on the phenyl rings had a significant effect on the inhibitory potency.

[3] Multicomponent reaction for the synthesis of new 1,3,4-thiadiazole-thiazolidine-4-one molecular hybrids as promising antidiabetic agents through α-glucosidase and α-amylase inhibition. Bioorganic Chemistry, 2021, 115, 105210.

Jose Suarez, Celera Dynamics, UK

Smart opto-chemical platform for online protein measurement and digital twin development of biopharmaceutical processes

The efficiency of biotechnological production of biomolecules is critical for new biopharmaceuticals development. Product quality is determined by process development and the availability of high standard analytics, especially for very high-volume production. The biological products can be at very different concentration levels along the production chain, in different matrix environments and undergoing degradation. In all these stages, from cell culture, fermentation and down to ultrafiltration steps, there is a need to control the production through protein quantification without compromising the media sterility. We have devised a sensing system that allows the measurement of important parameters in real time and with unprecedented sensitivity. Our system also incorporates the capacity to provide real-time analytics through modeling and forecasting to determine product yield and purity, which is the ultimate goal of online monitoring systems.

Swetha Mundanattu, Lekshmi R. Nath, Ravi Shankar Lankalapalli, Noah Isakov, Rheal Towner, Deepa Sathyaseelan, Ruby John Anto, Rajiv Gandhi Centre for Biotechnology and National Institute for Interdisciplinary Sciences and

Technology, India, Ben-Gurion University of the Negev, Israel, and University of Oklahoma Health Sciences Centre, USA

Uttroside B, a US FDA-approved "Orphan drug" is more potent than sorafenib against HCC

Hepatocellular carcinoma (HCC) is associated with a high rate of mortality. Uttroside B (Utt-B), a furostanol saponin isolated in our laboratory from the leaves of Solanum nigrum Linn, exhibits exceptional cytotoxicity toward liver cancer cells compared to sorafenib, the first-line treatment option against HCC. Utt-B drastically inhibits the growth of HepG2 xenografts and exhibits better antitumor efficacy than sorafenib in NOD-SCID mice. It induces vacuolated structures indicative of autophagy and activates autophagy markers such as LC3 II, Beclin 1 & Atg proteins. Inhibition of autophagy by BafA1, 3-MA, and beclin siRNA enhances the Utt-B-mediated apoptosis. The anti-malarial drug, Chloroquine, which is a well-known autophagy inhibitor, when used in combination with Utt-B significantly enhances the chemotherapeutic efficacy of Utt-B. Utt-B is pharmacologically safe and does not induce hemolysis. Recently, US FDA granted "Orphan Drug" designation to Utt-B against HCC.

Eseoghene Helen Umukoro, Obafemi Awolowo University, Nigeria

Photo-assisted Electrochemical Oxidation of Organic Pollutant at a WO3-Exfoliated Graphite Electrode

In the quest for novel and efficient electrochemical materials as electrodes for the degradation of organic pollutants in water treatment, tungsten trioxide (WO3) and exfoliated graphite (EG) nanocomposite were synthesized, characterized, and fabricated into an electrode for the electrochemical and photo-assisted electrochemical degradation of organic pollutants in water treatment process. The nanocomposite was synthesized by a modified facile chemical method and then characterized using cyclic voltammetry, electrochemical impedance spectroscopy, X-ray diffractometry, Raman spectroscopy, Fourier transformed infrared spectroscopy, scanning and transmission electron microscopy, and energy dispersive X-ray spectrometry. The applicability of the WO3–EG as photoanode material was tested by the degradation of 4-nitrophenol and orange II dye as target organic pollutants in 0.1 M Na2SO4 solution at a current density of 10 mAcm-2 and the extent of mineralization was investigated.

GPSR Compliance

The European Union's (EU) General Product Safety Regulation (GPSR) is a set of rules that requires consumer products to be safe and our obligations to ensure this.

If you have any concerns about our products, you can contact us on ProductSafety@springernature.com

In case Publisher is established outside the EU, the EU authorized representative is:

Springer Nature Customer Service Center GmbH
Europaplatz 3
69115 Heidelberg, Germany

Batch number: 10726135

Printed by Printforce, the Netherlands